ORTHOGONAL FREQUENCY DIVISION MULTIPLEXING FOR WIRELESS COMMUNICATIONS

Edited by

YE (GEOFFREY) LI
GORDON STÜBER
Georgia Institute of Technology

Springer

Editors:

Ye (Geoffrey) Li
Georgia Institute of Technology
School of Electrical & Computer Engineering
777 Atlanta Drive
Atlanta, GA 30332-0250

Gordon L. Stüber
Georgia Institute of Technology
School of Electrical & Computer Engineering
777 Atlanta Drive
Atlanta, GA 30332-0250

Orthogonal Frequency Division Multiplexing for Wireless Communications

Library of Congress Control Number: 2005935341

ISBN 0-387-29095-8 e-ISBN 0-387-30235-2
ISBN 978-0387-29095-9

Printed on acid-free paper.

Printed in the United States of America.

9 8 7 6 5 4 3 2 1 SPIN 11546566

springer.com

CONTENTS

PREFACE

Orthogonal frequency division multiplexing (OFDM) has been shown to be an effective technique to combat multipath fading in wireless channels. It has been and is going to be used in various wireless communication systems. This book gives a comprehensive introduction on the theory and practice of OFDM for wireless communications. It consists of seven chapters and each has been written by experts in the area. Chapter 1, by G. Stüber, briefly motivates OFDM and multicarrier modulation and introduces the basic concepts of OFDM, Chapter 2, by Y. (G.) Li, presents design of OFDM systems for wireless communications, various impairments caused by wireless channels, and some other types of OFDM related modulation. Chapters 3 to 6 address different techniques to mitigate the impairments and to improve the performance of OFDM systems. Chapter 3, by J. Cioffi and L. Hoo, focuses on system optimization techniques, including channel partitioning, loading of parallel channels, and optimization through coding. Chapter 4, by S. K. Wilson, addresses timing and frequency offset estimation in OFDM systems. It also briefly discusses sampling clock offset estimation and correction. Chapter 5, by Y. (G.) Li, deals with pilot aided and decision-directed channel estimation for OFDM systems. Chapter 6, by C. Tellambura and M. Friese, discusses various techniques to reduce the peak-to-average power ratio of OFDM signals. Chapter 7, by G. Stüber and A. Mody, presents recent results on synchronization for OFDM systems with multiple transmit and receive antennas for diversity and multiplexing. To facilitate the readers, about 300 subject indexes and 300 references are given at the end of the book.

This book is designed for engineers and researchers who are interested in learning and applying OFDM for wireless communications. The readers are expected to be familiar with technical concepts of communications theory, digital signal processing, linear algebra, probability and random processes.

It can be also used as a textbook for graduate courses in advanced digital communications. Nevertheless, to accommodate readers having a variety of technical backgrounds, most of the key concepts in our book are developed with detailed derivations and proofs.

Even through each chapter is written by different people, we have tried to make symbols, notations, writing styles in different chapters consistent.

The editors of the book would like to thank L. Cimini, Jr. for initiating the book project, discussing skeleton of the book, identifying potential chapter contributors, and providing insight comments on the first draft of almost every chapter. The editors are also deeply indebted to J. Cioffi, L. Hoo, S. Wilson, P. O. Börjesson, P. Ödling, J. J. van de Beek, C. Tellambura, M. Friese, and A. Mody, who not only have done important and crucial work in OFDM related research topics but also contributed chapters in this book.

In particular, Y. (G.) Li would like to thank Professor S.-X. Cheng of Southeast University, P. R. China, who first introduced the concept of parallel modem (an OFDM related modulation) to him about 20 years ago. He wishes to thank some of his pervious colleagues at AT&T Labs - Research, including, L. Cimini, Jr., N. Sollenberger, J. Winters, and J. Chuang, for their technical advising and help in his OFDM related research. He also thanks his wife, Rena, for constant support and his sons, Frank and Micheal, for understanding while carrying out the book project.

We wish to thank the National Science Foundation, the U.S. Army Research Lab, Bell Labs of Lucent Technologies, Hughes Network Systems, Nortel Networks, Nokia Research Center, Mitsubishi Electric Research Labs, Motorola Labs, and Yamacraw Program of Georgia for their support of related research and educational activities of the editors.

Finally, A. N. Greene and M. Guasch of Springer deserve our special thanks for their tireless efforts in editing and promoting this book that we would otherwise have been unable to complete.

Ye (Geoffrey) Li and Gordon Stüber
Atlanta, Georgia

Chapter 1

INTRODUCTION

Gordon Stüber

Digital bandpass modulation techniques can be broadly classified into two categories. The first is single-carrier modulation, where data is transmitted by using a single *radio frequency* (RF) carrier. The other is multi-carrier modulation, where data is transmitted by simultaneously modulating multiple RF carriers. This book is concerned with a particular type of multi-carrier modulation known as *orthogonal frequency division multiplexing* (OFDM). OFDM has gained popularity in a number of applications including digital subscriber loops, wireless local area networks. It is also a strong contender for fourth generation cellular land mobile radio systems.

OFDM transmits data in parallel by modulating a set of orthogonal sub-carriers. OFDM is attractive because it admits relatively easy solutions to some difficult challenges that are encountered when using single-carrier modulation schemes on wireless channels. Simplified frequency domain equalization is often touted as a primary advantage of OFDM over single-carrier modulation with conventional time-domain equalization. However, frequency domain equalization can be applied just as easily to single-carrier modulation techniques as it can to OFDM. Perhaps the greatest benefit of using OFDM is that the modulation of closely-spaced orthogonal sub-carriers partitions the available bandwidth into a collection of narrow sub-bands. Motivated by the water-pouring capacity of a frequency selective channel, adaptive transmission techniques can be readily used to increase the overall bandwidth efficiency. One such possibility is to use adaptive bit loading techniques, where the modulation alphabet size on each sub-carrier is adjusted according to channel conditions. A larger signal constellation is used on sub-carriers where the received signal-to-noise ratio is large, and vice versa. As will be shown later in this book, OFDM waveforms are resilient to timing errors, yet highly sensitive to frequency offsets and phase noise in the transmitter and receiver RF and sampling clock oscillators. These characteristics are opposite those of single-carrier modulation, which is more sensitive to timing errors and less sensitive to frequency offsets. This is a manifestation of the long OFDM modulated symbol duration and the closely-spaced orthogonal sub-carriers. Hence, OFDM has its own set of unique implementation challenges that are not present in single-carrier systems. This book provides a comprehensive treatment of these challenges and their solutions.

1.1 High Rate Wireless Applications

The demand for high speed wireless applications and limited RF signal bandwidth has spurred the development of power and bandwidth efficient air interface schemes. Cellular telephone systems have gone through such a growth process. After the introduction of the first analog cellular systems in the early 1980s, subscriber growth for basic cellular voice services increased dramatically. This lead to the introduction of several second generation digital cellular standards in the early 1990s, such as the *Global System for Mobile communication* (GSM) and *Code Division Multiple Access* (CDMA), with the objective of providing greater system capacity so that the growing demand for voice services could be accommodated with scarce bandwidth resources.

The 1990s also seen a tremendous growth of Internet related services and applications, mostly using a wired Internet Protocol (IP) infrastructure. With the growing demand for wireless data and multimedia applications, cellular telephony and the Internet have become convergent technologies. This has lead to the development of third generation cellular standards, such as *Wideband CDMA* (WCDMA) and cdma2000, that support wireless voice, data, and multimedia applications. With the pervasiveness of the Internet, the cellular telephone network is evolving from a circuit switched to a packet switched IP-based core network. Such an infrastructure can support not only delay insensitive applications such as mobile data, but delay sensitive applications such as *voice over IP* (VoIP) as well.

The growth of the Internet also led to the development of various *wireless local area network* (WLAN) standards, such as those developed under IEEE802.11, to permit mobile connectivity to the Internet. Such services typically operate in unlicensed bands. With a surging demand for wireless Internet connectivity, new WLAN standards have been developed including IEEE802.11b, popularly known as Wi-Fi that provides up to 11 Mb/s raw data rate, and more recently IEEE802.11a/g that provides wireless connectivity with speeds up to 54 Mb/s. IEEE802.11b uses a signaling technique

Data Rate	6, 9, 12, 18, 24, 36, 48, 54 Mb/s
Modulation	BPSK, QPSK, 16-QAM, 64-QAM
Coding	1/2, 2/3, 3/4 CC
Number of subchannels	52
Number of pilots	4
OFDM symbol duration	4 μs
Guard interval	800 ns
Subcarrier spacing	312.5 kHz
3 dB bandwidth	16.56 MHz
Channel spacing	20 MHz

Table 1.1. Key parameters of the IEEE 802.11a OFDM standard, from [1].

based on *complementary code keying* (CCK), while IEEE802.11a uses OFDM which is the subject of this book. The main physical layer parameters of the IEEE 802.11a OFDM standard are summarized in Table 1.1. Dual mode radio access devices have been developed allowing access both public cellular networks and private WLANs, to provide a more ubiquitous and cost efficient connectivity.

More recent developments such as IEEE802.16 *wireless metropolitan area network* (WMAN) standard address *broadband fixed wireless access* (BFWA), that provides a last mile solution to compete with wireline technologies such as *Asymmetric Digital Subscriber Loop* (ADSL), coaxial cable, and satellite. Similar to IEEE802.11a, IEEE802.16 uses OFDM. The emerging *Mobile Broadband Wireless Access* (MBWA) IEEE802.20 standard extends IEEE802.16 to mobile environments. Once again, IEEE802.20 is based on OFDM. OFDM is also being considered in the IEEE802.11n standard that considers *Multiple-Input Multiple-Output* (MIMO) systems , where multiple antennas are used at the transmitter for the purpose of spatial multiplexing or to provide increased spatial diversity. Finally, OFDM has also found application in *Digital Audio Broadcast* (DAB) and *Digital Terrestrial Video Broadcast* (DVBT) standards in Europe and Japan.

1.2 Wireless Channel

To comprehend the benefits and drawbacks of OFDM, we must first understand the basic characteristics of the radio propagation environment. Radio signals generally propagate according to three mechanisms; reflection, diffraction, and scattering. The appropriate model for radio propagation depends largely on the intended application, and different models are used for the different applications such as cellular land mobile radio, WMANs, and indoor WLANs. In general, however, radio propagation can be roughly characterized by three nearly independent phenomenon; path loss attenuation with distance, shadowing, and multipath-fading. Each of these phenomenon is caused by a different underlying physical principle and each must be accounted for when designing, evaluating, and deploying any wireless system to ensure adequate coverage and quality of service.

1.2.1 Path Loss and Shadowing

It is well known that the intensity of an electromagnetic wave in free space decays with the square of the radio path length, d, such that the received power at distance d is

$$\Omega_p(d) = \Omega_t k \left(\frac{\lambda_c}{4\pi d} \right)^2 \tag{1.2.1}$$

where Ω_t is the transmitted power, λ_c is the wavelength, and k is a constant of proportionality. Although it may seem counter-intuitive, path loss is essential in high capacity frequency reuse systems, the reason being that a rapid attenuation of signal strength with distance permits the bandwidth to be reused within a close physical proximity without excessive interference. Such principles form the basis for cellular mobile radio systems.

Free space propagation does not apply in a typical wireless operating environment, and the propagation path loss depends not only on the distance and wavelength, but also on the antenna types and heights and the local topography. The site specific nature of radio propagation makes theoretical prediction of path loss difficult, except for simple cases such as propagation over a flat, smooth, reflecting surface. A simple path loss model assumes that the received power is

$$\Omega_{p\ (\mathrm{dBm})}(d) = \mu_{\Omega_{p\ (\mathrm{dBm})}}(d_o) - 10\beta \log_{10}(d/d_o) + \epsilon_{(\mathrm{dB})} \quad (\mathrm{dBm}) \tag{1.2.2}$$

where $\mu_{\Omega_{p\ (\mathrm{dBm})}}(d_o) = \mathrm{E}[\Omega_{p\ (\mathrm{dBm})}(d_o)]$ is the average received signal power (in dBm) at a known reference distance. The value of $\mu_{\Omega_{p\ (\mathrm{dBm})}}(d_o)$ depends on the transmit power, frequency, antenna heights and gains, and other factors. The parameter β is called the path loss exponent and is a key parameter that affects the coverage of a wireless system. The path loss exponent lies in the range $3 \le \beta \le 4$ for a typical cellular land mobile radio environment. Usually, the path loss exponents are determined from empirical measurement campaigns.

The parameter $\epsilon_{(\mathrm{dB})}$ in (1.2.2) represents the error between the actual and estimated path loss. It is usually modelled as a zero-mean Gaussian random variable (in decibel units). This error is caused by large terrain features such as buildings and hills, and is sometimes known as shadowing or shadow fading. Shadows are generally modelled as being log-normally distributed, meaning that the probability density function of received power in decibel units, $\Omega_{(\mathrm{dBm})}(d)$, is

$$p_{\Omega_{p\ (\mathrm{dBm})}(d)}(x) = \frac{1}{\sqrt{2\pi}\sigma_\Omega} \exp\left\{ -\frac{(x - \mu_{\Omega_{p\ (\mathrm{dBm})}}(d))^2}{2\sigma_\Omega^2} \right\} \tag{1.2.3}$$

where

$$\mu_{\Omega_{p\ (\mathrm{dBm})}}(d) = \mu_{\Omega_{p\ (\mathrm{dBm})}}(d_o) - 10\beta \log_{10}(d/d_o) \ \ (\mathrm{dBm}) \ . \tag{1.2.4}$$

The parameter σ_Ω is the shadow standard deviation, and usually ranges from 5 to 12 dB, with $\sigma_\Omega = 8$ dB being a typical value for cellular land mobile radio applications. Shadows are spatially correlated, and sometimes modelled as having an exponential decorrelation with distance.

1.2.2 Multipath-Fading

A typical radio propagation environment exhibits multipath, where the plane waves incident on the receiver antenna arrive from many different directions with random amplitudes, frequencies and phases. Since the wavelength is relatively short (approximately 30 cm at 1 GHz), small changes in the location of the transmitter, receiver and/or scattering objects in the environment will cause large changes in the phases of the incident plane wave components. The constructive and destructive addition of plane waves combined with motion results in envelope fading, where the received envelope can vary by as much as 30 to 40 dB over a spatial distance equal to a fraction of a wavelength. Multipath-fading results in a doubly dispersive channel that exhibits dispersion in both the time and frequency domains. Time dispersion arises because the multipath components propagate over transmission paths having different lengths and, hence, they reach the receiver antenna with different time delays. Time dispersion causes intersymbol interference (ISI) that can be mitigated by using a time- or frequency domain equalizer in single-carrier systems, a RAKE receiver in CDMA systems, or frequency domain equalization in OFDM systems. Channel time variations due to mobility are characterized by Doppler spreading in the frequency domain. Such time-variant channels require an adaptive receiver to estimate and track channel the channel impulse response or parameters such as the signal-to-noise ratio that are related to the channel impulse response.

A multipath-fading channel can be modelled as a linear time-variant filter having the complex low-pass impulse response

$$h(t,\tau) = \sum_{n=1}^{N} C_n e^{j\phi_n(t)}\delta(\tau - \tau_n) \tag{1.2.5}$$

where $g(\tau, t)$ is the channel response at time t due to an impulse applied at time $t-\tau$, and $\delta(\,\cdot\,)$ is the dirac delta function. In (1.2.5), C_n, ϕ_n, and τ_n are the random amplitude, phase, and time delay, respectively, associated with the nth propagation path, and N is the total number of arriving multipath components. The time-variant phases $\phi_n(t)$ are given by [2]

$$\phi_n(t) = 2\pi(f_{D,n}t + \phi_n) \tag{1.2.6}$$

where ϕ_n is an arbitrary random phase uniformly distributed on the interval $[-\pi, \pi]$ and

$$f_{D,n} = f_m \cos\theta_n \tag{1.2.7}$$

is the Doppler frequency associated with the nth propagation path, where $f_d = v/\lambda_c$, λ_c is the carrier wavelength, and f_d is the maximum Doppler frequency occurring when the angle of arrival $\theta_n = 0$.

1.3 Interference and Noise

All communication systems are affected by thermal noise or *additive white Gaussian noise* (AWGN). However, wireless systems that employ frequency reuse are also affected by the more dominant *co-channel interference* (CCI). Co-channel interference arises when the carrier frequencies are spatially reused. In this case, the power density spectra of the co-channel signals overlap causing mutual interference. CCI places a limit on the minimum spatial separation that is required such that the carrier frequencies can be reused. CCI is the primary additive impairment in high capacity frequency reuse systems, such as cellular land mobile radio systems. Fig. 1.1 depicts the worst case forward channel co-channel interference situation in a cellular radio environment, which occurs when the mobile station is located at the corner of a cell at the maximum possible distance from its serving base station. With omni-directional antennas, there are six primary co-channel interferers; two at distance $D - R$, two at distance D, and two at distance of $D + R$, where R is the cell radius. Using the simple path loss model in (1.2.4) and neglecting shadowing, the worst case *carrier-to-interference ratio* is

$$\begin{aligned}\frac{C}{I} &= \frac{1}{2}\frac{R^{-\beta}}{(D-R)^{-\beta}+D^{-\beta}+(D+R)^{-\beta}} \\ &= \frac{1}{2}\frac{1}{\left(\frac{D}{R}-1\right)^{-\beta}+\left(\frac{D}{R}\right)^{-\beta}+\left(\frac{D}{R}+1\right)^{-\beta}}\end{aligned} \tag{1.3.1}$$

where β is the propagation path loss exponent. The parameter D/R is called the co-channel reuse factor. In a hexagonal cell deployment D/R is related to the reuse cluster size, N, by $D/R = \sqrt{3N}$. Clearly, the C/I increases with the cluster size, thereby providing better link quality. However, at the same time the available bandwidth (and number of channels) per cell decreases, thereby increasing the new call and handoff call blocking probabilities.

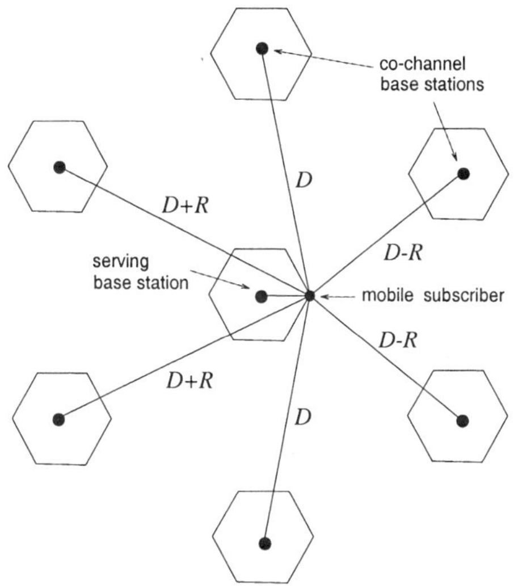

Figure 1.1. Worst case co-channel interference on the forward channel.

Frequency reuse also introduces adjacent channel interference (ACI). This type of interference arises when adjacent cells use channels that are spectrally adjacent to each other. In this case, the power density spectrum of the desired and interfering signals partially overlap. Although ACI degrades link quality it is less severe than CCI, since the interfering signals do not completely overlap in frequency.

1.4 Orthogonal frequency Division Multiplexing

OFDM is a multi-carrier modulation technique where data symbols modulate a parallel collection of regularly spaced sub-carriers. The sub-carriers have the minimum frequency separation required to maintain orthogonality of their corresponding time domain waveforms, yet the signal spectra corresponding to the different sub-carriers overlap in frequency. The spectral overlap results in a waveform that uses the available bandwidth with a very high bandwidth efficiency. OFDM is simple to use on channels that exhibit time delay spread or, equivalently, frequency selectivity. Frequency selective channels are characterized by either their delay spread or their channel coherence bandwidth which measures the channel decorrelation in frequency. The coherence bandwidth is inversely proportional to the *root-mean-square* (rms) delay spread. By choosing the sub-carrier spacing properly in relation to the channel coherence bandwidth, OFDM can be used to convert a frequency selective channel into a parallel collection of frequency flat sub-channels. Techniques that are appropriate for flat fading channels can then be applied in a straight forward fashion.

1.4.1 OFDM Concept

An OFDM modulator can be implemented as an N-point *inverse discrete Fourier transform* (IDFT) on a block of N information symbols followed by *digital-to-analog converter* (DAC) on the IDFT samples [3]. In practice, the IDFT can be implemented with the computationally efficient *inverse fast Fourier transform* (IFFT) as shown in Fig. 1.2. Let $\{s_k, k = 1, \ldots, N\}$ represent a block of N complex data symbols chosen from an appropriate signal constellation such as *quadrature amplitude modulation* (QAM) or *phase shift keying* (PSK). The IDFT of the data block is

$$S_n = \sum_{k=0}^{N-1} s_k \exp\left\{\frac{j2\pi nk}{N}\right\} , \quad n = 0,\ 1,\ \ldots,\ N-1 \ , \tag{1.4.1}$$

yielding the time-domain sequence $\{S_n, n = 1, \ldots, N\}$. To mitigate the effects of ISI caused by channel delay spread, each block of N IFFT coefficients is typically preceded by a *cyclic prefix* (CP) or a guard interval consisting of N_g samples, such that the length of the CP is at least equal to the channel length N_h in samples, where $\mu = \frac{T_h}{T_s} N$, T_h is the length of (continuous) channel, and T_s is the duration of a OFDM block or symbol. The cyclic prefix is simply a repetition of the last N_g IFFT coefficients. Alternatively, a cyclic suffix can be appended to the end of a block of N IFFT coefficients, that is a repetition of the first N_g IFFT coefficients. The guard interval of length N_g is an overhead that results in a power and bandwidth penalty, since it consists of redundant symbols. However, the guard interval is useful for implementing time and frequency synchronization functions in the receiver, since the guard interval contains repeated symbols at a known sample spacing. The time duration of an OFDM symbol is $N + N_g$ times larger than the modulated symbol in a single-carrier system.

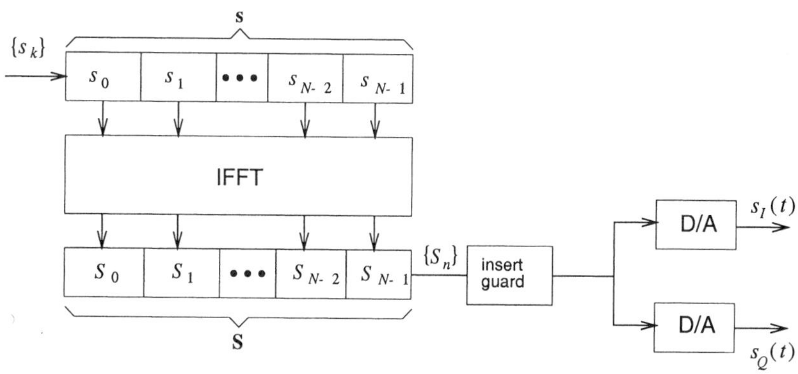

Figure 1.2. Block diagram of basic OFDM transmitter, from [2].

At the receiver, the received complex baseband signal is sampled with an *analog-to-digital converter* (ADC), usually with a sampling interval, $\Delta_T = \frac{T_s}{N}$. Sometimes fractional sampling is used, where the sample period is $\frac{1}{M}\Delta_T$, where M is an integer greater than one. For simplicity, assume here that $M = 1$. Then the combination of the DAC in the transmitter, the waveform channel, and the ADC in the receiver creates an overall discrete-time channel with tap spacing Δ_T. After ADC, the N_g samples received during the guard interval of each OFDM symbol are discarded in the case of a cyclic prefix; in case of a cyclic suffix the N_g received samples at the beginning of an OFDM symbol are replaced with the μ received samples at the end of the OFDM symbol. Under the condition that $N_g \geq N_h$, the *linear* convolution of the transmitted sequence of IFFT coefficients with the discrete-time channel is converted into a *circular* convolution. As a result, the effects of the ISI are completely and easily removed. After removal of the guard interval, each block of N received samples is converted back to the frequency domain using an FFT as shown in Fig. 1.3. The FFT operation performs baseband demodulation. The N frequency domain samples are each processed with a simple one-tap Frequency Domain Equalizer (FDE) and applied to a decision device to recover the data symbols or to a metric computer if error correction coding is used. The one-tap FDE simply multiplies each FFT coefficient by a complex scalar.

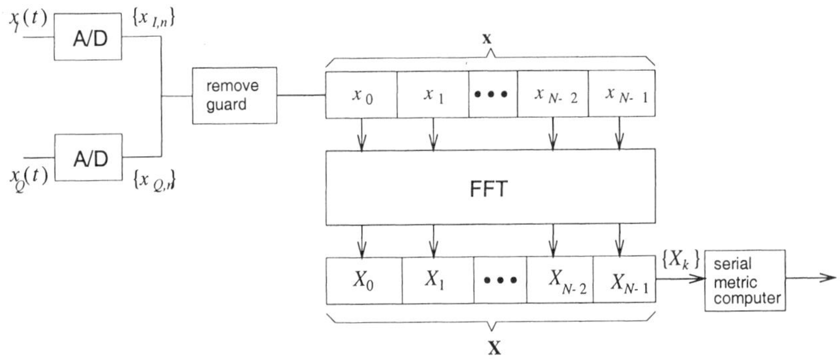

Figure 1.3. Block diagram of basic OFDM receiver, from [2].

Ease of equalization is often touted as the primary advantage of OFDM. However, as mentioned earlier, similar equalization techniques can be applied to single-carrier systems as well. Such a technique is called *single-carrier frequency domain equalization* (SC-FDE). Similar to OFDM, SC-FDE systems transmit data in blocks of N symbols at a time. Each block of N data symbols is preceded by a cyclic prefix of length N_g that is simply a repetition of the last N_g samples in each length-N block. Alternatively, a length-N_g cyclic suffix can be appended to each length-N block that is a repetition of the first G samples in the block. The length-$N + N_g$ block is then applied to a DAC, upconverted to RF, and transmitted over the waveform channel. The received waveform is downconverted to complex baseband and applied to an ADC. The receiver then removes the guard interval in exactly same fashion as an OFDM receiver. Afterwards, the length-N time-domain sample sequence is converted to the frequency domain using an N-point FFT. Frequency domain equalization (FDE) is then applied to the N FFT coefficients. Similar to OFDM, the FDE simply multiplies each FFT coefficient by a complex scalar to perform zero-forcing or minimum mean square error equalization. Afterwards, the equalized samples are converted back to the time-domain using an N-point IFFT and applied to a decision device or metric computer. The overall system complexity of SC-FDE is comparable to OFDM. The main difference is that OFDM uses an IFFT in the transmitter and an FFT in the receiver, while SC-FDE does not perform any transformation in the transmitter but employs an FFT/IFFT pair in the receiver.

1.4.2 Channel Capacity and OFDM

Consider a time-invariant frequency selective channel with transfer function $H(f)$, such that the amplitude response $|H(f)|$ varies across the channel bandwidth W. The power spectral density of the additive Gaussian noise is $S_{nn}(f)$ and may not be flat either. Shannon [4] proved that the capacity of such a non-ideal additive Gaussian noise channel is achieved when the transmitted power $\Omega_t(f)$ is adjusted across the bandwidth W according to

$$\Omega_t(f) = \left\{ \begin{array}{ll} K - S_{nn}(f)/|H(f)|^2 \ , & f \in W \\ 0 \ , & f \notin W \end{array} \right\} \tag{1.4.2}$$

where K is a constant chosen to satisfy the constraint

$$\int_W \Omega_t(f) df \leq \Omega_{\text{av}} \tag{1.4.3}$$

with Ω_{av} being the average available power to the transmitter. One method for approaching the channel capacity is to divide the bandwidth W into N sub-bands of width $W/\Delta f$, where $\Delta f = 1/T_s$ is chosen small enough so that $|H(f)|^2/S_{nn}(f)$ is essentially constant within each sub-band. The signals in each sub-band may then be transmitted with the optimum power allocation $\Omega_t(f)$, while being individually coded to achieve capacity. From (1.4.2), the transmitter power should be high when $|H(f)|^2/S_{nn}(f)$ is large and small when $H(f)/S_{nn}(f)$ is small. In a practical system with a target bit error rate, this implies the use of a larger signal constellation and/or higher rate error correction code in sub-bands where $|H(f)|^2/S_{nn}(f)$ is larger. The technique of using adaptive modulation and coding on the different OFDM sub-carriers requires knowledge of the channel at the transmitter. Such channel knowledge at the transmitter is readily available in OFDM systems that employ *time-division duplexing* (TDD), where the same set of sub-carrier frequencies are alternately used for transmission and reception in each direction of a full-duplex link. Reciprocity ensures that the channel in each direction has the same impulse response provided that the channel time-variations are slow enough. Adaptive bit loading is more complicated in OFDM systems that employ *frequency division duplexing* (FDD) since the reciprocity principle does not apply due to the significant frequency decorrelation of the forward and reverse channels. With FDD, the channel must be estimated and, afterwards, full or partial information of the channel is relayed back to the transmitter for adaptation purposes.

1.5 Synchronization and Channel Estimation

An OFDM receiver operating in the acquisition mode must perform time synchronization, RF and sample clock frequency offset estimation and correction, and initial channel estimation. For systems that transmit information in a packetized or burst mode, these synchronization processes are usually aided by a synchronization preamble consisting of a training sequence or the concatenation of several training sequences. Training sequences in the synchronization preamble have length $N_P = N/I$, where N is the OFDM block length and I is integer. Each training sequence must have an appropriate cyclic guard interval. The synchronization preamble is periodically inserted into the stream of OFDM symbols containing the transmitted data. Synchronization algorithms can also exploit the cyclic guard interval of the OFDM symbol, since the guard interval consists of repeated symbols separated in time. The guard interval can be used to estimated the change in phase due to the channel and oscillator frequency offsets.

After acquisition has been achieved, the receiver enters the data mode and tracks the drift in the RF and sample clock oscillators, and variations in the channel. For applications characterized large Doppler spreads, such as land mobile radio, the channel coefficients are usually tracked by inserting additional OFDM pilot symbols or by using pilot sub-carriers followed by time and frequency interpolation. In WLAN and WMAN applications, the channel is relatively static since the user terminals are usually stationary or slowly moving. However, even in this case channel time variations are expected due to the presence of frequency offsets between transmitter and the receiver local RF and sample clock oscillators. Generally, the components used in the customer premises equipment are low cost and have low tolerances. A typical drift of 10-20 *parts per million* (ppm) is expected in the oscillators. Therefore, an OFDM signal with a bandwidth of 4 MHz, for example, may produce a sampling offset of 80 samples for every one second of transmission. Such RF and sampling frequency offsets cause phase rotation, amplitude distortion and may result in a complete loss of synchronization.

1.6 Peak-to-Average Power Ratio

Consider again the time-domain IFFT coefficients in (1.4.1). For purpose of illustration, suppose data symbols are chosen from *binary phase shift keying* (BPSK), such that $s_k \in \{-1, +1\}$. In this case (1.4.1) can be rewritten as

$$\begin{aligned} S_n &= S_n^{(R)} + jS_n^{(I)} \\ &= \sum_{k=0}^{N-1} s_k \cos\frac{2\pi nk}{N} + j\sum_{n=0}^{N-1} s_k \sin\frac{2\pi nk}{N} , \qquad n = 0,\ 1,\ \ldots,\ N-1\ . \end{aligned} \tag{1.6.1}$$

When N is large, the central limit theorem can be invoked such that $S_n^{(R)}$ and $S_n^{(I)}$ can be treated as independent zero-mean Gaussian random variables with variance $\sigma^2 = N/2$. Under this assumption, the $|S_n|^2, n = 0, \ldots, N-1$ are exponentially distributed random variables and $\mathrm{E}[|S_n|^2] = 2\sigma^2 = N$. Treating the $|S_n|^2$ as independent exponential random variables and applying order statistics, the peak value of $|S_n|^2$, denoted as $S_{\max}^2 = \max_{0 \le n \le N-1} |S_n|^2$, is a random variable with cumulative distribution function $F_{S_{\max}^2}(y) = (1 - e^{-y/N})^N$. The peak-to-average power ratio (PAPR) can be defined as

$$\mathrm{PAPR} = \frac{S_{\max}^2}{\mathrm{E}[|S_n|^2]} = \frac{S_{\max}^2}{N} \tag{1.6.2}$$

Note that the PAPR is a random variable due to the random data $\{s_k, k = 0, \ldots, N-1\}$, and the probability that the PAPR exceeds a specified level z is

$$\begin{aligned} \mathrm{Prob}(\mathrm{PAPR} > z) &= \mathrm{Prob}(S_{\max}^2 > zN) \\ &= 1 - F_{S_{\max}^2}(zN) \\ &= 1 - (1 - e^{-z})^N\ . \end{aligned} \tag{1.6.3}$$

Observe that $\mathrm{Prob}(\mathrm{PAPR} > z)$ increases with the number of sub-carriers, N, for any level $z \ge 1$. We can also compute the mean PAPR using the probability density function $f_{S_{\max}^2}(y) = (1 - e^{-y/N})^{N-1} e^{-y/N}$. The result is

$$\begin{aligned} \overline{\mathrm{PAPR}} &= \int_0^\infty y f_{S_{\max}^2}(y)dy \\ &= N \sum_{k=0}^{N-1} \binom{N-1}{k} \frac{(-1)^k}{(k+1)^2}\ . \end{aligned} \tag{1.6.4}$$

Again, observe that the average PAPR increases with N. For $N = 16$, $\overline{\text{PAPR}} = 3.38$ and for $N = 32$, $\overline{\text{PAPR}} = 4.06$. Note that the above analysis of PAPR is based on a simple central limit theorem approximation. A more accurate and detailed analysis of PAPR is considered in Chapter 6 of this book.

The high PAPR of OFDM signals is a fundamental drawback when compared to single-carrier modulation. Practical power amplifiers are linear only over a finite range of input amplitudes. In order to prevent saturation and clipping of the OFDM signal peaks, the amplifiers must be operated with sufficient "back-off" or head room. The required back-off increases with the PAPR and, hence, the number of sub-carriers N. However, increased back-off reduces the efficiency of the power amplifier. Generally, there are two solutions to the high PAPR problem of OFDM signals. The first is to reduce the PAPR of the transmitted signals through such methods as clipping and filtering, constrained coding, and selective mapping. The second is to use linearization techniques to increase the range of linearity of the power amplifier. Such PAPR reduction methods, however, reduce PAPR while sacrificing complexity and/or bandwidth efficiency.

1.7 MIMO OFDM

Wireless communications systems with multiple transmit and receiving antennas can exploit a dense scattering propagation environment to increase the channel capacity [5], [6]. Generally, there are two categories of MIMO techniques. One category improves the power efficiency by maximizing spatial diversity. Such techniques include delay diversity, and space-time block- and trellis-coding (STBC and STTC). The other category uses linear processing to increase data rate, typically under conditions where full spatial diversity would not be not achieved. Such techniques include Bell Labs Layered Space-Time (BLAST) [7].

The first transmit diversity approach for MIMO systems was delay diversity. Multiple transmit antennas send delayed copies of same signal, and maximum likelihood sequence estimation (MLSE) is used at the receiver to estimate the transmitted sequence. Decision feedback equalizers can also be used in such systems. The delay-diversity approach is simple and can be considered as a particular space-time code.

STBC and STTC can provide full spatial diversity for MIMO systems. STBC utilizes the orthogonality property of the code to achieve full diversity; however, it cannot achieve full-rate transmission when the number of transmit antennas is greater than two. STTC uses enough trellis coding to guarantee full diversity; but the decoding complexity increases exponentially with the number of transmit antennas. Both STBC and STTC lack scalability with number of transmit antennas. As the number of transmit antennas is changed, different space-time codes are needed. STBC and STTC was developed for quasi-static flat fading channels. To apply STBC and STTC to frequency selective fading channels, they must be used in conjunction with other techniques such as equalization or orthogonal frequency division multiplexing (OFDM), that effectively generate one or more flat faded coding channels.

MIMO-OFDM arrangements have been suggested for frequency selective fading channels, where either STBC or STTC is used across the different antennas in conjunction with OFDM. Such approaches can provide very good performance on frequency selective fading channels. However, the complexity can be very high, especially for for a large number of transmit antennas. Another approach uses delay diversity together with OFDM on flat fading channels. For frequency selective fading channels, a cyclic delay diversity approach can be used with OFDM something we call multi-carrier delay diversity modulation (MDDM). Our work has shown that full spatial diversity can be achieved for MDDM on flat fading channels, provided that the minimum Hamming distance of the outer (pre-IFFT) error correcting code either equals or exceeds the number of transmit antennas, and to obtain good coding gain, a simple block interleaving will do [8]. Unlike STBC and STTC the scheme is scalable; the number of transmit antennas can be changed without changing the error correcting code. MDDM can easily handle frequency selective fading using MRC, or other type of combining depending on the environment.

1.8 Outline of This Book

Chapter 2 introduces the basic concepts of OFDM including FFT implementation, comparison with single-carrier modulation, and basic system design. Chapter 2 also analyzes the impact of the various impairments introduced by the wireless channel on OFDM performance. These impairments include frequency offsets in the RF and sample clock oscillators, channel time variations, sample timing offsets, multipath delay spread, and amplifier non-linearities. Finally, Chapter 2 briefly considers other approaches to multi-carrier modulation.

Chapter 3 is concerned with performance optimization of OFDM systems with the objective of maximizing bandwidth efficiency. These optimization methods often rely on full or partial information of the channel transfer function at the transmitter. This information is used, for example, to adaptively load the OFDM sub-carriers, a technique commonly referred to as discrete multitone modulation (DMT). Chapter 3 discusses issues of partitioning the available bandwidth, (adaptive) loading of the parallel sub-channels, and optimization through error correction coding.

Chapter 4 addresses the synchronization of OFDM signals, including details of time and frequency synchronization. The time and frequency synchronization methods considered in Chapter 4 use either pilot-based methods or non-pilot-based methods that exploit the redundant symbols in the cyclic guard interval. Time and frequency synchronization processes can be implemented in a separate or joint fashion. Generally, the synchronization functionality has two modes, acquisition and tracking. Finally, the problem of sampling clock offset estimation and correction is considered.

Chapter 5, is concerned with channel estimation. With OFDM the channel transfer function must be estimated at each sub-carrier. There are three basic methods for channel estimation. The first is pilot-symbol aided estimation, where known pilot symbols are transmitted in the 2-D time-frequency grid. The second is decision directed estimation, where decisions on data symbols are used to update the channel estimates. The third category are blind channel estimation techniques that do not rely on the transmission of known data symbols.

Chapter 6 considers the important issue of OFDM peak-to-average power ratio (PAPR) reduction techniques. As mentioned earlier, multi-carrier modulation techniques such as OFDM exhibit a high PAPR. One method for reducing the PAPR is to intentionally limit or clip the OFDM waveform prior to amplification. Such methods distort the OFDM waveform that may result in a bit error rate degradation. Another class of methods uses distortionless methods such as constrained coding and selective mapping. Such methods do not distort the OFDM signal, but at the cost of bandwidth or computational complexity.

Finally, Chapter 7 considers MIMO-OFDM with an emphasis on synchronization. A complete suite of signal acquisition and tracking algorithms is presented for MIMO-OFDM systems. Our algorithms use a preamble to perform signal acquisition, which consists of time synchronization, RF and sampling frequency offset estimation, and channel estimation. This is followed by open loop tracking of the RF and sampling frequency offsets, and the channel. We suggest a preamble structure and pilot matrix design for MIMO-OFDM systems with any number of transmit antennas that enable our algorithms to work efficiently. Simulation results are presented that account for all required signal acquisition and tracking functions in a system very similar to IEEE 802.16a.

1.9 Summary and Further Reading

This introductory chapter briefly introduced the main concepts of OFDM, while provided the motivation and outlining the scope of the remainder of the book.

Several systems have previously used OFDM or other multicarrier techniques [3], [9], [10], [11], [12]. In particular, in the early 1960's, this multicarrier techniques were used in several high-frequency military systems, such as KINEPLEX [13], ANDEFT [14], KATHRYN [15], [16], where fast fading was not a problem. Similar modems have found applications in voice bandwidth data communications [17] to alleviate the degradations caused by an impulsive noise environment. More information on multicarrier related research in 1960's and 1970's can be found in [18], [19], [20] and the references therein. In 1985, Cimini first investigated OFDM for mobile wireless communications systems in [21]. In [22], Casas and Leung discussed the application of multicarrier techniques on mobile radio FM channels. Willink and Wittke [23] and Kalet [24] investigated the theoretical capacity of multicarrier systems. In 1990, Bingham [25] studied the performance and implementation complexity of OFDM and concluded that the time for OFDM had come. The application of original OFDM, clustered OFDM, and MC-CDMA in mobile wireless systems can be found in [26], [27], [28], [29]. For MIMO and BLAST, see the work of Foschini, [5], [7], [6]. Recently, serval books [30], [31], [32] on OFDM have been published.

Chapter 2

BASIC CONCEPTS

Ye (Geoffrey) Li

In this chapter, we first introduce the basic concepts of *orthogonal frequency division multiplexing* (OFDM), discuss the advantages and disadvantages compared single-carrier modulation, and present an implementation example. We then address various impairments of wireless channels on OFDM systems. Finally, we briefly describe other forms of multicarrier modulation.

2.1 Basic OFDM

High data-rate is desired in many applications. However, as the symbol duration reduces with the increase of data-rate, the systems using single-carrier modulation suffer from more severe *intersymbol interference* (ISI) caused by the dispersive fading of wireless channels, thereby needing more complex equalization. OFDM modulation divides the entire frequency selective fading channel into many narrow band flat fading subchannels[1] in which high-bit-rate data are transmitted in parallel and do not undergo ISI due to the long symbol duration. Therefore, OFDM modulation has been chosen for many standards, including *Digital Audio Broadcasting* (DAB) and terrestrial TV in Europe, and *wireless local area network* (WLAN). Moreover, it is also an important technique for high data-rate transmission over mobile wireless channels. Here we introduce the basic concepts of OFDM.

2.1.1 OFDM

OFDM was first introduced in [3], which is the form used in all present standards. It can be regarded as a time-limited form of multicarrier modulation.

Let $\{s_k\}_{k=0}^{N-1}$ be the complex symbols to be transmitted by OFDM mod-

[1]Subchannel is sometimes also called subcarrier or tone.

With the cyclic extension, the actual OFDM symbol duration is increased from T_s to $T = T_s + T_g$. In the following discussion, the cyclic suffix extension in Fig. 2.1 (b) is assumed. However, the results can be also applied to the other types of cyclic extension.

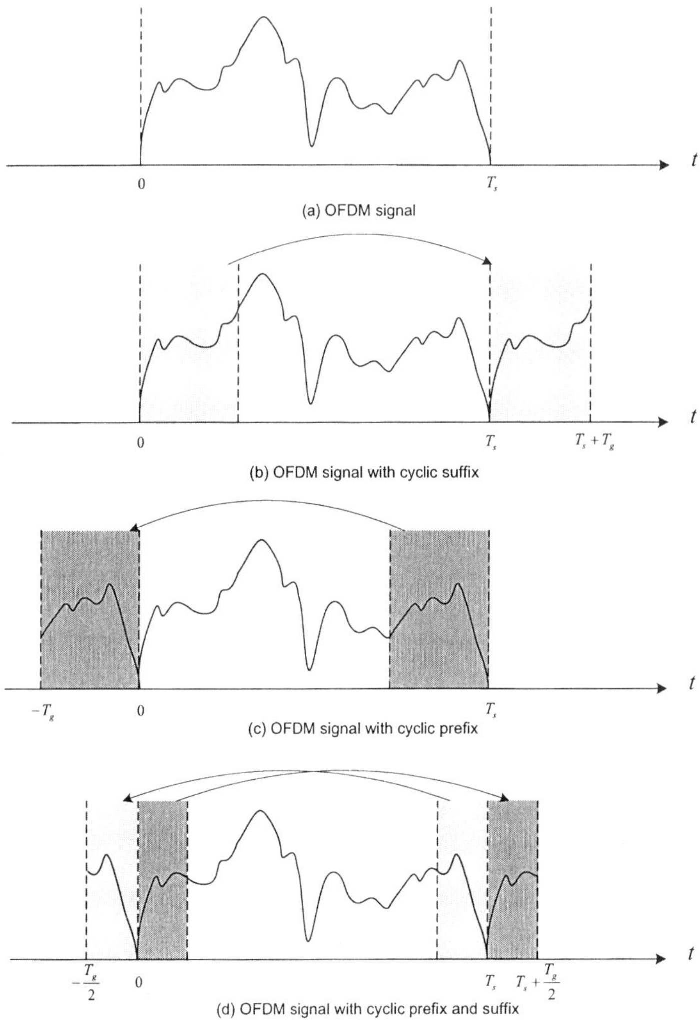

Figure 2.1. OFDM signal with different cyclic extensions.

Because $s(t)$ in (2.1.1) is a summation of truncated complex exponential functions with different frequencies, the power density spectrum of $s(t)$ consists of $|\sin(f)/f|^2$-shaped spectra, as sketched in Fig. 2.2.

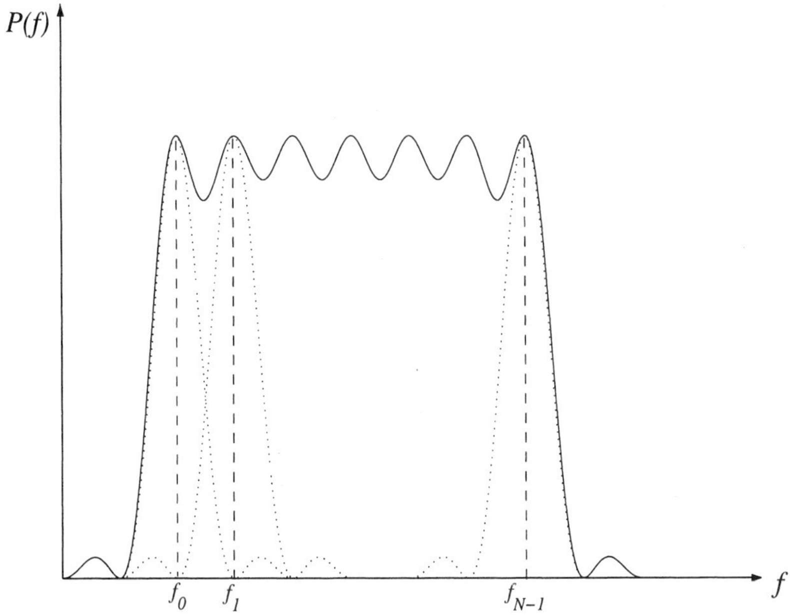

Figure 2.2. Power spectrum of OFDM Signal.

Fig. 2.2 shows that, for an OFDM signal consisting of N subchannels, the signal bandwidth is about $(N+1)\Delta f$. Since the transmission rate of each subchannel is $\frac{1}{T}$ symbols/sec., the *total transmission rate* of OFDM signal is $\frac{N}{T}$ symbol/sec. Therefore, the bandwidth efficiency of the OFDM system is

$$\begin{aligned}
\eta &= \frac{N/T}{(N+1)\Delta f} \\
&= \frac{N/(T_s+T_g)}{(N+1)/T_s} \\
&= \frac{1}{1+\frac{1}{N}}\frac{1}{1+\frac{T_g}{T_s}}, \qquad (2.1.7)
\end{aligned}$$

in symbols/sec/Hz. For most practical OFDM systems, N is much larger than 1 and the guard interval or cyclic extension is much smaller than the

OFDM symbol duration, so $W \approx 1$. If each symbol carries k bit information, the bandwidth efficiency will be k bits/sec/Hz.

2.1.4 Comparison with Single-Carrier

As indicated in [33], the dispersive Rayleigh fading in wireless channels limits the highest data rate of single-carrier systems. To reduce the effect of ISI in unequalized systems, the symbol duration must be much larger than the delay spread of wireless channels. In OFDM, the entire channel is divided into many narrow subchannels, which are transmitted in parallel, thereby increasing the symbol duration and reducing the ISI. Therefore, OFDM is an effective technique for combating multipath fading and for high-bit-rate transmission over mobile wireless channels.

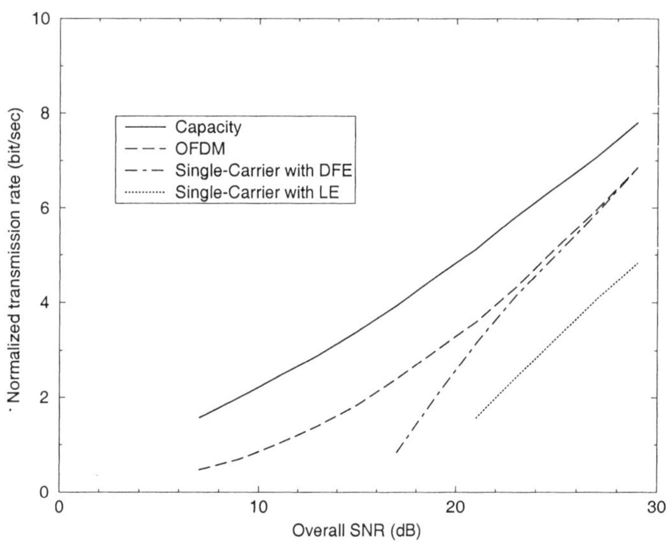

Figure 2.3. Transmission rate of OFDM and single-carrier systems for AWGN channel.

Figure 2.3, from [23], compares the transmission rate of an OFDM system with that of a single-carrier system using a *decision-feedback equalizer* (DFE) or a *linear equalizer* (LE). Note that the curve for the DFE in the figure is obtained by assuming that the feedback symbols at the DFE are error free, so it is in fact an upper bound for the transmission rate of the DFE. From the figure, for the same overall SNR, the normalized transmission rate of the OFDM system is much higher than that of the single-carrier system.

2.1.5 Design Example

Here, we present a simple example to demonstrate the design of an OFDM system. Consider a sample system that is required to transmit 1.2 Mbits/sec using *quadrature PSK* (QPSK) over an 800 kHz bandwidth in a wireless environment with a maximum delay span up to 40 μsec. Note that from the results in [33], for a channel with a 40-μsec delay span, the maximum symbol rate is only 5 kbaud. It is obvious that the required transmit rate can not be obtained by a single-carrier system. However, it can be easily achieved using OFDM modulation.

To construct the OFDM signal, we assume the entire channel bandwidth, 800 kHz, is divided into N=128 subchannels or tones. Thus, the subchannel or subchannel space is 6.25 kHz. Let the 4 subchannels on each end be used as guard tones to facilitate filtering, and the rest (120 tones) are used to transmit data. To make the tones orthogonal to each other, the symbol duration is $T_s = 160$ μsec. An additional $T_g = 40$ μsec cyclic extension is used to provide protection from intersymbol interference due to channel multipath delay spread. This results in a total block length $T = 200$ μsec and a subchannel symbol rate $r_b = 5$ kbaud. For QPSK, each symbol carries 2 bit information; consequently, the data transmission rate of the OFDM system is

$$R_b = \frac{120 \times 2 \text{ bits}}{200 \ \mu\text{sec}} = 1.2 \text{ Mbits/sec.}$$

2.1.6 Baseband versus Passband

In Sections 2.1.1-2.1.5, the OFDM signals are complex baseband signals. However, in wireless communication systems, *complex* baseband signals must be converted into *real* passband signals. In this section, we briefly introduce the baseband and passband representations.

The baseband signal, $s(t)$, is usually a complex function of time. Therefore, it can be written into rectangular form as

$$s(t) = s_I(t) + \jmath s_Q(t),$$

where the real part, $s_I(t)$, is called *in-phase component* of the baseband signal; and the imaginary part, $s_Q(t)$, is called *quadrature component.* For the baseband OFDM signal in (2.1.1), we have

$$\begin{aligned} s(t) &= \sum_{k=0}^{N-1} \left(\Re\{s_k\} \cos(2\pi f_k t) - \Im\{s_k\} \sin(2\pi f_k t) \right) \\ &\quad + \jmath \sum_{k=0}^{N-1} \left(\Im\{s_k\} \cos(2\pi f_k t) + \Re\{s_k\} \sin(2\pi f_k t) \right), \end{aligned}$$

and therefore,

$$s_I(t) = \sum_{k=0}^{N-1} \left(\Re\{s_k\} \cos(2\pi f_k t) - \Im\{s_k\} \sin(2\pi f_k t) \right),$$

and

$$s_Q = \sum_{k=0}^{N-1} \left(\Im\{s_k\} \cos(2\pi f_k t) + \Re\{s_k\} \sin(2\pi f_k t) \right),$$

where $\Re\{s_k\}$ and $\Im\{s_k\}$ denote the real and imaginary parts of the complex symbol s_k, respectively.

Figure 2.4 shows conversion between baseband and passband signals. From the figure, the passband signal can be expressed as

$$\begin{aligned} s_p(t) &= \Re\{s(t) e^{\jmath 2\pi f_c t}\} \\ &= s_I(t) \cos(2\pi f_c t) - s_Q(t) \sin(2\pi f_c t), \end{aligned}$$

where f_c is the carrier frequency of a communication system. It is assumed that the variations of the signal are much slower than the carrier frequency. For OFDM, the passband signal can be further simplified as

$$\begin{aligned} s_p(t) &= s_I(t) \cos(2\pi f_c t) - s_Q(t) \sin(2\pi f_c t) \\ &= \sum_{k=0}^{N-1} \left(\Re\{s_k\} \cos(2\pi (f_c + f_k) t) - \Im\{s_k\} \sin(2\pi (f_c + f_k) t) \right). \end{aligned}$$

If we denote the magnitude and the phase of complex symbol, s_k, as d_k and θ_k, respectively, that is, $s_k = d_k e^{\jmath \theta_k}$, then

$$s_p(t) = \sum_{k=0}^{N-1} d_k \cos(2\pi (f_c + f_k) t + \theta_k).$$

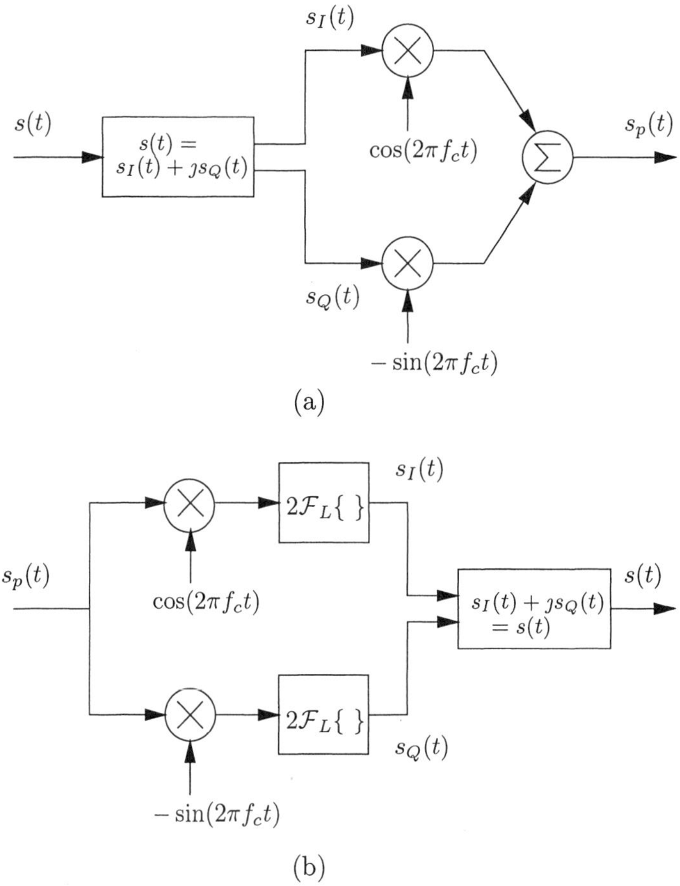

Figure 2.4. Baseband versus passband: (a) baseband to passband conversion, and (b) passband to baseband conversion.

At the receiver, the baseband signal can be obtained from the passband signal. Figure 2.4 (b) shows conversion from the baseband signal to the passband signal, where $\mathcal{F}_L$ represents a low-pass filter operation. From Fig. 2.4 (b), we have

$$\begin{aligned} s(t) &= 2\mathcal{F}_L\{s_p(t)\cos(2\pi f_c t) - js_p(t)\sin(2\pi f_c t)\} \\ &= 2\mathcal{F}_L\{s_p(t)e^{-j2\pi f_c t}\}. \end{aligned}$$

Similarly, the linear distortion of any physical channel can be also equivalent to a *baseband (complex) channel*, $h(t)$, so that the baseband channel output is the convolution of the baseband signal and the baseband channel impulse response, *i.e.*

$$x(t) = s(t) \circledast h(t) = \int_{-\infty}^{+\infty} s(\tau)h(t-\tau)d\tau.$$

More detailed information about baseband and passband conversion can be obtained from Proakis [34].

2.2 Impairments of Wireless Channels to OFDM Signals

In this section, we introduce the impairments in OFDM systems, including Doppler shift, dispersive fading, timing and frequency offsets, sampling clock offset, and nonlinear distortion due to large *peak-to-average-power ratio* (PAPR) of the OFDM signal.

2.2.1 Time-Varying Impairments

Both Doppler shift and frequency offset can be modelled as time-varying impairments. Here we first derive a general expression for the effect of the time-varying impairments and then discuss the effect of Doppler shift and frequency offset, respectively.

Consider an OFDM signal,

$$s(t) = \sum_k s_k e^{j2\pi f_k t}, \quad 0 \le t \le T_s,$$

where $f_k = f_o + k\Delta f$ and s_k is the signal transmitted over the k-th subchannel. If there is a multiplicative time-varying distortion, $\gamma(t)$, that is caused by either frequency offset or Doppler spread, the received signal will be

$$x(t) = \gamma(t)s(t). \tag{2.2.1}$$

The demodulated signal will be [2]

$$\begin{aligned} X_m &= \frac{1}{T_s}\int_0^{T_s} x(t)e^{-\jmath 2\pi f_m t}dt \\ &= \frac{1}{T_s}\int_0^{T_s} \gamma(t)\sum_k s_k e^{\jmath 2\pi f_k t}e^{-\jmath 2\pi f_m t}dt \\ &= \sum_k \left\{\frac{1}{T_s}\int_0^{T_s} \gamma(t)e^{-\jmath 2\pi (f_m - f_k)t}dt\right\} s_k \\ &= a_0 s_m + \underbrace{\sum_{k\neq m} a_{m-k}s_k}_{ICI}, \end{aligned} \tag{2.2.2}$$

where a_l is defined as

$$a_l \triangleq \frac{1}{T_s}\int_0^{T_s} \gamma(t)e^{-\jmath 2\pi l\Delta f t}dt. \tag{2.2.3}$$

a_0 is usually a complex number, whose magnitude and phase represent the attenuation and the phase shift on the desired signal, respectively. a_l's, for $l \neq 0$, are complex gains of the *interchannel interference* (ICI). If $\gamma(t)$ is not a constant, then $a_l \neq 0$ for some $l \neq 0$, and ICI exists.

Effect of Frequency Offset

If there is a frequency offset, δf, between the transmitter and the receiver, then $\gamma(t)$ in (2.2.1) is a deterministic function and can be expressed as

$$\gamma(t) = e^{\jmath 2\pi \delta f t} = e^{\jmath 2\pi\alpha\Delta f t},$$

where $\alpha = \frac{\delta f}{\Delta f}$. From (2.2.3), we have

$$\begin{aligned} a_l &= \frac{1}{T_s}\int_0^{T_s} e^{\jmath 2\pi\alpha\Delta f t}e^{-\jmath 2\pi l\Delta f t}dt \\ &= \frac{\sin[\pi(l-\alpha)]}{\pi(l-\alpha)}e^{-\jmath\pi(l-\alpha)} \\ &= -\frac{\sin(\pi\alpha)}{\pi(l-\alpha)}e^{\jmath\pi\alpha}. \end{aligned} \tag{2.2.4}$$

[2]For simplicity, an integral is used here instead of the DFT. However, the integration is almost the same as the DFT for systems with a large number of carriers.

Let $\alpha = k_o + \epsilon$, where k_o is an integer and ϵ is a fractional number with $|\epsilon| \leq 1/2$, then

$$a_l = -\frac{\sin(\pi\epsilon)}{\pi(l - k_o - \epsilon)} e^{+\jmath\pi\epsilon}. \qquad (2.2.5)$$

When $\alpha \leq 1/2$ ($k_o = 0$ and $\epsilon = \alpha$), $0 < |a_l| \leq |a_0|$. The desired signal is the dominant component in the demodulated signal. However, there is also ICI since $a_l \neq 0$ for $l \neq 0$. When α is an integer ($k_o = \alpha$ and $\epsilon = 0$), $a_{k_o} = 1$, $a_l = 0$ for $l \neq k_o$, and $X_l = s_{l-k_o}$. Therefore, the frequency offset causes a simple tone shift and there is no ICI. In general, neither k_o nor ϵ is zero; consequently, tone shift, attenuation, phase shift, and ICI all exist. However, the signal distortion caused by frequency offset is deterministic. Furthermore, once the frequency offset is known, its effect can be corrected. Chapter 4 will present techniques for frequency offset estimation and correction in OFDM systems, where coarse and fine synchronization is used to cancel the effects of k_o and ϵ, respectively.

Effects of Doppler Shift

For channels with Doppler spread, $\gamma(t)$ is can be modelled as a zero-mean and narrow-band *wide-sense stationary* (WSS) stochastic process. For the classical Doppler spectrum [35], the spectral density of $\gamma(t)$ is

$$P_J(f) = \begin{cases} \frac{1}{\pi f_d} \frac{1}{\sqrt{1-(\frac{f}{f_d})^2}}, & \text{if } |f| < f_d, \\ 0, & \text{otherwise}, \end{cases} \qquad \text{(Classical)}$$

where f_d is the maximum Doppler frequency. Two extreme cases of the Doppler spectrum are the uniform and the two-path models, which have been studied in [36]. For these two models, the spectral densities are

$$P_u(f) = \begin{cases} \frac{1}{2f_d} & \text{if } |f| < f_d, \\ 0, & \text{otherwise}, \end{cases} \qquad \text{(Uniform)}$$

and

$$P_t(f) = \frac{1}{2}[\delta(f + f_d) + \delta(f - f_d)], \quad \text{(Two-path)}$$

respectively. The correlation function of $\gamma(t)$, defined as $r(\tau) = E\{\gamma(t + \tau)\gamma^*(t)\}$, is easily obtained as

$$r(\tau) = \mathcal{F}^{-1}\{P(f)\}.$$

The correlation functions for the three models given above are

$$r_J(\tau) = J_0(2\pi f_d \tau), \quad \text{(Classical)}$$

$$r_u(\tau) = \text{sinc}(f_d \tau), \quad \text{(Uniform)}$$

and

$$r_t(\tau) = \cos(2\pi f_d \tau), \quad \text{(Two-path)}$$

respectively, where $J_0(x)$ is the zero-order Bessel function of the first kind and

$$\text{sinc}(x) \triangleq \frac{\sin(\pi x)}{\pi x}.$$

It should be noted that the two-path model corresponds to an OFDM system with a fixed frequency offset of f_d Hz.

Since $\gamma(t)$ is a stochastic process, from (2.2.3), a_l is a random variable. Furthermore, it is proved in [37] that a_l is zero mean and with variance

$$\sigma_l^2 \triangleq E\{|a_l|^2\} = \int_{-1}^{1} r(T_s x)(1-|x|)e^{-j2\pi l x}dx,$$

and the total ICI power due to Doppler spread is

$$\begin{aligned} P_{ICI} &\triangleq E\left|\sum_{l\neq 0} a_l s_{m-l}\right|^2 \\ &= \int_{-1}^{1} (1-|x|)\left(1-r(T_s x)\right)dx \\ &= 1-\int_{-f_d}^{f_d} P(f)\text{sinc}^2(fT_s)df. \end{aligned} \quad (2.2.6)$$

Once the time-domain correlation or the Doppler spectral density of the time-varying channel is known, the ICI power can be calculated. For the classical model, we have

$$P_{ICI} = 1 - \int_{-1}^{1} (1-|x|)J_0(2\pi f_d T_s x)dx, \quad (2.2.7)$$

which was first derived by Russell and Stüber [38]. For the uniform and two-path models, we have

$$P_{ICI} = 1 - \frac{1-\cos(2\pi f_d T_s) - 2\pi f_d T_s \text{Si}(2 f_d T_s)}{2(\pi f_d T_s)^2}, \quad (2.2.8)$$

Table 2.1. α_1's and α_2's for different time-varying models

model	α_1	α_2
Classical	1/2	3/8
Uniform	1/3	1/5
Two-path	1	1

and

$$P_{ICI} = 1 - \text{sinc}^2(f_d T_s), \tag{2.2.9}$$

respectively, where

$$\text{Si}(x) = \pi \int_0^x \frac{\sin(\pi u)}{\pi u} du = \pi \int_0^x \text{sinc}(u) du.$$

The expressions in (2.2.8) and (2.2.9) were first derived by Robertson and Kaiser [36].

Using the expressions derived above, the ICI power can be exactly calculated. However, the exact expressions are complicated and do not easily provide much insight. Furthermore, in many instances, the exact time-domain correlation or power spectrum is not available. Here, we introduce several bounds on the ICI power, which are derived in [37]. These bounds are less complicated and the insight is more readily obtained.

It has been proved in [37] that the ICI power has the following lower and upper bounds:

$$P_{ICI} \geq \frac{\alpha_1}{12}(2\pi f_d T_s)^2 - \frac{\alpha_2}{360}(2\pi f_d T_s)^4, \tag{2.2.10}$$

and

$$P_{ICI} \leq \frac{\alpha_1}{12}(2\pi f_d T_s)^2, \tag{2.2.11}$$

where α_i's, for $i = 1,\ 2$, are defined as

$$\alpha_k \triangleq \frac{1}{f_d^{2k}} \int_{-f_d}^{f_d} f^{2k} P(f) df = \frac{2}{f_d^{2k}} \int_0^{f_d} f^{2k} P(f) df.$$

The constants α_1 and α_2 are easy to calculate and are given in Table 2.1 for the three models introduced before.

It should be indicated that without knowing the Doppler spectrum, α_1 and α_2 can be also evaluated using other approaches. For example, it can be proved that

$$E\left\{\left|\gamma^{(k)}(t)\right|^2\right\} = (2\pi)^{2k} \int_{-f_d}^{f_d} f^{2k} P(f) df,$$

where $\gamma^{(k)}(t) = \frac{d^k \gamma(t)}{dt^k}$. Then, α_k can be evaluated by

$$\alpha_k = \frac{E\left\{\left|\gamma^{(k)}(t)\right|^2\right\}}{(2\pi f_d)^{2k}},$$

which is much simpler than obtaining by the Doppler spectrum.

From the definition of α_1, it is clear that $\alpha_1 \leq 1$. Using this fact, together with (2.2.11), we can obtain a *universal upper bound* on the ICI power, which depends only on $f_d T_s$,

$$P_{ICI} \leq \frac{1}{12}(2\pi f_d T_s)^2. \tag{2.2.12}$$

This universal upper bound can be used in OFDM systems with any Doppler spectra, including with frequency offset. Since α_1 is no more than 1, the above bound is looser than the bound in (2.2.11). However, (2.2.12) is much easier to calculate since it only depends on $f_d T_s$. For the two-path model, $\alpha_1 = 1$, and the universal bound is also the tight bound.

In the above discussions, we have introduced tight and universal bounds for time-varying flat fading channels. For OFDM systems with a proper cyclic extension, the exact expressions for the ICI power and the various bounds are also applicable to time-varying dispersive channels.

In Figure 2.5, we compare the upper and the lower tight bounds and the universal bound with the exact value of the ICI power for the classical Doppler spectrum. From Fig. 2.5, the tight bounds are very close to the exact ICI power. When designing OFDM systems, if symbol duration, T_s, is chosen so that $f_d T_s$ is very small, then the ICI due to Doppler spread will be negligible. In the design example in Section 2.1.1, $T_s = 160$ μsec, then $P_{ICI} < 0.168\%$(-27.7 dB) when $f_d = 200$ Hz. Therefore, the P_{ICI} is much less than the noise or co-channel interference level.

Though Doppler shift has only minor effect on the ICI of OFDM signals when $f_d T_s$ is very small. It *does* make the channel parameters to vary from one OFDM block to another. Therefore, when channel parameters are used for coherent detection or adaptive antenna arrays in mobile wireless systems, channel tracking is still essential in most environments. In Chapter 5, we describe different channel parameter estimation approaches.

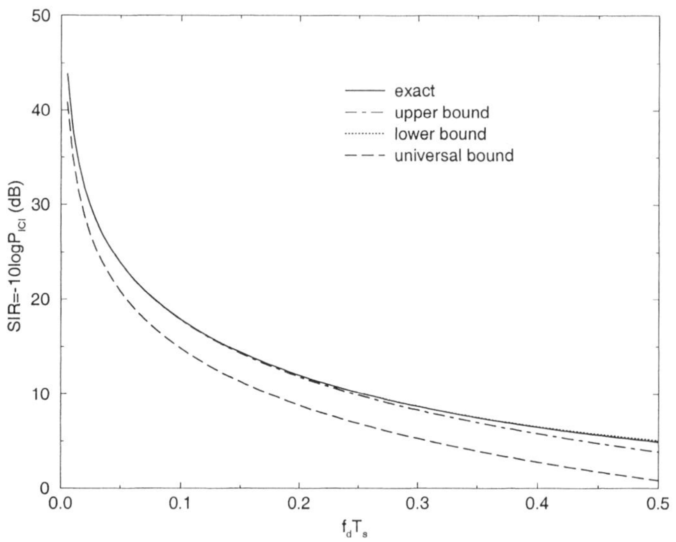

Figure 2.5. Comparison of exact P_{ICI}, its upper and lower bounds, and universal bound for the classical model.

2.2.2 Effect of Sampling Clock Offset

In practice, the sampling clock at the receiver is often different from that at the transmitter. The sampling clock difference will degrade the performance of the systems. As discussed in Section 2.1.2, OFDM signal can be simply demodulated by performing DFT to the samples if the continuous signal $s(t) = \sum_{k=0}^{N-1} s_k e^{j2\pi f_k t}$ is sampled at a sampling interval of $T_{sa} = \frac{T_s}{N}$ at the receiver. However, if the sampling interval at the receiver is $T'_{sa} = T_{sa} + \beta T_{sa}$, other than T_{sa}, then the samples will be $\{s(nT'_{sa})\}_{n=0}^{N-1}$. If DFT is still used for OFDM demodulation [39], then we will have

$$\begin{aligned} X_m &= \frac{1}{N}\sum_{n=0}^{N-1} s(nT'_{sa})e^{-\jmath 2\pi \frac{nm}{N}} \\ &= a_{m,m}s_m + \sum_{k\neq m} a_{m,k}s_k, \end{aligned} \tag{2.2.13}$$

where

$$a_{m,k} = \frac{\sin[\pi(k-m+\beta k)]}{N\sin[\frac{\pi}{N}(k-m+\beta k)]}e^{\jmath\pi\frac{N-1}{N}(k-m+\beta k)}.$$

From (2.2.13), the demodulated signal, X_m, consists of the desired symbol component, s_m, and ICI. The desired symbol is modified by

$$\begin{aligned} a_{m,m} &= \frac{\sin(\pi\beta m)}{N\sin(\frac{\pi}{N}\beta m)}e^{\jmath\pi\frac{N-1}{N}\beta m} \\ &\approx e^{\jmath\pi\beta m}, \end{aligned}$$

which is a subchannel-dependent phase rotation. In the above approximation, we have assumed that $N \gg 1$ and $\beta N \ll 1$, which is usually true in practice. It can be also shown in [39] that the average ICI power at the m-th subchannel is

$$\begin{aligned} P_{ICI}[m] &= E\left|\sum_{k\neq m} a_{m,k}s_k\right|^2 \\ &\approx \frac{\pi^2}{3}\beta^2 m^2, \end{aligned}$$

which is also subchannel-dependent.

In Section 4.4, we will briefly discuss sampling clock offset estimation and correction.

2.2.3 Effect of Timing Offset

In Figure 2.6, the effect of timing offset on an OFDM signal is shown. When there is a timing offset, $\tau > 0$[3], between the transmitter and the receiver, the observed signal will be

$$\bar{s}(t,\tau) = \begin{cases} s(t+\tau), & \text{if } 0 \le t \le T_s - \tau, \\ e(t-T_s+\tau), & \text{if } T_s - \tau \le t \le T_s. \end{cases}$$

[3]Even through we assume that $\tau > 0$ here, the derived result can be also used to the case with $\tau < 0$

where $e(t)$ denotes interference due to the timing offset. For OFDM systems without a guard interval or cyclic extension, $e(t)$ is a part of next OFDM block, as shown in Fig. 2.6 (a). For OFDM systems with a guard interval larger than τ, $e(t) = 0$, as shown in Fig. 2.6 (b). For OFDM systems with a cyclic extension period larger than τ, $e(t) = s(t)$, as shown in Fig. 2.6 (c). For OFDM systems with a cyclic extension period or guard interval less than τ, $e(t)$ is a mixture of the above two or three signals.

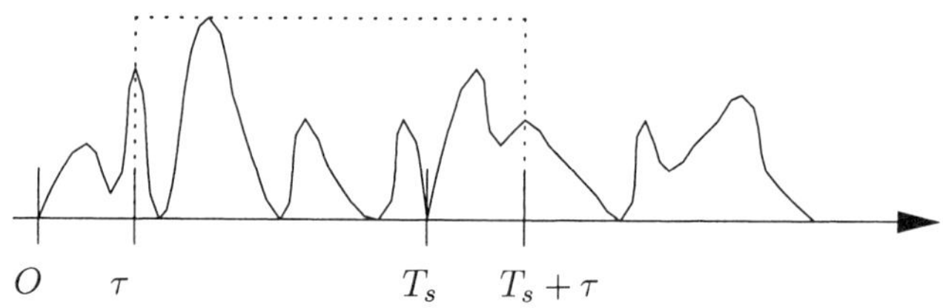

(a) $e(t)$ is a part of the next block

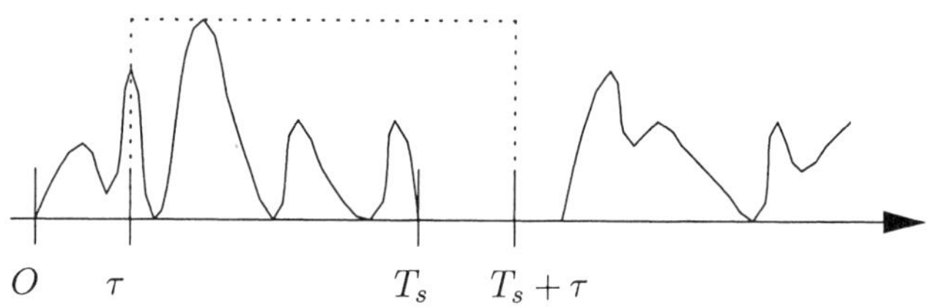

(b) $e(t)$ is a part of guard interval

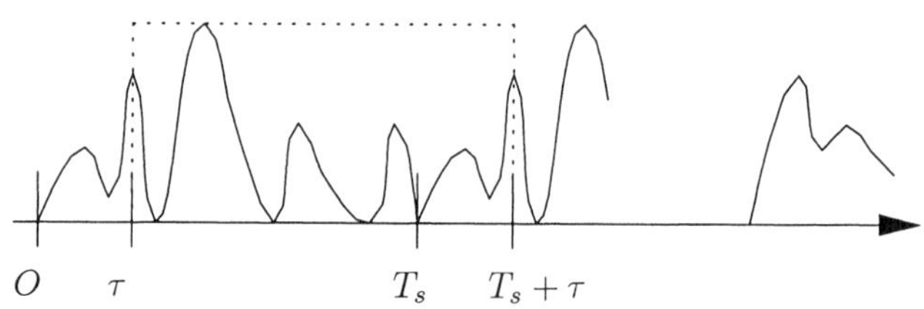

(c) $e(t)$ is a part of cyclic extension

Figure 2.6. Effect of timing offset on OFDM signal.

When there is a timing offset, the demodulated signal at the receiver is

$$\begin{aligned}
X_m &= \frac{1}{T_s}\int_0^{T_s} \bar{s}(t,\tau)e^{-\jmath 2\pi f_m t}dt \\
&= \frac{1}{T_s}\int_0^{T_s-\tau} s(t+\tau)e^{-\jmath 2\pi f_m t}dt + \frac{1}{T_s}\int_{T_s-\tau}^{T_s} e(t-T_s+\tau)e^{-\jmath 2\pi f_m t}dt \\
&= \frac{1}{T_s}\int_{\tau}^{T_s} s(t)e^{-\jmath 2\pi f_m (t-\tau)}dt + \frac{1}{T_s}\int_0^{\tau} e(t)e^{-\jmath 2\pi f_m (t-\tau)}dt \\
&= \frac{1}{T_s}\int_0^{T_s} s(t)e^{-\jmath 2\pi f_m (t-\tau)}dt - \frac{1}{T_s}\int_0^{\tau} s(t)e^{-\jmath 2\pi f_m (t-\tau)}dt \\
&\quad + \frac{1}{T_s}\int_0^{\tau} e(t)e^{-\jmath 2\pi f_m (t-\tau)}dt \\
&= s_m e^{\jmath 2\pi f_m \tau} + \frac{1}{T_s}\int_0^{\tau} [e(t)-s(t)]e^{-\jmath 2\pi f_m (t-\tau)}dt. \qquad (2.2.14)
\end{aligned}$$

From (2.2.14), timing offset introduces a phase shift to the desired symbol component and an additive interferer depending on whether a cyclic extension or a null interval is used. When the system has no guard interval or cyclic extension, then as shown in Fig. 2.6, $e(t)$ is a part of the next OFDM block, which is independent of $s(t)$. Therefore, the resulting average interference power is the summation of the powers $\frac{1}{T_s}\int_0^{\tau} s(t)e^{\jmath 2\pi f_m (t-\tau)}dt$ and $\frac{1}{T_s}\int_0^{\tau} e(t)e^{\jmath 2\pi f_m (t-\tau)}dt$. When the system has a guard interval that is larger than τ, then $e(t)=0$, only single term remains. However, if a proper cyclic extension is used as in Fig. 2.6 (c), then $e(t)=s(t)$ and there is no interference. Therefore, a proper cyclic extension can effectively cancel additive interference caused by timing offset.

In Chapter 4, timing offset estimation is introduced. With estimated timing offset, the sampling time and integration region can be adjusted to reduce the induced phase shift on the desired symbol and to mitigate additive interference.

2.2.4 Effect of Delay Spread

For a channel with multipath delay spread, the received signal is a summation of the transmitted signal with different (complex) gains and delays, as shown in Fig. 2.7. Here we assume that the transmitted signal has a proper cyclic suffix extension and the length of the cyclic extension, T_g, is larger than the delay span or channel length, T_h, of the multipath fading channel.

Furthermore, we assume that the starting time of integration/observation is between T_h and T_g, that is, $T_h \leq \tau \leq T_g$.

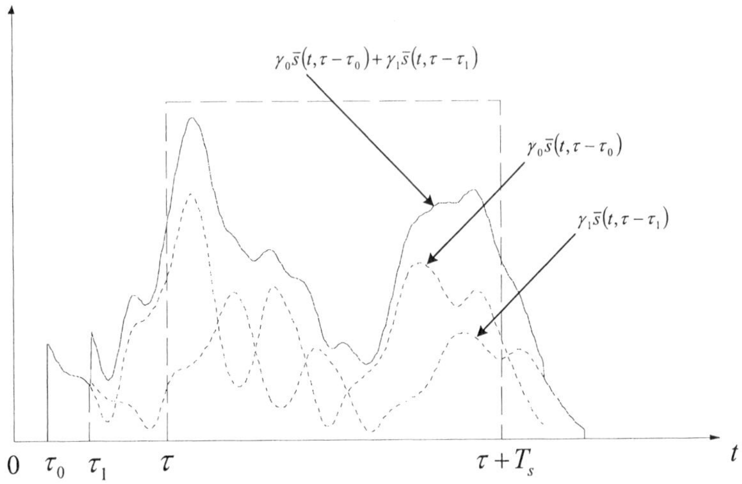

Figure 2.7. Effect of delay spread on OFDM signal.

Let the gain and delay of each path be γ_i and τ_i, respectively. As shown in Figure 2.7, the received signal can be expressed as

$$x(t) = \sum_i \gamma_i \bar{s}(t, \tau - \tau_i).$$

Therefore, the demodulated signal at the receiver is

$$\begin{aligned} X_m &= \frac{1}{T_s}\int_0^{T_s} r(t)e^{-\jmath 2\pi f_m t}dt \\ &= \sum_i \gamma_i \frac{1}{T_s}\int_0^{T_s} \bar{s}(t, \tau - \tau_i)e^{-\jmath 2\pi f_m t}dt. \end{aligned}$$

When $\tau_{max} \leq \tau \leq T_g$, from (2.2.14), we have

$$\frac{1}{T_s}\int_0^{T_s} \bar{s}(t, \tau - \tau_i)e^{-\jmath 2\pi f_m t}dt = s_m e^{\jmath 2\pi f_m (\tau - \tau_i)}.$$

Hence,

$$\begin{aligned} X_m &= \sum_i \gamma_i s_m e^{j2\pi f_m(\tau-\tau_i)} \\ &= s_m e^{j2\pi f_m \tau} \sum_i \gamma_i e^{-j2\pi f_m \tau_i} \\ &= H(f_m) e^{j2\pi f_m \tau} s_m, \end{aligned} \tag{2.2.15}$$

where $H(f)$ is the frequency response of the multipath channel defined as

$$H(f) = \sum_i \gamma_i e^{-j2\pi f \tau_i}.$$

From (2.2.15), the received symbol is the original symbol with a phase shift determined by the timing offset, and multiplicative distortion determined by the frequency response at each subchannel, which makes signal detection very simple and is also a crucial difference between OFDM and single-carrier modulation. For single-carrier modulation, delay spread or frequency selectivity of wireless channels will cause ISI, which makes signal detection very complicated.

In Chapters 4 and 5, we will introduce various timing offset estimation and channel estimation approaches, respectively. Once the timing offset and channel parameters are estimated, the phase shift and the multiplicative distortion can be corrected.

2.2.5 System Nonlinearity

Nonlinear devices in wireless systems will distort OFDM signals. In this section, we will briefly discuss the impact of system nonlinearity on OFDM signals.

In Chapter 1.6, the PAPR of OFDM signals has been discussed. It can be easily checked that for an OFDM signal with N subchannels the peak power can be as large as N^2 while the average power is N when $\mathrm{E}\{|s_k|^2\} = 1$; consequently, the largest PAPR will be

$$\mathrm{PAPR} = N.$$

For an OFDM signal with 128 subchannels, PAPR=21 dB, while it is about 6 dB for single carrier modulation. It should be noted that the probability for an OFDM signal to have a large PAPR is very small even though the largest possible PAPR is very large.

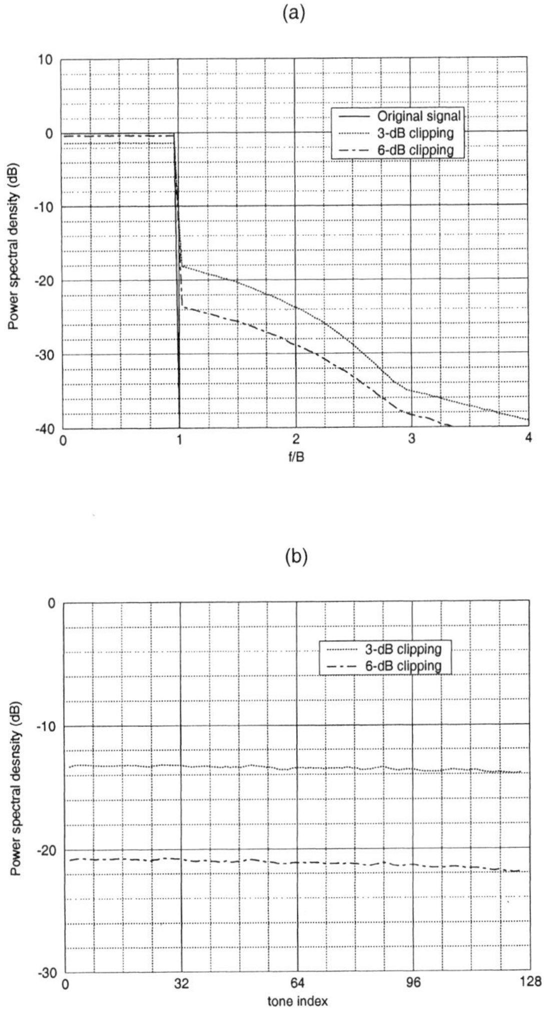

Figure 2.8. (a) Spectral spread and (b) inband distortion caused by clipping.

When an OFDM signal is passed through a nonlinear device, such as a transmitter power amplifier, it will suffer significant nonlinear distortion, which generates spectral spreading and in-band noise. Figure 2.8 demonstrates inband distortion and spectral spread due to the nonlinearity of an amplifier. If the amplifier is modeled as a 3-dB clipper, then it will cause about 14 dB inband noise, about 22 dB adjacent channel interference. As indicated in [40], adjacent channel interference or spectral spread can be mitigated by a clipping-and-filtering algorithm. However, inband noise still degrades the system performance. In Chapter 6, we will present different PAPR reduction techniques for OFDM systems.

2.3 Other Multicarrier Modulation

In the previous sections, we have introduced OFDM modulation, which belongs to a more general group of multicarrier modulation,. In this section, we briefly present two band-limited multicarrier approaches: orthogonal and filter modulation.

2.3.1 Orthogonal Approach

The orthogonal approach was proposed by Chang in [9]. It is shown in Figure 2.9, where $s_{n,k}$ and $X_{n,k}$ are the transmitted and demodulated symbols, respectively, at the k-th subchannel of the n-th OFDM block. The filters in the figure are band-limited and with overlap in the frequency domain. Therefore, their frequency responses must satisfy certain conditions so that the transmitted signal can be demodulated at the receiver.

Denote $\Phi_k(f) = |\Phi_k(f)| \exp\{\jmath\theta_k(f)\}$ the frequency response of the k-th filter in Fig. 2.9. The receiver can recover the transmitted signal without any ISI and ICI at the receiver if the amplitude and the phase of the frequency response satisfy the following conditions [9].

- *Amplitude condition*: The amplitudes for different subchannels have the same shape, *i.e.*,

  $$|\Phi_k(f)|^2 = C + Q(f - f_k), \text{ for } |f - f_k| < \Delta f,$$

  for $k = 0, 1, \cdots, N-1$, where $f_k = f_o + k\Delta f$ is the center frequency of the k-th subchannel. Furthermore, $Q(f)$ is a symmetric real function that is zero outside $[-\Delta f, \Delta f]$ and satisfies

  $$Q(f) = Q(-f), \text{ for all } |f| \leq \Delta f,$$

  and

  $$Q(f) = -Q(\Delta f - f) \text{ for all } 0 \leq f \leq \Delta f,$$

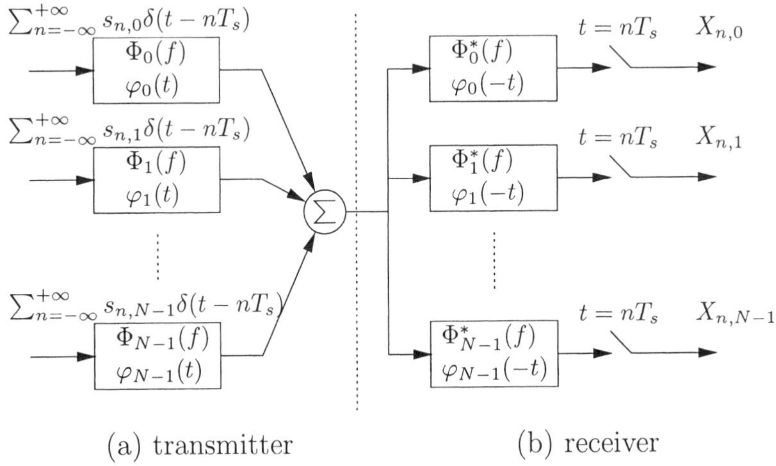

Figure 2.9. An orthogonal approach based band-limited multicarrier scheme.

and C in the above is a constant such that $C + Q(f)$ is nonnegative for $|f| \leq \Delta f$.

- *Phase condition*: The phases for different subchannels also have the same shape, *i.e.*,

$$\theta_k(f) = \theta(f - f_k), \text{ for } |f - f_k| < \Delta f.$$

The $\theta(f)$ in the above equation satisfies

$$\theta(f) = m\pi\frac{f}{2\Delta f} + \alpha_0 + \sum_n \alpha_{2n+1} \cos\left((2n+1)\pi\frac{f}{\Delta f}\right) + \sum_n \alpha_{2n} \sin(2n\pi\frac{f}{\Delta f}),$$

where m is any odd integer and α_n's for all integer n's are arbitrary (real) numbers.

One of example in this category of multicarrier modulation is to choose

$$Q(f) = \frac{1}{2}\cos\left(\pi\frac{f}{\Delta f}\right),$$

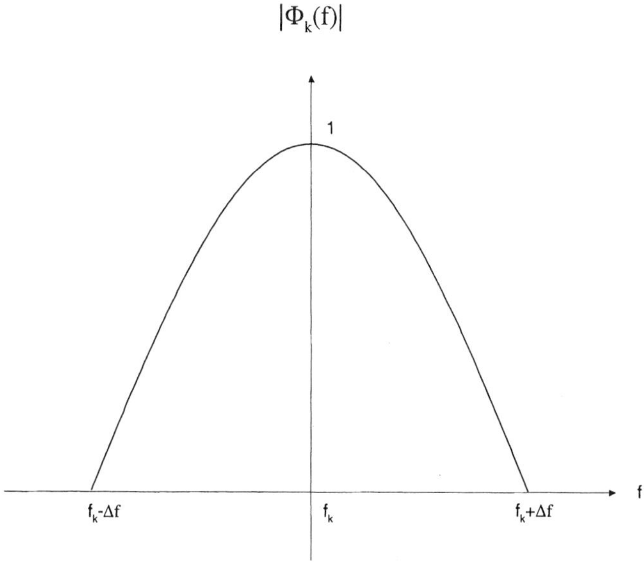

Figure 2.10. An example of amplitude shaping

and $C = 1/2$. Then

$$|\Phi_k(f)|^2 = \frac{1}{2} + \frac{1}{2}\cos\left(\pi\frac{f-f_k}{\Delta f}\right)$$

and

$$|\Phi_k(f)| = \cos(\pi\frac{f-f_k}{2\Delta f}),$$

which is shown in Figure 2.10

2.3.2 Filter Approach

Filter bank based multicarrier modulation was proposed by Saltzberg in [10]. The modulator and demodulator of this scheme are shown in Figure 2.11. All filters $F(f)$'s in the transmitter and receiver are identical and are assumed to be real. It is demonstrated in [10] that the use of the same filtering at transmitter and receiver assures optimum performance for the AWGN channels. In order to eliminate the possibility of interference between any channels that are not immediately adjacent, the filters are bandlimited to the subchannel space, Δf, that is

$$F(f) = 0, \quad |f| \geq \Delta f,$$

and the transmit and receive filters in tandem have a Nyquist roll-off,

$$F^2(f) + F^2(\Delta f - f) = 1, \quad 0 \leq f \leq \Delta f. \tag{2.3.1}$$

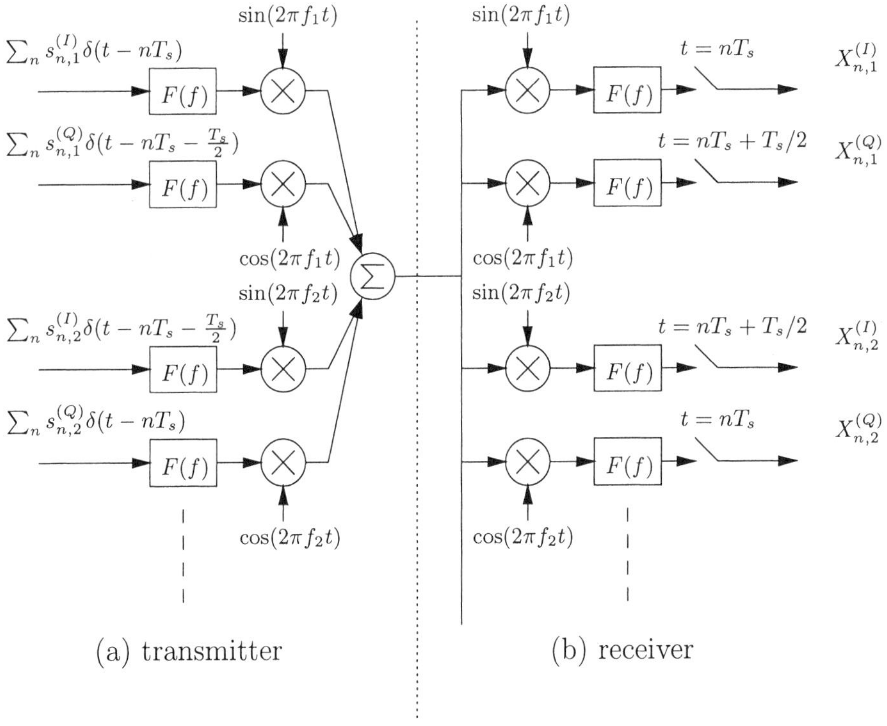

Figure 2.11. A filtering based multicarrier scheme.

The symbol duration for each stream is $T_s = 1/\Delta f$, and the timing of the two streams for the same subchannel are staggered by $T_s/2$. Adjacent channels are staggered oppositely so that the data streams that modulate the cosine carriers of the even numbered channels and the sine carriers of the odd numbered channels are in phase. Consequently, the pass-band expression of the modulated signal from the transmitter is the following:

$$\begin{aligned} s(t) &= \sum_{k \text{ odd}} \sum_{n=-\infty}^{+\infty} s_{n,k}^{(I)} f(t - nT_s) \cos(2\pi f_k t) \\ &+ \sum_{k \text{ odd}} \sum_{n=-\infty}^{+\infty} s_{n,k}^{(Q)} f(t - nT_s - T_s/2) \sin(2\pi f_k t) \\ &+ \sum_{k \text{ even}} \sum_{n=-\infty}^{+\infty} s_{n,k}^{(I)} f(t - nT_s - T_s/2) \cos(2\pi f_k t) \\ &+ \sum_{k \text{ even}} \sum_{n=-\infty}^{+\infty} s_{n,k}^{(Q)} f(t - nT_s) \sin(2\pi f_k t), \end{aligned}$$

where $f(t)$ is the inverse Fourier transform of $F(f)$, $f_k = f_o + k\Delta f$, and $s_{n,k}^{(I)}$ and $s_{n,k}^{(Q)}$ are the information-bearing real random variables.

It can be shown that for the ideal channel (distortionless and noiseless), the demodulator can recover the transmitted sequences $a_{n,k}$ and $b_{n,k}$ without the ISI and ICI, if $F(f)$ satisfies (2.3.1).

For channels with linear distortion, the ISI and ICI are analyzed in [10]. The equalization for the filter based approach to reduce ISI and ICI is studied in [41].

2.3.3 General Multicarrier Modulation

We have discussed two special multicarrier modulations: time-limited form (OFDM) and band-limited form. Here we describe general multicarrier modulations.

To facilitate the basic concept of orthogonal multicarrier modulations, we first introduce *(T_s-shift) orthogonal complex functions.* A set of N complex functions, $\varphi_k(t)$ for $k = 0, 1, \cdots, N-1$, are called T_s-shift orthogonal if for any k, l and n, m,

$$\int_{-\infty}^{+\infty} \varphi_k(t - nT_s)\varphi_l^*(t - mT_s)dt = c\delta[k - l]\delta[n - m], \tag{2.3.2}$$

where c is a positive constant. For simplicity, we may assume that the set is normalized, *i.e.* $c = 1$.

Based on a T_s-shift orthogonal complex function set, a general multicarrier transmitter and receiver can be designed. Let $s_{n,k}$ represent the information to be transmitted at the k-th subchannel of the n-th block. Then, the complex baseband signal transmitted over the channel can be expressed as

$$s(t) = \sum_{n=-\infty}^{+\infty} \sum_{k=0}^{N-1} s_{n,k}\varphi_k(t - nT_s). \tag{2.3.3}$$

In an AWGN channel, the received signal is

$$x(t) = s(t) + n(t), \tag{2.3.4}$$

where $n(t)$ is additive white complex Gaussian noise with zero mean and variance N_o. The transmitted sequence can be detected by

$$\begin{aligned} X_{n,k} &= \int_{-\infty}^{+\infty} x(t)\varphi_k^*(t - nT_s)dt \\ &= \int_{-\infty}^{+\infty} (s(t) + n(t))\varphi_k^*(t - nT_s).dt \end{aligned} \tag{2.3.5}$$

Because of the orthogonality of $\{\varphi_k(t - nT_s)\}$, (2.3.5) reduces to

$$X_{n,k} = s_{n,k} + N_{n,k}, \tag{2.3.6}$$

where

$$N_{n,k} = \int_{-\infty}^{+\infty} n(t)\varphi_k^*(t - nT_s)dt. \tag{2.3.7}$$

Since $\{\varphi_k(t-nT_s)\}$'s are orthonormal for different n's or k's, the noise $N_{n,k}$ is zero-mean and with variance N_o, and is uncorrelated (therefore independent) for different n's and k's.

OFDM introduced in Section 2.1 is clearly a special case when $\varphi_k(t) = \frac{1}{\sqrt{T_s}} e^{\jmath \frac{2\pi kt}{T_s}}$.

Chapter 3

PERFORMANCE OPTIMIZATION

John M. Cioffi and Louise M. C. Hoo

Optimization of an OFDM signal enables the highest data rates and/or most reliable performance in the bandwidth of transmission. Optimization is important simply because it leads to greater spectral efficiency, i.e., more bits/second/Hz in the given bandwidth allocated to the OFDM signal. Optimization can thus lead to better use of the scarce spectrum available for wireless transmission. As this chapter will show, use of every frequency is not always best and that shift of information and energy from channel frequencies of poor SNR to frequencies of comparatively higher SNR can be good and lead to higher performance. Optimized OFDM transmission systems then often carry information unequally among the subchannels or tones. An OFDM system with optimization is often given a special name, *Discrete Multitone* (DMT), and is often known better by that name in applications where optimization is used. For example, DMT is a well-known name in digital subcriber line systems. Optimization methods have not yet been used in practical wireless applications of OFDM, but are particularly well suited to the emerging area known as spatial-temporal wireless transmission, which is briefly described in the conclusion.

Optimization includes any or all of the component subproblems of *partitioning*, *loading*, and *coding* that are addressed in Sections 3.2, 3.3, and 3.4 respectively. Optimization through partitioning determines the best set of modulation basis functions as a function of channel characteristics. Optimization through loading basically consists of the best allocation of bits and energies to each of the OFDM subchannels. Coding additionally improves performance and some caution is necessary when using coding with loading. All optimization depends on the channel and thus estimation of the channel through handshaking is a crucial prerequisite. Since channels can change with time, each of the optimization subproblems (and especially loading) need to have an ability to vary with channel changes, as discussed individually in each of the relevant subsections.

This chapter specifically investigates the areas of partitioning, loading, and coding, with the intent of enabling the reader to design a system that can take advantage of the optimization of any or all. The objective will be the reader's facility with optimization, enabling their study of the feasibility of optimization for various specific applications of OFDM.

3.1 History of OFDM Optimization

Shannon's classic 1948 work [4] found that the highest data rate on a bandlimited channel is achieved when the sum of the channel noise energy (normalized by the channel attenuation) and the transmit energy at each frequency is constant, a classic result now known as *water-filling*. Shannon's construction of this optimized transmission used a multicarrier system with an infinitely dense set of tones. Indeed, transmission engineers know today that no amount of conventional coding alone, without OFDM's water-fill energy distribution, can achieve the best performance. Shannon thus laid a foundation for future practical realizations and thus motivated the loading and partitioning areas of multitone transmission. Holsinger [18] verified Shannon's work with some experimental prototypes where he measured bandlimited channels and then altered the energy and information distribution of a rudimentary OFDM transmitter, observing the improvements of the method. His thesis [18] revisited the whole area of bandlimited channel capacity and formalized the concepts, which were subsequently popularized in Gallager's Information Theory text book [42], including the popular nomenclature of "water-filling." The work of these three does not appear to be known to several at Bell Laboratories in the late 1960s [9] [10] who independently studied the channel partitioning or generation of a finite set of parallel independent subchannels, but who subsequently did not optimize bit and energy distributions.

A series of voiceband modems over a 30-year period each independently used multitone transmission to achieve data rates approximately double (or more) those of competing standardized modems, including a 3000 bps modem in 1958 by Collins Radio (now Rockwell), a 9.6 kbps modem in the early 1970's by Gandolf, and a 20 kbps modem in the early 1980's by Telebit (and again IMC). IBM independently studied optimized OFDM for disk-drives and voiceband modems in the early 1980's, including the work of Ruiz [20], who later took his experience with him to Stanford University, where he initiated a more complete study of the area [43]. Each of the incompatible modem products was sold into niche markets for highest-speed private-line data transfer. A concerted effort to standardize the technique did not occur until 1992 when Amati Communications Corporation (now Texas Instruments, and directly related, as a Stanford spin-out, to Ruiz' early work) convinced the American National Standards Institute to standardize on the optimized OFDM (called DMT there) for digital subscriber line transmission (the next generation of data modems running at speeds from .5 to 50 Mbps on ordinary phone lines). Today, optimization of OFDM is taken for granted in wireline transmission, but not in wireless. Perhaps, this chapter will help motivate the transcension of some of the more ominous practical hurdles in such wireless application, which will be discussed as this development progresses.

3.2 Channel Partitioning

All multicarrier modulation techniques are based on the concept of *channel partitioning*. Channel partitioning methods divide a wideband, spectrally-shaped transmission channel into a set of parallel and ideally independent narrowband subchannels to which the loading algorithms of Section 3.3 can be applied. Loading algorithms optimize the amount of energy and also the number of bits in each subchannel to maximize the overall performance over a given channel.

Ideal OFDM transmission, also known as *multitone transmission* (MT) here, is an example of channel partitioning where each subchannel has a frequency index, and all subchannels are independent because each subchannel exists only over a finite band of continuous frequencies and is zero outside that band. The ideal OFDM basis functions $\{\varphi_n(t) = \frac{1}{\sqrt{T}}\text{sinc}\left(\frac{t}{T}\right) e^{\jmath\frac{2\pi nt}{N}}\}$ form an orthonormal basis while also exhibiting the desirable property that the set of channel-output functions $\{h(t) * \varphi_n(t)\}$ remains orthogonal for any $h(t)$. The ideal OFDM basis functions, when combined with continuous-time water-filling, and $N \to \infty$, are optimum. For a finite N, the ideal OFDM basis functions are not optimum because the channel shaping induced by $h(t)$ in each frequency band is not flat in general. Furthermore, the ideal OFDM basis functions use sinc functions and exist over an infinite time interval, which is not suitable for practical implementation. Thus, ideal OFDM transmission makes a useful example for the introduction of multichannel modulation, but is not a particularly good initial choice for channel

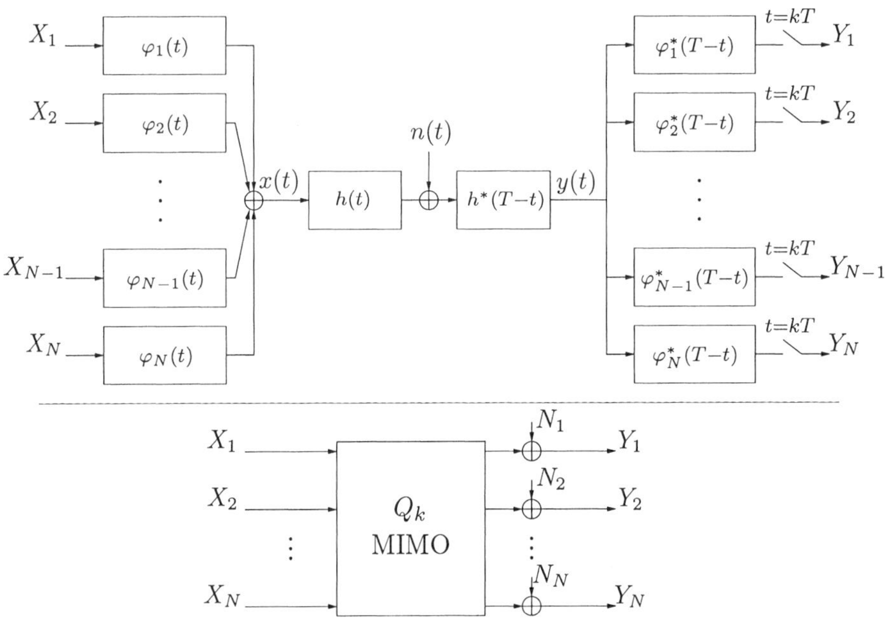

Figure 3.1. Transmission system using optimum channel partitioning.

partitioning unless N is large.

This section introduces optimum basis functions for modulation based on a finite time interval of T of channel inputs and outputs, and presumes input basis functions also only exist over this symbol-period interval, presuming that preceding and succeeding transmission of additional symbols over the channel is desired. Subsection 3.2.1 show that the optimum basis functions become channel-dependent and are the eigenfunctions of a certain channel covariance function. Subsection 3.2.2 investigates the interference between subchannels, and its elimination with guard periods. Subsection 3.2.3 introduces discrete-time channel partitioning and the discrete-time equivalent of modulation with the continuous-time eigenfunctions, known as *Vector Coding* (VC). Subsection 3.2.4 determines OFDM to be a special case of Vector Coding.

3.2.1 Eigenfunction Transmission

Figure 3.1 depicts a transmission system that uses optimum channel partitioning.

Define the pulse response as

$$p_n(t) \stackrel{\Delta}{=} \varphi_n(t) * h(t), \ n = 1, \ldots, N \quad . \tag{3.2.1}$$

The set of sampled outputs of the N matched filters $p_n^*(-t)$ at all symbol instants kT forms a sufficient statistic or complete representation of the transmitted signal as shown in Figure 3.1. The transmission system can be compactly represented as a *multiple-input multiple-output* (MIMO) system that is characterized by an $N \times N$ square matrix $\boldsymbol{Q}(t)$ whose entries are given by

$$q_{n,m}(t) \stackrel{\Delta}{=} \frac{p_n(t) * p_m^*(-t)}{\|p_n\| \cdot \|p_m\|}, \ n, m = 1, \ldots, N \quad . \tag{3.2.2}$$

Its sampled-time representation is thus

$$\boldsymbol{Q}_k = \boldsymbol{Q}(kT) \quad . \tag{3.2.3}$$

Generalization of the Nyquist Criterion

A generalization of the Nyquist Criterion for no intersymbol interference, which is also known as inter-block interference, nor intra-symbol interference, which is also called inter-carrier or inter-subchannel interference, is for

$$\boldsymbol{Q}_k = \mathbf{I}\delta_k \quad , \tag{3.2.4}$$

that is, the sampled MIMO channel matrix is an identity matrix. Thus, to avoid inter-symbol and intra-symbol interference, the designer could choose basis functions so that (3.2.4) holds. In general, there are many choices for such a function. Although there are many approaches to the selection and development of these functions, all other studies that the authors are currently aware of, essentially restrict attention to the situation where $h(t) = \delta(t)$. Bandlimitations in the form of a channel impulse response where $h(t) \neq \delta(t)$ for PAM or QAM transmission cause loss of the orthogonality of what would otherwise be Nyquist transmit pulse shapes. So then too will bandlimited channel filters cause loss of orthogonality of generalized Nyquist functions, except in some well-known special cases to which this development will now progress.

Choice of transmit basis

The channel impulse response forms a *channel autocorrelation function* according to

$$r(t) = h(t) * h^*(-t) \quad . \tag{3.2.5}$$

This channel autocorrelation function defines a possibly countably infinite-size set of orthonormal *eigenfunctions*, $\{\varphi_n(t)\}$, possibly nonzero only over the time interval $[-T/2, T/2)$, that satisfy the relation[1]

$$\rho_n \varphi_n(t) = \int_{-T/2}^{T/2} r(t-\tau)\varphi_n(\tau)d\tau \quad n = 1, ..., \infty \quad . \tag{3.2.6}$$

That is, a filter $\varphi_n(t)$ convolved with the matched-filtered channel over the interval $[-T/2, T/2)$ produces that same filter, scaled by a constant ρ_n, called the *eigenvalue*. The set of eigenvalues is unique to the channel autocorrelation $r(t)$, while the eigenfunctions are not necessarily unique. The eigenfunctions are generally difficult to compute (unless the symbol period is infinite), and are used here for theoretical purposes. Each eigenfunction represents a "mode" of the channel through which data may be transmitted independently of the other modes with gain ρ_n. The use of the channel-dependent orthonormal set of eigenfunctions (the N with largest eigenvalues) as a basis for modulation is called *Modal Modulation* (MM).

From Figure 3.1,

$$x(t) = \sum_{n=1}^{N} X_n \varphi_n(t) \quad . \tag{3.2.7}$$

The receiver's matched-filter-output signal is then

$$y(t) = \sum_{n=1}^{N} X_n \left[r(t) * \varphi_n(t)\right] + \tilde{n}(t) \quad , \tag{3.2.8}$$

where $\tilde{n}(t)$ is the matched-filtered noise with autocorrelation function $\frac{N_0}{2} r(t)$. Through (3.2.6), the output signal $y(t)$ is

$$y(t) = \sum_{n=1}^{N} (\rho_n X_n)\, \varphi_n(t) \quad , \tag{3.2.9}$$

[1]Most developments of eigenfunctions develop the functions over the causal time interval $[0, T)$. We instead non-causally center this interval about the origin at $[-T/2, T/2)$ in this theoretically abstract case so that taking the limit as $T \to \infty$ produces a true two-sided convolution integral, which will then have eigenvalues as the Fourier transform.

which is a sum of the orthonormal basis functions, weighted by the channel input samples, $\rho_n x_n$. This summation has the same form as any modulated waveform where each symbol element x_n is now replaced by $\rho_n x_n$.

The MAP/ML receiver is thus also the same and recovers $\rho_n x_n$ by processing each filter output with a matched-filter/sampler for each of the basis functions, as shown in Figure 3.1. The noise samples at the output of these matched filters are independent for different n because the basis functions are orthogonal. Equivalently,

$$E[N_i N_j^*] = \frac{\mathcal{N}_0}{2}\int_0^T\int_0^T r(t-s)\varphi_i(t)\varphi_j^*(s)dtds \tag{3.2.10}$$

$$= \frac{\mathcal{N}_0}{2}\int_0^T \rho_i\varphi_i(t)\varphi_j^*(t)dt \tag{3.2.11}$$

$$= \rho_i\frac{\mathcal{N}_0}{2}\delta_{ij} \quad . \tag{3.2.12}$$

Thus, the joint-probability density function $p_{\boldsymbol{y}/\boldsymbol{x}}$ factors into N independent component distributions, and an ML/MAP detector is equivalent to an independent detector on each sample output. Therefore, an MM transmission system generates a set of parallel channels whose SNR's are $\text{SNR}_n = (\rho_n^2\mathcal{E}_n)/(\rho_n\frac{\mathcal{N}_0}{2}) = \rho_n\mathcal{E}_n/\frac{\mathcal{N}_0}{2}$.

The functions $p_n(t)$ defined by MM clearly satisfy the generalized Nyquist criterion because of the independence of the subchannels created and because they are zero outside the interval $[-T/2, T/2)$, thus averting inter-symbol and intra-symbol interference.

MM is optimum given observation of a finite time interval $[-T/2, T/2)$. It is then natural to expect that MM converges to MT modulation as both allow the number of dimensions N to go to infinity while also allowing the time interval to span $(-\infty, \infty)$. In other words, for sufficiently large N and T, the designer can use either the MM method or the MT method and be assured that the design is as good as can be achieved on a linear ISI channel with Gaussian noise.

Periodic Channels

A channel with periodic $r(t) = r(t+T)$ will always have eigenfunctions $\varphi_n(t) = e^{\jmath\frac{2\pi}{T}nt}$ for the interval $[-T/2, T/2)$. Periodic channels do not exist in practice, but designers often use extra bandwidth in the design of a transmission problem to make the finite-duration channel appear as if it were periodic. In this case, MT and MM would again be the same, subject to any loss incurred in creating the periodic channel.

3.2.2 Overlap, Excess Bandwidth, and Guard Period

A realistic transmission channel has most or all its nonzero impulse response confined to a finite time interval T_h. Thus, convolution of the (necessarily causal) channel impulse response with nonzero input basis functions existing over $[0, T)$ produces an output with duration longer than the symbol period T[2]. The first T_h seconds of any symbol are thus possibly corrupted by intersymbol interference from the last symbol if that previous symbol had nonzero transmit energy over $(T - T_h, T)$. Thus, the receiver then usually ignores the first T_h seconds of any symbol. If $T_h << T$, then this overhead penalty, or "excess bandwidth" penalty, is small and no equalization (see Section 3.2.5 and 3.2.6) need be used. For a fixed channel and thus fixed T_h, as $T \to \infty$, this excess bandwidth becomes negligible. This time period where the receiver ignores channel outputs is often called a *guard period* in multichannel modulation. For a finite symbol period, the multichannel system then has an excess bandwidth given by the factor $\alpha = T_h/(T + T_h)$.

3.2.3 Discrete-Time Channel Partitioning

While MM restricts attention to a finite time interval for construction of the transmit basis functions, these functions remain continuous-time and a function of the channel. Such functions would be difficult to implement exactly in practice. *Discrete-time channel partitioning* partitions a discrete-time description of the channel. This description is a linear matrix relationship between a finite number of output samples (presumably sampled at a rate $1/\Delta_T$ exceeding twice the highest frequency to be transmitted) and a corresponding finite set of channel input samples at the same rate that constitute the transmitted symbol. It is assumed that the combined effect of transmit filters, transmission channel and received filters can be approximated by a finite impulse response (FIR) filter. Such a description can never be exact in practice, but with a sufficient number of input and output samples included, this description can be very close to the exact action of the ISI channel. Figure 3.2 illustrates general N-dimensional discrete-time channel partitioning.

[2]This section will now begin to pursue realizable transmission systems and not theoretical abstractions. Thus, the basis functions will necessarily have to be causal in the remainder of this section.

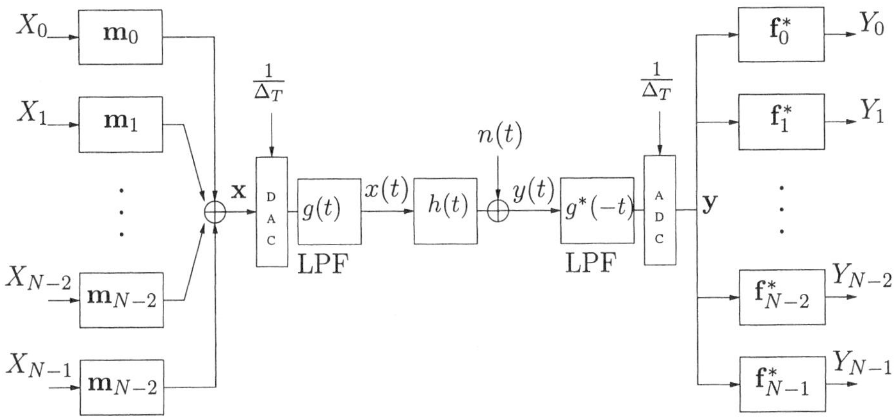

Figure 3.2. Digital realization of channel partitioning.

Discrete-time *basis vectors*, $\boldsymbol{m}_n$ replace the continuous basis functions of MM or MT. These basis vectors have a finite length of $N+\nu$ samples over a duration of time T. The sampled pulse response, $p(k\Delta_T)$, of the channel is assumed to span a time interval of no more than $\nu+1$ sample periods, where $(\nu+1)\Delta_T = T_h$. Each basis vector, $\boldsymbol{m}_n$, multiplies a subsymbol element X_n before being summed to form the transmit symbol vector $\boldsymbol{x}$. Then, $\boldsymbol{x}$ is passed through a Digital-to-Analog Converter (DAC). Some transmit filtering of the output DAC occurs (typically analog lowpass anti-aliasing) before the signals pass through the ISI channel. The combined filtering and channel form a pulse response $\varphi(t) * h(t)$. The sampling rate $1/\Delta_T = (N+\nu)/T$ is higher than twice the highest frequency that the designer hopes to use for the transmitted signals. The modulator attempts to use basis vectors, $\boldsymbol{m}_n$, that will remain orthogonal after undergoing the dispersive effect of the ISI channel, just as orthogonality is maintained with MM.

The receiver low-pass filters (with gain $1/\sqrt{\Delta_T}$ to maintain noise per dimension at $\frac{\mathcal{N}_0}{2}$), and then samples at rate $1/\Delta_T$ the received modulated signal. The matched filters $\mathbf{f}_n^*$ are discrete-time and also finite length. The matched filters are only complex if the entire set of modulated signals had been passband modulated by an external passband modulator that is not shown in Figure 3.2, and would in this case correspond to a complex baseband-equivalent channel. There are $N+\nu$ input samples that lead to N matched-filtered output samples, with ν samples lost to the guard period to avoid intersymbol interference. The samples can be real or complex, which in the latter case tacitly implies a doubling of the number of real dimensions.

Optimal Partitioning Vectors - Vector Coding

The last N ADC channel-output symbol samples in Figure 3.2 have vector representation, with time indexed within a symbol from $k = -\nu$ to $N-1$,

$$\begin{bmatrix} y_{N-1} \\ y_{N-2} \\ \vdots \\ y_0 \end{bmatrix} = \begin{bmatrix} p_0 & p_1 & \dots & p_\nu & 0 & \dots & 0 \\ 0 & p_0 & \ddots & p_{\nu-1} & p_\nu & \ddots & 0 \\ 0 & \ddots & \ddots & \ddots & \ddots & \ddots & 0 \\ 0 & \dots & 0 & p_0 & p_1 & \dots & p_\nu \end{bmatrix} \begin{bmatrix} x_{N-1} \\ \vdots \\ x_0 \\ x_{-1} \\ \vdots \\ x_{-\nu} \end{bmatrix} + \begin{bmatrix} n_{N-1} \\ n_{N-2} \\ \vdots \\ n_0 \end{bmatrix} , \tag{3.2.13}$$

or more compactly,

$$\boldsymbol{y} = \mathbf{P}\boldsymbol{x} + \boldsymbol{n} \quad . \tag{3.2.14}$$

The $N \times (N+\nu)$ matrix $\mathbf{P}$ has a *singular value decomposition* (SVD) given by

$$\mathbf{P} = \mathbf{F}\left[\mathbf{\Lambda} \vdots \mathbf{0}_{N,\nu}\right]\mathbf{M}^* \quad , \tag{3.2.15}$$

where $\mathbf{F}$ is an $N \times N$ unitary ($\{\mathbf{FF}^* = \mathbf{F}^*\mathbf{F} = \mathbf{I}$) matrix, $\mathbf{M}$ is an $(N+\nu) \times (N+\nu)$ unitary ($\mathbf{MM}^* = \mathbf{M}^*\mathbf{M} = \mathbf{I}$) matrix, $\mathbf{0}_{N,\nu}$ is an $N \times \nu$ matrix of zeros and $\mathbf{\Lambda}$ is an $N \times N$ diagonal matrix with *singular values* λ_n, $n = 1, ..., N$ along the diagonal. The vector $\boldsymbol{x}$ is a data symbol vector.

Vector Coding creates a set of N parallel independent channels by using the transmit basis vectors, $\boldsymbol{m}$, that are the first N columns of $\mathbf{M}$. In other words, the transmit vector $\boldsymbol{x}$ is obtained from the N vector components X_n, $n = 1, \dots, N$ according to

$$\boldsymbol{x} = \mathbf{M}\begin{bmatrix} \boldsymbol{X} \\ 0 \\ \vdots \\ 0 \end{bmatrix} = [\boldsymbol{m}_{N-1}\, \boldsymbol{m}_{N-2} \dots \boldsymbol{m}_1\, \boldsymbol{m}_0 \dots \boldsymbol{m}_{-\nu}] \begin{bmatrix} X_{N-1} \\ X_{N-2} \\ \vdots \\ X_0 \\ 0 \\ \vdots \\ 0 \end{bmatrix} = \sum_{n=0}^{N-1} X_n \boldsymbol{m}_n \; . \tag{3.2.16}$$

The last ν columns of $\mathbf{M}$ do not affect $\mathbf{P}$ because they are multiplied by zeros in (3.2.16). These ν columns does not contribute to the channel output and can be ignored in implementation. The corresponding vector-coding receiver uses discrete "matched filters" as the N rows of $\mathbf{F}^*$, forming

$$\boldsymbol{Y} = \mathbf{F}^* \boldsymbol{y} = \begin{bmatrix} \mathbf{f}_{N-1}^* \boldsymbol{y} \\ \vdots \\ \mathbf{f}_0^* \boldsymbol{y} \end{bmatrix} \quad . \tag{3.2.17}$$

The mathematical description of the resultant N parallel channels is

$$\boldsymbol{Y} = \boldsymbol{\Lambda} \boldsymbol{X} + \boldsymbol{N} \quad . \tag{3.2.18}$$

Each independent channel or entry in $\boldsymbol{Y}$ is represented by

$$Y_n = \lambda_n X_n + N_n \quad . \tag{3.2.19}$$

Since $\mathbf{F}$ is unitary, the noise vector $\boldsymbol{N}$ is also additive white Gaussian with variance per real dimension given by $\frac{\mathcal{N}_0}{2}$, which is identical to that of $\boldsymbol{n}$.

Theorem 3.2.1: Optimality of Vector Coding
Vector Coding has the maximum multichannel SNR, $SNR_{m,u}$ (see Section 3.3.2), for any discrete channel partitioning.

Proof: See [44].

Thus, vector coding is optimum for the N-dimensional channel of (3.2.13). This optimality is only strictly true when capacity-achieving codes with "Gaussian" distributions are used on each of the subchannels. The restriction to independent symbols, requiring the receiver to ignore the first ν samples of any symbol also restricts the optimality. If this latter restriction were relaxed or omitted, then, it is possible that a better design exists. For $N >> \nu$, such a potentially better method would offer only a small improvement. Thus, only the case of $\nu > .1N$ would be of interest in further investigation of this restriction.

In a discrete-time system as modelled in (3.2.13), the noise is unlikely to be white. To convert any channel into an equivalent white-noise channel, the designer may add receiver preprocessing by a canonical noise-whitening filter with the equivalent channel frequency response becoming the ratio of the channel to the square-root of the power-spectral density of the noise. Since this filter may be difficult to realize in practice, the discrete-time processing of factoring the noise covariance to

$$E\left[\boldsymbol{n}\boldsymbol{n}^*\right] = \mathbf{R}_{nn}\sigma^2 = \mathbf{R}_{nn}^{1/2}\mathbf{R}_{nn}^{-1/2}\sigma^2 \quad , \tag{3.2.20}$$

can be used. Then, a discrete white-noise equivalent channel becomes

$$\boldsymbol{y} \leftarrow \mathbf{R}_{nn}^{-1/2}\boldsymbol{y} = \left(\mathbf{R}_{nn}^{-1/2}\mathbf{P}\right)\boldsymbol{x} + \tilde{\boldsymbol{n}} \quad . \tag{3.2.21}$$

The extra matrix multiplication is absorbed into the receiver processing as $\mathbf{F}^*\boldsymbol{y} \leftarrow \mathbf{F}^*\mathbf{R}_{nn}^{-1/2}\boldsymbol{y}$. The preceding development then continues with $\mathbf{P}$ replaced by the matrix $\mathbf{R}_{nn}^{-1/2}\mathbf{P}$, the latter of which then undergoes SVD. The matrix $\mathbf{P}$ will not be Toeplitz any longer, but will be very close to Toeplitz if the duration of the noise whitening filter is much less than the symbol period.

3.2.4 Partitioning for OFDM

OFDM is a common form of Vector Coding that adds a restriction to reduce complexity of implementation. The OFDM channel partitioning forces the transmit symbol to have $x_{-k} = x_{N-k}$ for $k = 1, ..., \nu$. Such repeating of the last ν samples at the beginning of the symbol is called a *cyclic prefix*. The cyclic prefix is simply a structured guard period. Its use eliminates both inter-symbol and inter-subchannel (intra-symbol) interference, and results in tremendous processing simplification. The use of any guard period that is not in the form of a cyclic prefix and that spans the length of the channel impulse response will only eliminate inter-symbol or inter-block interference.

With the cyclic prefix, the matrix description of the channel has a square $N \times N$ circulant $\mathbf{P}$ matrix:

$$\begin{bmatrix} y_{N-1} \\ y_{N-2} \\ \vdots \\ y_0 \end{bmatrix} = \begin{bmatrix} p_0 & p_1 & \cdots & p_\nu & 0 & \cdots & 0 \\ 0 & p_0 & \ddots & p_{\nu-1} & p_\nu & \ddots & 0 \\ 0 & \ddots & \ddots & \ddots & \ddots & \ddots & 0 \\ 0 & \cdots & 0 & p_0 & p_1 & \cdots & p_\nu \\ p_\nu & 0 & \cdots & 0 & p_0 & \cdots & p_{\nu-1} \\ \vdots & \ddots & \ddots & \ddots & \ddots & \ddots & \vdots \\ p_1 & \cdots & p_\nu & 0 & \cdots & 0 & p_0 \end{bmatrix} \begin{bmatrix} x_{N-1} \\ \vdots \\ x_0 \end{bmatrix} + \begin{bmatrix} n_{N-1} \\ n_{N-2} \\ \vdots \\ n_0 \end{bmatrix}$$

$$\boldsymbol{y} = \mathbf{P}\boldsymbol{x} + \boldsymbol{n} \quad . \tag{3.2.22}$$

Singular value decomposition produces unique singular values if those values are restricted to be nonnegative real values (even when $\mathbf{P}$ is complex). For transmission, this restriction is superfluous and a variant of SVD will be simpler to compute. For the case of the cyclic prefix, the SVD may be replaced by the eigen-decomposition or "spectral factorization" of the circulant matrix $\mathbf{P}$,

$$\mathbf{P} = \mathbf{M}\mathbf{\Lambda}\mathbf{M}^* \tag{3.2.23}$$

where $\mathbf{M}\mathbf{M}^* = \mathbf{M}^*\mathbf{M} = \mathbf{I}$ and $\mathbf{\Lambda}$ is a square diagonal matrix with eigenvalues λ_n on the anti-diagonal. The magnitudes of the eigenvalues are equal to the magnitudes of the singular values. Furthermore, effectively, M and F become the same matrix for this special case. The product SNR for this alternative factorization in (3.2.23) is the same as for SVD-factorization because the magnitude of each eigenvalue equals the magnitude of the corresponding singular value[3].

The DFT of an N-dimensional vector $\mathbf{x} = [x_{N-1} \dots \ x_0]^T$ is also an N-dimensional vector $\mathbf{X} = [X_{N-1} \dots \ X_0]^T$ whose components are given by,

$$X_k = \frac{1}{\sqrt{N}} \sum_{n=0}^{N-1} x_n e^{-j\frac{2\pi}{N}kn}, \quad k = 0, \dots, N-1 \quad . \tag{3.2.24}$$

The corresponding IDFT is given by

$$x_n = \frac{1}{\sqrt{N}} \sum_{k=0}^{N-1} X_k e^{j\frac{2\pi}{N}kn}, \quad n = 0, \dots, N-1 \quad . \tag{3.2.25}$$

In matrix form, the DFT and IDFT are

$$\boldsymbol{X} = \mathcal{Q}\boldsymbol{x} \tag{3.2.26}$$

$$\boldsymbol{x} = \mathcal{Q}^*\boldsymbol{X} \quad , \tag{3.2.27}$$

where $\mathcal{Q}$ is the orthonormal matrix given by

$$\begin{aligned} \mathcal{Q} &= \frac{1}{\sqrt{N}} \begin{bmatrix} e^{-j\frac{2\pi}{N}(N-1)(N-1)} & \dots & e^{-j\frac{2\pi}{N}2(N-1)} & e^{-j\frac{2\pi}{N}(N-1)} & 1 \\ e^{-j\frac{2\pi}{N}(N-1)\cdot(N-2)} & \dots & e^{-j\frac{2\pi}{N}2(N-1)} & e^{-j\frac{2\pi}{N}(N-2)} & 1 \\ \vdots & \ddots & \vdots & \vdots & \vdots \\ e^{-j\frac{2\pi}{N}(N-1)} & \dots & e^{-j\frac{2\pi}{N}2} & e^{-j\frac{2\pi}{N}} & 1 \\ 1 & \dots & 1 & 1 & 1 \end{bmatrix} \\ &= [\boldsymbol{q}_{N-1}, \dots, \boldsymbol{q}_0] \quad . \end{aligned} \tag{3.2.28}$$

[3]One should recall that this section always whitens noise first before proceeding with developments, and white noise always has a circulant covariance.

Both $\boldsymbol{x}$ and $\boldsymbol{X}$ can be presumed to be one period of a periodic sequence to generate DFT/IDFT values for indices k and/or n outside the interval $(0, N-1)$.

For baseband modulation schemes where the time-domain signals are real, the restriction of $X_{N-n} = X_n^*$ for $n = 1, ..., N-1$ must hold. This restriction implies that there are $\bar{N} = N/2$ independent complex dimensions when N is even[4].

Lemma 3.2.1 (DFT is M for circularly prefixed VC) *The circulant matrix* $\mathbf{P} = \mathbf{M}\boldsymbol{\Lambda}\mathbf{M}^*$ *has a eigen-decomposition with* $\mathbf{M} = \mathcal{Q}$. *Furthermore, the diagonal values of* $\boldsymbol{\Lambda}$ *are* $\lambda_n = P_n$ *- that is, the eigenvalues of this circulant matrix are found through the DFT.*

Proof: The eigenvalues and corresponding eigenvectors of a matrix are found according to

$$\lambda_n \boldsymbol{q}_n = \mathbf{P}\boldsymbol{q}_n \quad , \tag{3.2.29}$$

which has matrix equivalent $\mathcal{Q}\boldsymbol{\Lambda} = \mathbf{P}\mathcal{Q}$. By investigating the effect of the bottom element of the product $\mathbf{P}\boldsymbol{q}_n$, where $\boldsymbol{q}_n$ is the column of $\mathcal{Q}$ corresponding to the nth point in the DFT, one finds

$$p_0 + p_1 e^{-\jmath\frac{2\pi}{N}n} + ... + p_{N-1} e^{-\jmath\frac{2\pi(N-1)}{N}n} = P_n \quad . \tag{3.2.30}$$

The next row up is then

$$p_0 e^{\jmath\frac{2\pi}{N}n} + p_1 + ... + p_{N-1} e^{-\jmath\frac{2\pi(N-2)}{N}n} = P_n e^{+\jmath\frac{2\pi}{N}n} \quad . \tag{3.2.31}$$

Clearly, repeating this procedure, one finds

$$\mathbf{P}\boldsymbol{q}_n = P_n \boldsymbol{q}_n \quad , \tag{3.2.32}$$

making $\boldsymbol{q}_n$ an eigenvector of $\mathbf{P}$. Thus, a choice of the eigenvectors is simply the columns of the DFT matrix $\mathcal{Q}$. For this choice, the eigenvalues must exactly be the DFT values of the rows of $\mathbf{P}$. **QED.**

[4]Actually, there are $\bar{N}-1$ complex dimensions and two real dimensions corresponding to $n = 0$ and $n = \bar{N}$. The extra two real dimensions are often grouped together and considered to be one "complex" dimension, with possibly unequal variances in the 2 dimensions. Usually, practical systems use neither of the real dimensions, leaving $\bar{N} - 1$ subchannels

Thus, the equivalent of the **F** and **M** matrices for VC with a cyclic prefix becomes the well-known DFT operation. Compared to the N^2 operations required by a full matrix multiplication, the DFT operation can be implemented with $N \log_2(N)$ operations, thereby offering a very large computational reduction for large N. Clearly, since an additional restriction of cyclic prefix is applied only to the case of OFDM channel partitioning, then, it can only perform as well or worse than VC without this restriction. When $N >> \nu$, the difference will be small simply because the difference in $\mathbf{P}$ is small. The designer should take care to note that the total transmit energy available for OFDM when executing loading algorithms is reduced by the factor $N/(N+\nu)$ because of the "wasted energy" in the cyclic prefix, the price for the computational simplification.

For the case of $N \to \infty$, the cyclic prefix becomes negligible and vector coding becomes OFDM. This is consistent with our earlier finding that eigenvector coding in the limit of large symbol size and infinite number of subchannels becomes MT. VC is a discrete form of eigenvector coding while OFDM is a discrete form of MT. In the limit, with sufficiently high sampling rate, OFDM, MT, VC and eigenvector coding all have the same performance. However, only OFDM is easily implemented. No other channel partitioning can perform better as long as N is sufficiently large.

An alternative view of OFDM

If the input sequence is periodic with a period of N samples, then, the corresponding output sequence from an FIR channel is also periodic. The output of the channel may be computed by taking the IDFT of the sequence $Y_n = P_n X_n$ (where P_n are the samples of the DFT of the channel pulse response samples). When the cyclic prefix is used, however, the input appears periodic as far as the last N samples of any symbol, which are all the receiver processes. OFDM thus can also be viewed as a circular convolution "trick," although this obscures the optimality and connections to VC.

Noise-Equivalent Channel Interpretation for OFDM

Colored Gaussian noise generally has power spectral density $\frac{\mathcal{N}_0}{2} R_n(f)$. The channel pulse response of the equivalent AWGN is the pulse response of the original channel adjusted by the inverse square-root autocorrelation of the Gaussian noise:

$$P(f) \to \frac{P(f)}{\sqrt{R_n(f)}} \quad . \tag{3.2.33}$$

The receiver uses a "noise-whitening" filtering $1/\sqrt{R_n(f)}$ as a preprocessor to the matched-filter and sampling device. The noise-equivalent viewpoint construes this whitening filter as part of the channel, thus altering the channel pulse response according to (3.2.33).

For OFDM, the noise equivalent view discussed earlier leads to a set of subchannels with signal gains $\bar{\mathcal{E}}_{\boldsymbol{x}}\frac{|P(n/T)|^2}{R_n(n/T)}$ and AWGN variance $\frac{\mathcal{N}_0}{2}$, as long as R_{nn} is also circulant. In practice, it is not unless $N \to \infty$; however, for reasonable noises, the autocorrelation matrix can be close to circulant for moderate values of N. A more common view, which avoids the use of the whitening filter, is that each of the subchannels has gain $\bar{\mathcal{E}}_{\boldsymbol{x}}|P(n/T)|^2$ and noise $\frac{\mathcal{N}_0}{2}R_n(n/T)$. The set of SNR's remains the same in either case, but the need for the whitening filter is eliminated in the second viewpoint. The second viewpoint has the minor flaw that the noise variances per subchannel are not completely correct unless the noise $n(t)$ is cyclo-stationary with period T or $N\Delta_T$ is large with respect to the duration of the noise autocorrelation function. In practice, the values used for N are almost always more than sufficiently large that implementations corresponding to the second viewpoint are ubiquitously found.

Partitioning using Filter Banks

By lifting the requirement on the basis functions of the multicarrier scheme to be DFT-based, better spectral containment properties or a greater reduction of out-of-band spectral energy can be obtained. To this end, Filtered MultiTone (FMT) modulation in the context of filter banks is argued in [45]. As discussed in [45], FMT-based scheme allows for efficient spectrum management while transferring the transmission overhead from the time to the frequency domain.

3.2.5 Stationary Equalization for Finite-length Partitioning

A potential problem in the implementation of OFDM and VC channel-partitioning methods is the use of the cyclic prefix or guard period of extra ν samples, where $\nu+1$ is the length in sampling periods of the channel pulse response. The required excess-bandwidth is then $\nu/(N+\nu)$. On many practical channels, ν can be large - values of several hundred occur in both wireless and wired transmission problems, especially when equivalent channels with narrowband noise are considered. To minimize the excess bandwidth, N needs to be very large, potentially 10,000 samples or more. Complexity is still minimal with OFDM methods with large N, when measured in terms of operation per unit time interval. However, large N implies a large memory requirement (to store the bit tables, energy tables, FEQ coefficients, and intermediate FFT/IFFT results), often dominating the implementation complexity. Further, large N implies longer **latency** (delay from transmitter input to receiver output) in processing. Long latency can create problems with synchronization and can also be unacceptable with certain higher-level data protocols used and/or with certain types of carried applications signals (like voice signals).

One solution to the latency problem, invented and studied by Chow [46] (not to be confused with the Chow of Loading Algorithms, see also the continuing work of Al-Dhahir [47]), is to use a combination of short-length fixed equalization and OFDM (or VC) to reduce ν, and thereby the required N, to reach the highest performance levels with less memory and less latency. The equalizer used is known as a *time equalizer* (TEQ), and is studied in this section. Subsection 3.2.6 reviews the performance of the finite-length TEQ for a target ν that is potentially less than the channel pulse-response length.

The very stationarity of the TEQ renders it suboptimum because the OFDM and VC transmission systems are inherently cyclo-stationary with period $N + \nu$ samples. Thus, an optimum equalizer is likely also cyclo-stationary or periodic (see [44]). There are several "simplifications" in the following development of the TEQ that are disturbing:

1. The equalizer should be periodic and is not.

2. The input sequence x_k will be assumed to be "white" or i.i.d. in the analysis when it is not in a system that optimizes the channel input.

3. A scalar stationary mean-square error is minimized rather than $\text{SNR}_{m,u}$.

Nonetheless, TEQs are heavily used and often very effective in bringing performance very close to the infinite-length multitone/modal-modulation optimum.

3.2.6 Finite-Length TEQ

Like Decision Feedback Equalization [48], the finite-length MMSE-TEQ problem is characterized by the error

$$e_k(\Delta) = \boldsymbol{b}^* \boldsymbol{x}_{k-\Delta} - \boldsymbol{w}^* \boldsymbol{y}_k \tag{3.2.34}$$

where $\boldsymbol{x}_{k-\Delta} = [x_{k-\Delta}\ x_{k-\Delta-1} \dots x_{k-\Delta-\nu}]^T$, $\boldsymbol{b}^* = [b_0^*\ b_1^* \ \dots\ b_\nu^*]$ is not necessarily causal, monic nor minimum-phase, $\boldsymbol{w}^* = [w_0^*\ w_1^* \ \dots\ w_{L-1}^*]$, L is the number of equalizer coefficients and may be at another sampling rate for the fractionally spaced case[5], and $\boldsymbol{y}_k = [y_k\ y_{k-1}\ \dots\ y_{k-L+1}]^T$. The length ν is typically less than the length of the channel pulse response and equal to the value anticipated for use in a multichannel modulation system like OFDM or VC.

The input/output relationship of the channel is

$$\boldsymbol{y}_k = \mathbf{P}\boldsymbol{x}_k + \boldsymbol{n}_k \quad . \tag{3.2.35}$$

P need not be a convolution matrix and can reflect whatever guard bands or other constraints placed upon the transmit signal. However, in most developments, this would render the channel output cyclo-stationary, so we will assume P is a convolution matrix.

The MSE is minimized with a constraint of $||\boldsymbol{b}||^2 = c$ a constant, with c usually chosen as $c = ||p||^2$ to force the SNR_{out} of the channel to be the SNR_{out} of the TEQ output. Minimization first over $\boldsymbol{w}$ forces

$$E[\boldsymbol{e}(\Delta)\boldsymbol{y}_k^*] = 0 \quad . \tag{3.2.36}$$

The solution forces

$$\boldsymbol{w}^* = \boldsymbol{b}^* \boldsymbol{R}_{\boldsymbol{xy}}(\Delta)\boldsymbol{R}_{\boldsymbol{yy}}^{-1} \tag{3.2.37}$$

for any $\boldsymbol{b}$. Then, $\boldsymbol{e}(\Delta)$ becomes

$$\boldsymbol{e}(\Delta) = \boldsymbol{b}^*(\boldsymbol{x}_{k-\Delta} - \boldsymbol{R}_{\boldsymbol{xy}}(\Delta)\boldsymbol{R}_{\boldsymbol{yy}}^{-1}\boldsymbol{y}) \quad . \tag{3.2.38}$$

(3.2.38) has a vector of the minimized MSE finite-length linear equalizer (LE) errors within the parentheses. This MMSE-LE error vector has an autocorrelation matrix that can be computed as

$$\boldsymbol{R}_{\boldsymbol{ee}}(\Delta) = \mathcal{E}_{\boldsymbol{x}}\mathbf{I} - \boldsymbol{R}_{\boldsymbol{xy}}(\Delta)\boldsymbol{R}_{\boldsymbol{yy}}^{-1}\boldsymbol{R}_{\boldsymbol{yx}}(\Delta) \quad . \tag{3.2.39}$$

Tacit in (3.2.39) is the assumption that $\boldsymbol{R}_{\boldsymbol{yy}}$ is a fixed $L \times L$ matrix, which is equivalent to assuming that the input to the channel is stationary as is the noise. In the case of transmission methods like OFDM and VC, this stationarity is not true and the input is instead cyclo-stationary over a symbol

[5] The equalizer output should always be decimated to the channel-input sampling rate, which restricts this development to integer oversampling factors to maintain a fixed equalizer rather than one that adds yet another level of cyclo-stationarity.

period. Furthermore, the initial solution of the TEQ is often computed before transmit optimization has occurred so that $\boldsymbol{R_{xx}}$ is often assumed to be white or a diagonal matrix as will also be the case in this development. With these restrictions/assumptions, the overall MMSE is then the minimum of

$$\text{MMSE} = \min_{\boldsymbol{b},\Delta} \boldsymbol{b}^* \boldsymbol{R_{ee}}(\Delta)\boldsymbol{b} \quad . \tag{3.2.40}$$

with the constraint that $\|\boldsymbol{b}\|^2 = c$. The solution of this problem is easily shown to be the eigenvector of $\boldsymbol{R_{ee}}$ that corresponds to the minimum eigenvalue. Thus

$$\boldsymbol{b} = \sqrt{c}\boldsymbol{q}_{min}(\Delta) \quad . \tag{3.2.41}$$

Thus, the TEQ setting is determined by computing $\boldsymbol{R_{ee}}(\Delta)$ for all reasonable values of Δ and finding the MMSE (Δ) with the smallest value. The eigenvector corresponding to this Δ is the correct setting for $\boldsymbol{b}$ (with scaling by $\sqrt{c}$) and $\boldsymbol{w}$, the TEQ, is determined through (3.2.37).

The $L \times L$ matrix $\boldsymbol{R_{yy}}$ need only be inverted once, but the eigenvectors must be computed several times corresponding to all reasonable values of Δ. This computation can be intensive. The matrix $\boldsymbol{R_{yy}}$ is given by

$$\boldsymbol{R_{yy}} = \mathbf{P}\boldsymbol{R_{xx}}\mathbf{P}^* + \boldsymbol{R_{nn}} = \bar{\mathcal{E}}_{\boldsymbol{x}}\mathbf{P}\mathbf{P}^* + \boldsymbol{R_{nn}} \quad . \tag{3.2.42}$$

Any stationary $\boldsymbol{R_{xx}}$ can be assumed, but the simplification on the right in (3.2.42) occurs only for white $\boldsymbol{R_{xx}}$. If a non-stationary (perhaps an $\boldsymbol{R_{xx}}$ representing some known input optimization and cyclo-stationarity) $\boldsymbol{R_{xx}}$ is used, then $\boldsymbol{R_{yy}}$ becomes cyclo-stationary also and a function of both position within a symbol as well as delay Δ. The cross correlation matrix is

$$\bar{\mathbf{R}}_{xy}(\delta) = E\left[\boldsymbol{x}_{k-\Delta}\boldsymbol{y}_k\right] = E\left[\boldsymbol{x}_{k-\Delta}\boldsymbol{x}_k^*\right]\mathbf{P}^* \quad . \tag{3.2.43}$$

The cross-correlation matrix will have zeros in many positions determined by Δ unless an optimized input is being used and then again this matrix would also become a function of both Δ and the position within a symbol. Since it is hard to know which position within a symbol to optimize, or to know in advance the optimized spectrum used on the channel (which usually is affected by the TEQ), the TEQ is initialized first prior to channel identification, partitioning, and loading.

The TEQ performance increase can indeed be much greater than 4 dB on channels with more severe responses. Note the equalizer is not a DFE and does not maintain white noise nor channel magnitude in the finite-length case. Pal [49] has noted that in severely-constrained ISI channels that the element of fractional spacing of the TEQ for finite lengths can lead to significant performance gain with respect to implementation at the sampling rate of the transmitter, even when both have the same complexity.

3.3 Loading of Parallel Channels

Using the channel partitioning methods described in Section 3.2, an OFDM transmission system has N independent subchannels where each subchannel can transmit two-dimensional modulation schemes, such as QAM. Therefore, an OFDM system is seen to have a total number of $2N$ dimensions. To achieve the best performance, the amount of energy and the number of bits allotted to each subchannel should be optimized according to the channel and noise frequency responses. This form of optimization is commonly known as *loading*. Although extremely efficient loading algorithms abound, the challenge lies in the accurate estimation of the channel and noise characteristics within a relatively short period of time. Therefore, loading algorithms are most suitable for use in slow-fading scenarios, which are typical of fixed wireless applications, where the channels change at rates that are significantly lower than the OFDM symbol rate, and the parametrization of the channel and noise frequency responses can be communicated back to the transmitter at the expense of some information rate. An alternative transmission scheme is to assign equal number of bits and equal amount of energy to each subchannel, and implement powerful coding techniques described in Section 3.4 to compensate for the frequency selectivity of the channel.

This section describes the performance analysis and optimization of performance for the entire set of subchannels. Since the concept of the *SNR gap* for a single channel is crucial to some of the analysis, Subsection 3.3.1 first reviews a single channel. Then, Subsections 3.3.2, 3.3.3, and 3.3.4 use these single-channel results to establish results for parallel channels. Optimal and suboptimal loading algorithms, and their respective complexities are next presented in Subsections 3.3.5 to 3.3.10.

3.3.1 Single-Channel Gap Analysis

The so-called "gap" analysis of QAM is reviewed here briefly in preparation for its use in loading algorithms for OFDM transmission.

On an AWGN channel, with input energy per dimension $\bar{\mathcal{E}}$, gain H, and noise power per dimension σ^2, the signal-to-noise ratio is $\text{SNR} = \bar{\mathcal{E}}|H|^2/\sigma^2$.

$\bar{b}$	.5	1	2	3	4	5
SNR for $P_e = 10^{-6}$ (dB)	8.8	13.5	20.5	26.8	32.9	38.9
$2^{2\bar{b}} - 1$ (dB)	0	4.7	11.7	18.0	24.1	30.1
Γ (dB)	8.8	8.8	8.8	8.8	8.8	8.8

Table 3.1. Table of SNR Gaps for $P_e = 10^{-6}$.

This channel has a maximum data rate or capacity of

$$\bar{c} = \frac{1}{2}\log_2(1 + \text{SNR}) \quad \text{bits/dimension} \quad . \tag{3.3.1}$$

For passband transmission where the subchannels are complex, then the capacity of this channel is $2\bar{c}$ bits per transmission. Any reliable and implementable system must transmit at a data rate below capacity. Throughout the rest of this section, a bar on top of any variable indicates that the unit is quantified per dimension.

The gap is a convenient mechanism for analyzing systems that transmit at a rate of $\bar{b} < \bar{c}$. For any given coding scheme, and a given target probability of symbol error P_e, the *SNR gap* Γ is defined according to

$$\Gamma \triangleq \frac{2^{2\bar{c}} - 1}{2^{2\bar{b}} - 1} = \frac{\text{SNR}}{2^{2\bar{b}} - 1} \quad . \tag{3.3.2}$$

For many practical and used classes of coding schemes, this gap is constant for nearly all values of $\bar{b}$. When the gap $\Gamma = 1$ (0 dB), then $\bar{b} = \bar{c}$, that is, the code is so good that the maximum number of bits per dimension can be transmitted. As the gap increases, the coding scheme is less powerful. To illustrate such a constant gap, Table 3.1 lists achievable $\bar{b}$ for uncoded QAM schemes using square constellations. For $\bar{b} < 1$, this gap concept assumes additional mild hard coding in the "uncoded" system to obtain a small additional coding gain. The development of parallel channels will not include subchannel data rates with $\bar{b} <.5$.

For uncoded square QAM, which has the same gap as uncoded PAM, and $P_e = 10^{-6}$, the SNR gap Γ is constant at 8.8 dB (again, assuming 1.7 dB coding gain for $\bar{b} = .5$). For uncoded square QAM or PAM and $P_e = 10^{-7}$, the gap is fixed at about 9.5 dB. The use of codes, say trellis coding and/or forward error correction, reduces the gap. A very well-coded system may have a gap as low as 1 dB at $P_e \leq 10^{-6}$. Figure 3.3 plots $\bar{b}$ versus SNR for various gaps. The curve for $\Gamma = 9$ dB approximates uncoded PAM or QAM transmission at a probability of symbol error $P_e \approx 10^{-6}$. More powerful, but implementable, "turbo" codes [50] have gaps as small as 1 dB.

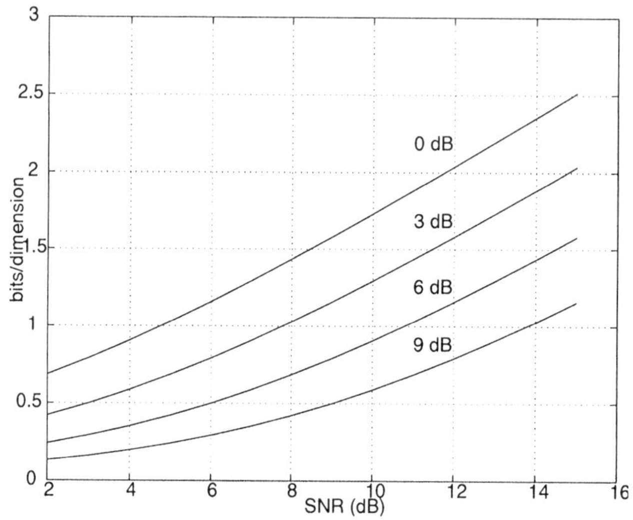

Figure 3.3. Illustration of achievable bit rates versus SNR for various gaps.

The gap Γ measures the SNR distance from capacity, as evident in the following rewritten definition of (3.3.2):

$$\bar{b} = \frac{1}{2}\log_2\left(1 + \frac{\text{SNR}}{\Gamma}\right) \quad . \tag{3.3.3}$$

Equivalently, (3.3.3) states that the achieved bit rate on the channel is reduced from capacity as if the channel had a lower SNR. That is, the achievable bit rate is the capacity of an AWGN channel with SNR $\leftarrow$ SNR/Γ. As the gap of the code goes to 0 dB ($\Gamma = 1$), capacity is achieved. The fact that this gap is constant for essentially all values of $\bar{b}$ is particularly useful with multichannel transmission methods, which may vary $\bar{b}$ over the constituent subchannels while using a coding scheme of constant gap.

For a given coding scheme, practical transmission designs often mandate a specific value for $\bar{b}$, or equivalently a fixed data rate. In this case, the design is not for $\bar{b}_{max} = \frac{1}{2}\log_2(1 + \text{SNR}/\Gamma)$, but rather for some $\bar{b} \leq \bar{b}_{max}$. The *margin* measures the excess SNR for that given bit rate.

Definition 3.3.1: *The* margin, γ_m, *for transmission on an AWGN (sub)channel with a given SNR, a given number of bits per dimension $\bar{b}$, and a given coding-scheme/target $\bar{P}_e$ with gap Γ is the amount by which the SNR can be reduced (increased for negative margin in dB) and still maintain a probability of error at or below the target P_e.*

Margin can be calculated through the use of the gap formula by

$$\gamma_m = \frac{2^{2\bar{b}_{max}} - 1}{2^{2\bar{b}} - 1} = \frac{\frac{\text{SNR}}{\Gamma}}{2^{2\bar{b}} - 1} \quad . \tag{3.3.4}$$

The margin is the amount by which the SNR on the channel may be lowered before performance degrades to a probability of error greater than the target P_e used in defining the gap. A negative margin in dB means that the SNR must improve by the magnitude of the margin before the target P_e is achieved. Rewriting the margin relation in (3.3.4) gives the bit rate for a channel with SNR, a coding scheme of gap Γ, and a margin γ_m:

$$\bar{b} = .5 \log_2 \left(1 + \frac{\text{SNR}}{\Gamma \gamma_m}\right) \quad \text{bits/dimension} \quad . \tag{3.3.5}$$

EXAMPLE 3.3.1: Assume an AWGN channel with an SNR of 20.5 dB. The capacity of this channel is then

$$\bar{c} = .5 \log_2 (1 + \text{SNR}) = 3.5 \ \text{bits/dimension} \ \ (\ 7 \ \text{bits/Hz}\ ) \quad .$$

With QPSK at a $P_e = 10^{-6}$, which has a gap Γ =8.8 dB,

$$\bar{b} = .5 \log_2 \left(1 + \frac{\text{SNR}}{10^{.88}}\right) = 2 \ \text{bits/dimension} \ \ (\ 4 \ \text{bits/Hz}\ ) \quad .$$

With concatenated trellis and forward error correction (or with "turbo codes") and a coding gain of 7 dB at $P_e = 10^{-6}$, the achievable data rate is

$$\bar{b} = .5 \log_2 \left(1 + \frac{\text{SNR}}{10^{.18}}\right) = 3 \ \text{bits/dimension} \ \ (\ 6 \ \text{bits/Hz}\ ) \quad .$$

Suppose a transmission application requires $\bar{b} = 2.5$ bits/dimension, then the margin for the coded system is

$$\gamma_m = \frac{2^{2 \cdot 3} - 1}{2^{2 \cdot 2.5} - 1} = \frac{63}{31} \approx 3 \ \text{dB} \quad .$$

This means the noise power would have to be increased (or the transmit power reduced) by more than 3 dB before the target probability of error of 10^{-6} is no longer met when $\bar{b} = 2.5$. Alternatively, suppose a design transmits 4-QAM over this channel and no coding is used, then the margin is

$$\gamma_m = \frac{2^{2\cdot 2} - 1}{2^{2\cdot 1} - 1} = \frac{15}{3} \approx 7 \text{ dB} \quad .$$

The gap is a powerful concept and will be used to facilitate the understanding of loading algorithms for a set of parallel channels.

3.3.2 A Single Performance Measure for Parallel Channels - Geometric SNR

In an *optimized* OFDM transmission system, it is usually desirable to have all subchannels with the same P_e. Otherwise, if one subchannel had significantly higher P_e than the others, then this high-P_e subchannel would dominate the bit error rate. Constant P_e occurs when all subchannels use the same coding scheme with a constant gap Γ. In this case, a single performance measure characterizes a multichannel transmission system. This measure is a *geometric SNR* that can be directly compared to the detection SNR of equalized transmission systems.

For a set of N parallel complex subchannels, where b_n denotes the number of bits assigned to the nth subchannel, the aggregate number of bits per dimension is

$$\bar{b} = \frac{1}{2N}\sum_{n=1}^{N} b_n = \frac{1}{2N}\sum_{n=1}^{N}\log_2\left(1 + \frac{\text{SNR}_n}{\Gamma}\right) \tag{3.3.6}$$

$$= \frac{1}{2N}\log_2\left(\prod_{n=1}^{N}\left[1 + \frac{\text{SNR}_n}{\Gamma}\right]\right) \tag{3.3.7}$$

$$= \frac{1}{2}\log_2\left(1 + \frac{\text{SNR}_{m,u}}{\Gamma}\right) \quad . \tag{3.3.8}$$

The variable $\text{SNR}_{m,u}$ denotes the *multichannel SNR* for a set of parallel channels and is defined by

$$\text{SNR}_{m,u} = \left[\left(\prod_{n=1}^{N}\left[1 + \frac{\text{SNR}_n}{\Gamma}\right]\right)^{\frac{1}{N}} - 1\right]\Gamma \quad . \tag{3.3.9}$$

When, as typical, the terms involving +1 can be ignored, the multichannel SNR is approximately the geometric mean of the SNRs on each of the subchannels

$$\text{SNR}_{m,u} \approx \text{SNR}_{geo} = \left(\prod_{n=1}^{N} \text{SNR}_n\right)^{\frac{1}{N}} \quad . \tag{3.3.10}$$

The bit rate given in (6.2.16) is as if the aggregate set of parallel channels were a single AWGN with $\text{SNR}_{m,u}$. Thus, the entire set can be replaced in analysis by a single AWGN channel with no intersymbol interference and an SNR $= \text{SNR}_{m,u}$, even if all the subchannels have different SNRs. If all the subchannels had the same SNR, then $\text{SNR}_{m,u}$, as given in (3.3.9), would trivialize to that SNR.

3.3.3 Water-Filling Optimization

To *maximize* the data rate, $R = b/T$, for a set of parallel subchannels when the symbol rate $1/T$ is fixed, requires maximization of the achievable $b = \sum_n b_n$ over b_n under a given total input energy and a target probability of error. Specifically, the number of bits allocated to the nth subchannel is

$$b_n = \log_2(1 + \frac{\mathcal{E}_n g_n}{\Gamma}) \quad , \tag{3.3.11}$$

where $g_n = |H_n|^2/2\sigma_n^2$ represents the subchannel signal-to-noise ratio when the transmitter applies a unit energy to that subchannel. The ratio g_n is a fixed function of the channel, but $\mathcal{E}_n$, which denotes the 2-D subsymbol energy allotted to the nth subchannel, can be optimized to maximize b, subject to a total transmit energy constraint of

$$\sum_{n=1}^{N} \mathcal{E}_n \leq \mathcal{E}_x \quad . \tag{3.3.12}$$

Since $\log(1+x)$ is a strictly increasing function of x, the total energy constraint of (3.3.12) will be binding, i.e., equality is met. Using Lagrange multipliers, the cost function becomes

$$\frac{1}{\ln 2} \sum_n \ln\left(1 + \frac{\mathcal{E}_n g_n}{\Gamma}\right) + \lambda\left(\mathcal{E}_x - \sum_{n=1}^{N} \mathcal{E}_n\right) \quad . \tag{3.3.13}$$

Differentiating with respect to $\mathcal{E}_n$ produces

$$\frac{1}{\ln 2} \frac{1}{\mathcal{E}_n + \frac{\Gamma}{g_n}} = \frac{\lambda \Gamma}{g_n} \quad . \tag{3.3.14}$$

Thus, the aggregate bit rate in b is maximized when the optimum subchannel transmit energies satisfy

$$\mathcal{E}_n + \frac{\Gamma}{g_n} = K = \text{ constant} \quad , \tag{3.3.15}$$

and K is chosen such that the total energy constraint given by (3.3.12) is met.

When $\Gamma = 1$ (0 dB), the maximum data rate or capacity of the parallel set of channels is achieved. Figure 3.4 illustrates water-filling for a multicarrier transmission system with 6 subchannels. The term "water-filling" arises from the analogy of the curve of Γ/g_n to a bowl into which water (energy) is poured, filling the bowl until there is no more energy to use. The water will rise to a constant flat level in the bowl, as shown by the shaded regions in Figure 3.4. The amount of water/energy in any subchannel is the depth of the water at the corresponding point in the bowl. When $\Gamma \neq 1$, the form of the water-filling optimization remains the same as long as Γ is constant over all subchannels. The scaling by $\Gamma > 1$ makes the inverse subchannel SNR curve, Γ/g_n appear steeper with n, thus leading to a narrower (fewer used subchannels) optimized transmit band than when capacity is achieved.

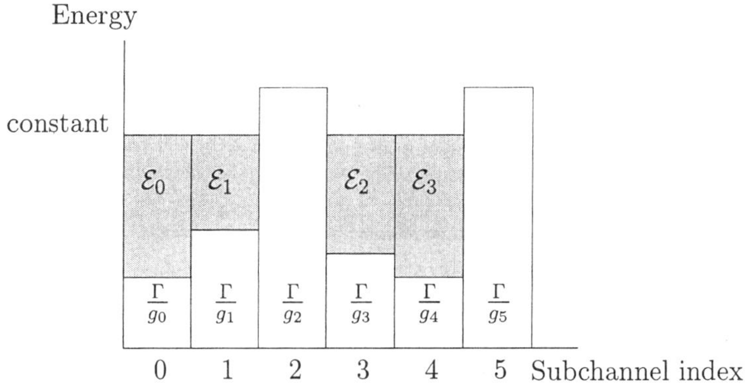

Figure 3.4. Illustration of discrete water-filling for 6 subchannels.

Equation (3.3.15) is solved for the constant K under the constraints that $\sum_n \mathcal{E}_n = \mathcal{E}_x$ and $\mathcal{E}_n \geq 0$. Note in Figure 3.4 that 4 of the six subchannels have positive energies, while 2 are eliminated for having negative energies, or equivalently for having normalized noise powers that exceed the constant line of water-filling. The 4 used subchannels have energies that make the sum of normalized noise powers and transmit energies constant for all subchannels.

The water-filling solution is unique because the rate function being maximized is concave. Therefore, there is a unique optimum energy distribution and a corresponding rate distribution among the subchannels with multichannel or multicarrier modulation.

3.3.4 Margin Maximization

For many transmission systems, a variable data rate is not desirable. In this case, the designer can instead maximize the performance margin at a given *fixed data rate.* Maximizing the fixed-rate margin is equivalent to minimizing the total energy allocated

$$\sum_{n=1}^{N} \mathcal{E}_n \quad , \tag{3.3.16}$$

subject to the data rate being fixed according to

$$\sum_{n=1}^{N} \log_2 \left(1 + \frac{\mathcal{E}_n g_n}{\Gamma}\right) \geq b \quad \text{bits per symbol} \quad . \tag{3.3.17}$$

Again, this fixed rate constraint is met at equality since $\log(1+x)$ is a strictly increasing function of x.

The maximum margin is then

$$\gamma_m = \frac{\mathcal{E}_x}{\sum_{n=1}^{N} \mathcal{E}_n} \quad . \tag{3.3.18}$$

Using Lagrange multipliers, the solution to the energy minimization problem is again the water-filling solution given by

$$\mathcal{E}_n + \frac{\Gamma}{g_n} = K_{ma} = \text{ constant} \quad . \tag{3.3.19}$$

However, in this case, the water/energy is poured only until the number of bits per symbol, which is computed by summing the bits carried by each subchannel according to (3.3.11), is equal to the given fixed rate b in (3.3.17). In other words, (3.3.19) is solved for the constant under the constraints that $\sum_n b_n = b$ and $\mathcal{E}_n \geq 0$. This solution is unique because the energy function being minimized is convex. Then, the maximum margin is computed according to (3.3.18). A $\gamma_m > 1$ implies that the total energy allocated is more than sufficient for providing the specified data rate b. A $\gamma_m < 1$ indicates that better coding or more transmit energy is necessary to transmit at a fixed rate of b bits per symbol at the target P_e over the channel. Scaling $\mathcal{E}_n$ of the used subchannels by γ_m in (3.3.18) alters the minimized total energy to be the correct transmitted limit of $\mathcal{E}_x$. Therefore, each used subchannel has the same maximum margin of γ_m.

The following sections discuss loading algorithms for computing the water-filling solutions for both the rate and margin maximization problems.

3.3.5 Loading Algorithm Classification

There are two types of *loading algorithms* – those that maximize data rate, and those that maximize performance margin at a given fixed data rate. Loading algorithms compute values for b_n and $\mathcal{E}_n$ for each subchannel in a parallel set of subchannels. Examples of loading algorithms are the optimum water-filling algorithms of Sections 3.3.6 and 3.3.7. These water-filling algorithms may produce b_n that have fractional parts or that may be very small. Such small or fractional b_n can complicate encoder and decoder implementation. To address this issue, Section 3.3.8 describes optimal and suboptimal discrete loading algorithms that constrain b_n to discrete information units.

Definition 3.3.2: Rate-adaptive (RA) loading criterion
A rate-adaptive *loading procedure maximizes (or approximately maximizes) the number of bits ber symbol subject to a fixed total transmit energy constraint:*

$$\begin{aligned} \max_{\mathcal{E}_n} \quad & b = \sum_{n=1}^{N} \log_2\left(1 + \frac{\mathcal{E}_n g_n}{\Gamma}\right) \\ \text{subject to}: \quad & \sum_{n=1}^{N} \mathcal{E}_n \le \mathcal{E}_x \\ & \mathcal{E}_n \ge 0, \quad n = 1, \ldots, N \end{aligned} \tag{3.3.20}$$

Definition 3.3.3: Margin-Adaptive (MA) loading criterion
A margin-adaptive *loading procedure minimizes (or approximately minimizes) the total energy used subject to a fixed total bits per*

symbol constraint:

$$\min_{\mathcal{E}_n} \quad \mathcal{E} = \sum_{n=1}^{N} \mathcal{E}_n$$

$$\text{subject to :} \quad \sum_{n=1}^{N} \log_2\left(1 + \frac{\mathcal{E}_n g_n}{\Gamma}\right) \geq b \qquad (3.3.21)$$

$$\mathcal{E}_n \geq 0, \quad n = 1, \ldots, N$$

The maximum margin is then

$$\gamma_{max} = \frac{\mathcal{E}_x}{\mathcal{E}} \quad . \qquad (3.3.22)$$

3.3.6 Computing Water Filling for RA Loading

From Section 3.3.3, the set of linear equations whose solution gives the water-filling distribution are

$$\mathcal{E}_1 + \Gamma/g_1 = K \qquad (3.3.23)$$

$$\mathcal{E}_2 + \Gamma/g_2 = K \qquad (3.3.24)$$

$$\vdots = \vdots$$

$$\mathcal{E}_N + \Gamma/g_N = K \qquad (3.3.25)$$

$$\mathcal{E}_1 + ... + \mathcal{E}_N = \mathcal{E}_x \quad . \qquad (3.3.26)$$

There are $N+1$ equations in $N+1$ unknowns, namely, $\mathcal{E}_n, n = 1, ..., N$ and K. Ignoring the nonnegativity constraints of $\mathcal{E}_n$ for now, the first N equations are summed to obtain the water-filling constant K

$$K = \frac{1}{N}\left[\mathcal{E}_x + \Gamma \sum_{n=1}^{N} \frac{1}{g_n}\right] \quad . \qquad (3.3.27)$$

Then, the subchannel energies are computed via:

$$\mathcal{E}_n = K - \frac{\Gamma}{g_n}, \quad \forall\, n = 1, ..., N \quad . \qquad (3.3.28)$$

It is clear that the first subchannel to violate the nonnegative energy constraint, that is $\mathcal{E}_n < 0$, is the one that has the smallest g_n. If this is true, the subchannel with the smallest g_n is eliminated from the subchannel set and its corresponding $\mathcal{E}_n$ should be zeroed. Then, (3.3.27) and (3.3.28) are solved again for the remaining subchannels with $N \leftarrow N-1$. This iteration process continues until the first solution with no negative energies occurs. The algorithm steps for RA water-filling are summarized in Table 3.2.

EXAMPLE 3.3.2: Let us assume that the complex baseband-equivalent channel is modelled by the discrete-time response of $H(D)^6 = 1 + (.3+j)D^{-1}$. Furthermore, let's assume that this channel is orthogonally subdivided into 4 subchannels (8 dimensions), where each subchannel is centered about $f = n/4$, $n = 0, 1, 2, 3$. The total available transmit energy is $\mathcal{E}_x = 8$, which translates to 1 unit of energy per dimension, and the total noise power per subchannel is $2\sigma^2 = 0.4$, which is equivalent to a noise power per dimension of 0.2.

n	$\|H_n\|$	$\|H_n\|^2$	$g_n = \frac{\|H_n\|^2}{2\sigma^2}$
0	1.64	2.69	6.725
1	2.02	4.09	10.225
2	1.22	1.49	3.725
3	0.3	0.09	0.225

The following outlines the steps for computing the water-filling solution with $\Gamma = 1$ for this channel.

With all subchannels used ($N = 4$), the water-filling level K is computed, using (3.3.27), to be

$$K = \frac{1}{4}\left(8 + \frac{1}{6.725} + \frac{1}{10.225} + \frac{1}{3.725} + \frac{1}{0.225}\right) = 3.2398 \quad .$$

Since this results in the last subchannel, which has the smallest g_n, to have negative energy, i.e., $\mathcal{E}_3 = 3.2398 - 1/.0225 = -1.205 < 0$, subchannel 3 should be eliminated from the optimization process. Using only the first 3 subchannels ($N = 3$),

$$K = \frac{1}{3}\left(8 + \frac{1}{6.725} + \frac{1}{10.225} + \frac{1}{3.725}\right) = 2.8383 \quad .$$

[6] $D = Z^{-1}$ for those not familiar with the coding/transmission theory's near ubiquitous use of D to denote a delay rather than the signal-processing theorist's Z^{-1}.

The subchannel energies, $\mathcal{E}_n$, for subchannels indexed from 0 through 2 are 2.6896, 2.7405, and 2.5699, which are all positive, indicating that the water-filling solution has been found. The corresponding number of bits carried by each subchannel, which is computed according to $b_n = \log_2(Kg_n)$, are 4.25, 4.86, and 3.40 bits respectively. This gives an aggregate number of bits per dimension of 1.56. The accuracy of the computation of capacity can be increased by further increasing N. The multichannel SNR for this 4-subchannel example is $\mathrm{SNR}_{m,u} = 2^{2(1.56)} - 1 = 8.9$ dB.

3.3.7 Computing Water-Filling for MA Loading

As described in Section 3.3.4, the MA water-filling solution will also have a constant K_{ma} that satisfies

$$\mathcal{E}_n = K_{ma} - \frac{\Gamma}{g_n} \tag{3.3.29}$$

for each *used* subchannel. The bit-rate constraint then becomes

$$\begin{aligned} b &= \sum_{n=1}^{N} \log_2\left(1 + \frac{\mathcal{E}_n g_n}{\Gamma}\right) && (3.3.30) \\ &= \sum_{n=1}^{N} \log_2\left(\frac{K_{ma} g_n}{\Gamma}\right) && (3.3.31) \\ &= \log_2\left(\prod_{n=1}^{N} \frac{K_{ma} g_n}{\Gamma}\right) && (3.3.32) \\ 2^b &= \left(\frac{K_{ma}}{\Gamma}\right)^N \prod_{n=1}^{N} g_n && (3.3.33) \end{aligned}$$

Thus, the water-filling constant for MA loading is

$$K_{ma} = \Gamma \left(\frac{2^b}{\prod_{n=1}^{N} g_n}\right)^{\frac{1}{N}} = \Gamma\, 2^{(b - \sum_{n=1}^{N} \log_2 g_n)/N} \quad , \tag{3.3.34}$$

and the subchannel energies are computed via:

$$\mathcal{E}_n = K - \frac{\Gamma}{g_n}, \ \forall\, n = 1, ..., N \ . \tag{3.3.35}$$

The same nonnegativity argument from Section 3.3.6 applies here as well. Table 3.2 summarizes the MA water-filling algorithm.

RA Water-filling Algorithm	**MA Water-filling Algorithm**
1. Sort g_n in a descending order such that $g_1 = \max_n g_n$ and $g_N = \min_n g_n$	
2. Set the number of used subchannels $N^* = N$	
3. Initialize $\mathcal{G} = \sum_{n=1}^{N} \frac{1}{g_n}$, $K = \frac{1}{N}(\mathcal{E}_x + \Gamma\mathcal{G})$	3. Initialize $\mathcal{G} = \sum_{n=1}^{N} \log_2 g_n$, $K = \Gamma 2^{(b-\mathcal{G})/N}$
4. While $(\mathcal{E}_{N^*} = K - \frac{\Gamma}{g_{N^*}}) < 0$: $\mathcal{G} \leftarrow \mathcal{G} - \frac{1}{g_{N^*}}$, $N^* \leftarrow N^* - 1$, $K = \frac{1}{N^*}(\mathcal{E}_x + \Gamma\mathcal{G})$	4. While $(\mathcal{E}_{N^*} = K - \frac{\Gamma}{g_{N^*}}) < 0$: $\mathcal{G} \leftarrow \mathcal{G} - \frac{1}{g_{N^*}}$, $N^* \leftarrow N^* - 1$, $K = \Gamma 2^{(b-\mathcal{G})/N^*}$
5. Compute $\mathcal{E}_n = \begin{cases} K - \frac{\Gamma}{g_n} & n=1,\ldots,N^* \\ 0 & n=N^*,\ldots,N \end{cases}$, $b_n = \log_2(1 + \frac{\mathcal{E}_n g_n}{\Gamma}) = \log_2(1 + \frac{K g_n}{\Gamma})$	5. Compute $\mathcal{E}_n = \begin{cases} K - \frac{\Gamma}{g_n} & n=1,\ldots,N^* \\ 0 & n=N^*,\ldots,N \end{cases}$, $b_n = \log_2(1 + \frac{\mathcal{E}_n g_n}{\Gamma}) = \log_2(1 + \frac{K g_n}{\Gamma})$

Table 3.2. The RA and MA water-filling algorithms

Note that the sum in the expression for K in Table 3.2 is always over the *used* subchannels. With N^* denoting the number of used subchannels, the aggregate bit rate in bits per dimension is

$$\bar{b} = \frac{1}{2N}\sum_{n=1}^{N^*} b_n = \frac{1}{2}\log_2\left(\frac{K^{\frac{N^*}{N}}\left(\prod_{n=1}^{N^*} g_n\right)^{\frac{1}{N}}}{\Gamma^{\frac{N^*}{N}}}\right) \tag{3.3.36}$$

$$= \frac{1}{2}\log_2\left(1 + \frac{\mathrm{SNR}_{m,u}}{\Gamma}\right) \quad . \tag{3.3.37}$$

and the SNR for the set of parallel channels is therefore

$$\mathrm{SNR}_{m,u} = \Gamma\left\{\left(\frac{K}{\Gamma}\right)^{\frac{N^*}{N}}\prod_{n=1}^{N^*} g_n^{\frac{1}{N}} - 1\right\} \quad . \tag{3.3.38}$$

The RA and MA algorithms in Table 3.2 each begins with an ordering of the subchannel signal-to-noise ratios g_n. As will be explained in Section 3.3.9, the complexities of these algorithms can be further reduced by recognizing the fact that an absolute ordering is not necessary.

3.3.8 Loading with Discrete Information Units

Water-filling algorithms produce bit distributions where b_n can be any real number. Realization of bit distributions with non-integer values can be difficult. Discrete loading algorithms allow the computation of bit distributions that are more amenable to implementation with a finite granularity.

Definition 3.3.4: Information Granularity
The granularity β *of a multichannel transmission system is the smallest incremental unit of information that can be transmitted. The number of bits on any subchannel is thus*

$$b_n = B_n\beta, \quad \forall n = 1, \ldots, N \quad , \tag{3.3.39}$$

where $B_n \geq 0$ *is an integer.*

Typically, β, takes on values such as .25, .5, .75, 1, or 2 bits.

There are two approaches to discrete (finite-granularity) loading. The first approach, which is based on greedy methods in mathematics, was first suggested by Hughes-Hartog [51]. Recently, Campello de Souza [52] and Levin [53] independently developed a complete and mathematically verifiable framework, and what are now known as **Levin-Campello (LC) Algorithms** that offer significant improvements as well as circumvent many drawbacks in the original Hughes-Hartog methods. The second approach, which computes the bit distribution by rounding of approximate water-filling results, is called **Chow's algorithm**.

Optimum Discrete Loading Algorithms

Optimum discrete loading algorithms recognize the discrete loading problem as an instance of what is known as "greedy optimization" in mathematics. The basic concept is that each increment of additional information to be transported by a multichannel transmission system is placed on the subchannel that requires the least incremental energy for its transport. Such algorithms are optimum for loading when the information granularity β is the same for all subchannels, which is usually the case.[7]

Several definitions are necessary to formalize discrete loading.

> **Definition 3.3.5: Bit distribution vector**
> *The* bit distribution vector *for a set of parallel subchannels that in aggregate carry b total bits of information is given by*
>
> $$\boldsymbol{b} = [b_1\ b_2\ ...\ b_N] \quad . \tag{3.3.40}$$

Discrete loading algorithms can use any monotonically increasing relation between transmit symbol energy and the number of bits transmitted on any subchannel. In general, the concept of incremental energy (essentially replacing the gap relation) is important to discrete loading:

> **Definition 3.3.6:** Incremental Energy *and* Symbol Energy
> *The symbol energy $\mathcal{E}_n$ for an integer number of information units $b_n = B_n\beta$ on each subchannel can be notationally generalized to the* energy function
>
> $$\mathcal{E}_n \leftarrow \mathcal{E}_n(b_n) \quad , \tag{3.3.41}$$

[7] With a subchannel granularity of β, a subchannel at DC (PAM) has a granularity of β bit per dimension while the other subchannels (QAM) each has a granularity of β bit per two dimensions, or equivalently $\beta/2$ bit per dimension. This anomaly rarely occurs in practice because DC is almost never passed in transmission channels and passband channels have two-dimensional complex QAM subchannels at the baseband-equivalent of DC.

where the dependence of the symbol energy on the number of information units transmitted, b_n, is explicitly shown. The incremental energy *needed to transmit b_n information units on a subchannel is the amount of additional energy required to send the B_n^{th} information unit with respect to the $(B_n - 1)^{th}$ information unit (that is, one more unit of β). The incremental energy is then defined as*

$$e_n(b_n) \stackrel{\Delta}{=} \mathcal{E}_n(b_n) - \mathcal{E}_n(b_n - \beta) \quad . \qquad (3.3.42)$$

EXAMPLE 3.3.3: Consider an uncoded SQuare (SQ) QAM with $\beta = 1$ bit, and d_n denoting the minimum input distance between points on each subchannel. Then, the energy function is given by

$$\mathcal{E}(b_n) = \begin{cases} \frac{2^{b_n}-1}{6} d_n^2, & b_n \text{ even} \\ \frac{2^{b_n+1}-1}{12} d_n^2, & b_n \text{ odd} \end{cases} \quad . \qquad (3.3.43)$$

The incremental energy is then

$$e_n(b_n) = \begin{cases} \frac{2^{b_n}-1}{12} d_n^2, & b_n \text{ even} \\ \frac{2^{b_n}+1}{12} d_n^2, & b_n \text{ odd} \end{cases} \quad . \qquad (3.3.44)$$

For large numbers of bits, the incremental energy for SQ QAM is approximately twice the amount of energy needed for the previous constellation size, or 3 dB per bit. A coded system with $\beta \neq 1$ may have a more complicated exact expression that perhaps is most easily represented by tabulation in general. For instance, the table for the uncoded SQ QAM is

b_n	0	1	2	3	4	5	6	7	8
$\mathcal{E}_n(b_n)$	0	$\frac{d_n^2}{4}$	$\frac{d_n^2}{2}$	$\frac{5d_n^2}{4}$	$\frac{5d_n^2}{2}$	$\frac{21d_n^2}{4}$	$\frac{21d_n^2}{2}$	$\frac{85d_n^2}{4}$	$\frac{85d_n^2}{2}$
$e_n(b_n)$	0	$\frac{d_n^2}{4}$	$\frac{d_n^2}{4}$	$\frac{3d_n^2}{4}$	$\frac{5d_n^2}{2}$	$\frac{11d_n^2}{4}$	$\frac{21d_n^2}{4}$	$\frac{43d_n^2}{4}$	$\frac{85d_n^2}{4}$

An energy function for QAM with $\beta = 1$ can also be defined via the gap approximation as

$$\mathcal{E}_n(b_n) = \frac{\Gamma}{g_n}\left(2^{b_n} - 1\right) \quad . \qquad (3.3.45)$$

The incremental energy is then

$$e_n(b_n) = \frac{\Gamma}{g_n} 2^{b_n - 1} = 2e_n(b_n - 1) \quad , \tag{3.3.46}$$

which is again 3 dB more than the incremental energy needed for the previous increase in constellation size by one bit. In general, the gap-based incremental energy function for QAM is given by

$$e_n(b_n) = \frac{\Gamma}{g_n} 2^{b_n} \left(1 - 2^{-\beta}\right) \quad . \tag{3.3.47}$$

Once an encoding system has been selected for the given fixed β and for each of the subchannels in a multichannel transmission system, then the incremental energy for each subchannel can be tabulated for use in discrete loading algorithms.

The optimum solution for the discrete MA loading problem is the bit distribution that requires the least amount of total energy to meet the fixed rate constraint of b bits. For the discrete RA loading problem, it is likely that there are many different bit distributions that achieve the maximum rate. Likewise, it is highly desirable to pick the bit distribution that requires the minimum total energy. Clearly, this minimum-energy property, called **efficiency**, is a necessary condition for a bit distribution to be optimum.

Definition 3.3.7: Efficiency of a Bit Distribution
A bit distribution vector $\boldsymbol{b}$ *is* **efficient** *for a given granularity* β *if*

$$\max_m \; e_n(b_n) \leq \min_n \; e_m(b_m + \beta), \quad n, m = 1, \ldots, N \tag{3.3.48}$$

Efficiency means that there is no movement of an information unit from one subchannel to another that reduces the multicarrier symbol energy.

Levin and Campello have independently formalized an iterative algorithm that will convert any bit distribution into an efficient bit distribution.

Levin-Campello (LC) Efficientizing (EF) Algorithm

1. $m \leftarrow \arg \min_{1 \leq i \leq N} e_i(b_i + \beta)$
2. $n \leftarrow \arg \max_{1 \leq j \leq N} e_j(b_j)$
3. While $e_m(b_m + \beta) < e_n(b_n)$ do

(a) $b_m \leftarrow b_m + \beta$

(b) $b_n \leftarrow b_n - \beta$

(c) $m \leftarrow \arg \min_{1 \le i \le N} e_i(b_i + \beta)$

(d) $n \leftarrow \arg \max_{1 \le j \le N} e_j(b_j)$

The EF algorithm swaps β information unit between the subchannels that violate the *efficiency* condition at each iteration and ends when no more such violations occur. The EF algorithm will produce an efficient bit distribution as long as the incremental energy function is monotonically increasing with information (bits), which is always the case in practical systems. Traditional greedy methods such as the Hughes-Hartog [51] algorithm only works with initial bit distributions that are efficient. Specifically, they achieve the optimum solution by starting with $\mathbf{b} = 0$ and performing, at each iteration, operations that preserve efficiency.

An additional property of *E-tightness* is necessary for the optimum solution to the discrete RA loading problem.

Definition 3.3.8: E-tightness
A bit distribution with granularity β is **E-tight** *if*

$$0 \le \mathcal{E}_x - \sum_{n=1}^{N} e_n(b_n) \le \min_{1 \le i \le N} e_i(b_i + \beta) \quad . \tag{3.3.49}$$

E-tightness implies that no additional unit of information can be carried without violation of the total energy constraint.

A bit distribution can be made E-tight by applying Levin-Campello E-tightening (ET) algorithm (see Table 3.3). The ET algorithm is efficiency-preserving. Given any bit distribution, it iteratively adds a β information unit to the least energy-consumptive subchannel when the total energy is sufficiently below the limit, and reduces a β information unit from the biggest energy-consumptive subchannel when the total energy exceeds the limit.

The optimality conditions that govern the solution to the discrete MA loading problem are efficiency and *B-tightness*, which is a dual of E-tightness.

Definition 3.3.9: B-tightness
A bit distribution $\boldsymbol{b}$ *with granularity β and total bits per symbol b is* B-tight *if*

$$b = \sum_{n=1}^{N} b_n \quad . \tag{3.3.50}$$

E-tightening (ET) algorithm

1. Set $S = \sum_{n=1}^{N} e_n(b_n)$;
2. While $\mathcal{E}_x - S \geq \min_{1 \leq i \leq N} e_i(b_i + \beta)$
   Or $(\mathcal{E}_x - S < 0)$

   If $(\mathcal{E}_x - S < 0)$
   1. $n \leftarrow \arg \max_{1 \leq i \leq N} e_i(b_i + \beta)$
   2. $S \leftarrow S - e_n(b_n)$
   3. $b_n \leftarrow b_n - \beta$

   Else
   1. $m \rightarrow \arg \min_{1 \leq i \leq N} e_i(b_i + \beta)$
   2. $S \leftarrow S + e_m(b_m + \beta)$
   3. $b_m \leftarrow b_m + \beta$

B-tightening (BT) algorithm

1. Set $b' = \sum_{n=1}^{N} b_n$
2. While $(b' \neq b)$

   If $(b' > b)$
   1. $n \leftarrow \arg \max_{1 \leq i \leq N} e_i(b_i)$
   2. $b' \leftarrow b' - \beta$
   3. $b_n \leftarrow b_n - \beta$

   Else
   1. $m \leftarrow \arg \min_{1 \leq i \leq N} e_i(b_i + \beta)$
   2. $b' \leftarrow b' + \beta$
   3. $b_m \leftarrow b_m + \beta$

Table 3.3. Levin-Campello Bit and Energy tightening algorithms

B-tightness simply guarantees that the fixed rate constraint in the discrete MA loading problem is met.

Except for a much simpler stopping criterion, the B-tightening (BT) algorithm (see Table 3.3) is essentially the same as the ET algorithm.

Combining the algorithms outlined, one can find the optimum solution to both the discrete RA and MA loading problems by first *efficientizing* and then *tightening* a candidate solution as outlined in Table 3.4.

The speed of convergence for both the RA and MA algorithms depends on the choice of the initial bit distribution $\boldsymbol{b}$. A natural choice for $\boldsymbol{b}$ is the water-filling solution, which has infinite granularity. The reasons are two-fold. First, it was proved in [52] that rounding of the water-filling solution, which is a simple operation, produces an efficient bit distribution. Second, the water-filling solution can be computed in expected $O(N)$ time, resulting in an overall expected complexity of $O(N)$ for both the optimum discrete RA and MA loading algorithms tabulated in Table 3.4. Further details on run-time issues are described in Section 3.3.9.

Levin-Campello General RA Algorithm	**Levin-Campello General MA Algorithm**
1. Choose any $\boldsymbol{b}$ 2. Make $\boldsymbol{b}$ efficient with the EF algorithm. 3. E-tighten the resultant $\boldsymbol{b}$ with the ET algorithm.	1. Choose any $\boldsymbol{b}$ 2. Make $\boldsymbol{b}$ efficient with the EF algorithm. 3. B-tighten the resultant $\boldsymbol{b}$ with the BT algorithm.

Table 3.4. Levin-Campello general optimum discrete RA and MA loading algorithms

The loading algorithms described in this section do not take into account any restrictions placed on the number of bits that each subchannel can carry. Some applications might mandate a minimum or/and a maximum allocation of bits for any of the used subchannels. These constraints are respectively called threshold and bit mask constraints. Adjustments that are required for the present loading algorithms to address these constraints will not be presented here but can be found in [52].

EXAMPLE 3.3.4: This example illustrates the steps for the optimum discrete RA loading with $\beta = 1$ on the $1 + (.3 + j)D^{-1}$ channel used in Example 3.3.2. By using the uncoded gap approximation at $P_e = 10^{-6}$, the QAM-subchannel energy is given by

$$\mathcal{E}_n(b_n) = \frac{10^{.88}}{g_n}\left(2^{b_n} - 1\right) \quad .$$

The incremental energies $e_n(b_n)$ then are:

n	0	1	2	3
$e_1(1)$	1.128	.742	2.036	33.715
$e_2(2)$	2.256	1.484	4.073	67.429
$e_3(3)$	4.512	2.968	8.146	134.858

Starting from a distribution of $\boldsymbol{b} = [1\ 0\ 1\ 0]$, the EF algorithm produces $\boldsymbol{b} = [1\ 1\ 0\ 0]$ which is efficient. The corresponding total energy is $\mathcal{E} = 1.87$. Since the total energy allocated is 8, the ET algorithm is applied to determine which subchannels are assigned the extra bits.

1. $\boldsymbol{b} = [1\ 2\ 0\ 0]$ with $\mathcal{E} = 3.354$
2. $\boldsymbol{b} = [1\ 2\ 1\ 0]$ with $\mathcal{E} = 5.39$
3. $\boldsymbol{b} = [2\ 2\ 1\ 0]$ with $\mathcal{E} = 7.646$
4. $\boldsymbol{b} = [2\ 3\ 1\ 0]$ with $\mathcal{E} = 10.614$

Clearly, the optimum bit distribution is $\boldsymbol{b} = [2\ 2\ 1\ 0]$, leading to a margin of $8/7.646 = 1.05$ (0.2 dB). If the initial bit distribution were instead $\boldsymbol{b} = [1\ 1\ 1\ 3]$, the sequence of steps for the EF algorithm would be:

$$b = [1\ 1\ 1\ 3] \xrightarrow{\text{EF}} b = [1\ 2\ 1\ 2] \xrightarrow{\text{EF}} b = [2\ 2\ 1\ 1] \xrightarrow{\text{EF}} b = [2\ 3\ 1\ 0]$$

and the sequence of steps for the ET algorithm would be:

$$b = [2\ 3\ 1\ 0] \xrightarrow{\text{ET}} b = [2\ 2\ 1\ 0]$$

As always, the EF algorithm retains the total number of bits and converges to the most energy-efficient solution for that number of bits from any initial distribution.

Chow's Algorithm

Chow [54] studied a number of transmission media with severe intersymbol interference, in particular the telephone-line digital transmission known generally as Digital Subscriber Line (DSL). Chow was able to verify that an "on/off" energy distribution, *as long as it uses the same or nearly the same transmission band as water-filling*, exhibits negligible loss with respect to the exact water-filling shape. The reader is cautioned not to misinterpret this result as saying that flat energy distribution is as good as water-filling. Rather an "on/off" energy distribution where "on" means flat energy distribution in the same used bands as water-filling and "off" means zero energy distribution where water-filling would be zero, gives an almost optimum performance.

The reason for the use of "on/off" energy distributions is that practical systems usually have a constraint on the maximum power spectral density (psd) along with a total energy or power constraint. Such psd constraints are typically flat over the region of bandwidth considered for transmission. Using this observation, Chow was able to devise loading algorithms that approximate water-filling with finite granularity. Both the RA and MA problems can be solved with Chow's algorithms, which in both cases begin with **Chow's "on/off" Loading Primer** below.

Chow's "on/off" Loading Primer

1. Sort the g_n such that $g_1 = \max_n g_n$ is the largest and $g_N = \max_n g_n$ is the smallest.

2. Let i denote the number of used subchannels and $b_{temp}(i)$ denote the tentative total number of bits. Initialize $i = N$ and $b_{temp}(N+1) = 0$.

3. Let $\mathcal{E}_n = \mathcal{E}_x / i$ for $n = 1, \ldots, i$ and $\mathcal{E}_n = 0$ for $n = i+1, \ldots N$. The subchannel SNR is then $\text{SNR}_n = g_n \mathcal{E}_n$. (When psd limits apply, the energy level is set at the psd limit on all subchannels to which it applies.)

4. Compute $b_{temp}(i) = \sum_{n=1}^{i} \log_2(1 + \frac{\text{SNR}_n}{\Gamma})$.

5. If $b_{temp}(i) < b_{temp}(i+1)$, then keep the bit distribution corresponding to $b_{temp}(i+1)$. Otherwise, $i \leftarrow i - 1$ and go to step 3.

The bit distribution produced by Chow's Primer will likely contain many subchannels with non-integer numbers of bits.

To solve the discrete *RA* problem, the following 3 steps adjust the result of the Chow's Primer (with steps 6 and 7 being executed for all used subchannels):

Chow's RA Algorithm

6. If $\frac{b_n}{\beta} - \lfloor \frac{b_n}{\beta} \rfloor < .5\beta$, then round b_n to the largest integer multiple of β contained in b_n, $B_n = \lfloor \frac{b_n}{\beta} \rfloor$. Reduce energy on this subchannel so that the performance is the same as on all other subchannels. That is, scale the energy by $\dfrac{2^{\beta B_n} - 1}{2^{b_{old,n}} - 1}$ such that the new b_n value then satisfies $b_n = \log_2(1 + \frac{\text{SNR}_n}{\Gamma})$.

7. If $\frac{b_n}{\beta} - \lfloor \frac{b_n}{\beta} \rfloor > .5\beta$, then round b_n to the next largest integer multiple of β, $B_n = \lceil \frac{b_n}{\beta} \rceil$. Increase the energy, $\mathcal{E}_n$, by the factor $\frac{2^{\beta B_n} - 1}{2^{b_{old,n}} - 1}$ so that the performance is the same as those of other subchannels and that the new b_n value satisfies $b_n = \log_2(1 + \frac{\text{SNR}_n}{\Gamma})$.

8. Compute the total energy used $\mathcal{E} = \sum_n \mathcal{E}_n$ and then scale all subchannels by the margin factor $\frac{\mathcal{E}_x}{\mathcal{E}}$.

This procedure produces a characteristic "sawtooth" energy distribution that may deviate from a flat distribution by as much as $\pm\frac{6\beta}{\tilde{N}}$ dB, where $\tilde{N} \in \{1, 2\}$ is the dimensionality of the subchannel. Figure 3.5 illustrates the sawtooth energy produced by Chow's algorithm with the $SNR(f)$ curve simultaneously displayed. The discontinuities are points at which the number of bits on a subchannel changes by the smallest information unit β. An increase in b_n by β means more energy is necessary, while a decrease in b_n by β means less energy is necessary. On average, the sawtooth curve approximates a flat straight line over the *used* tones, i.e., where energy is "on."

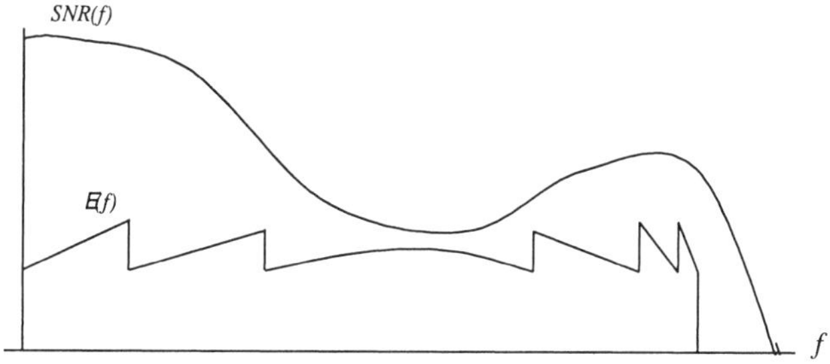

Figure 3.5. Illustration of sawtooth energy distribution characteristic of Chow's Algorithm.

EXAMPLE 3.3.5: Here, Chow's RA algorithm is applied to the $1+(.3+j)D^{-1}$ channel with $\Gamma = 0$ dB. The 3 passes necessary for determining the number of used subchannels are tabulated below:

	$i=4$	$i=3$	$i=2$
$\mathcal{E}_n$	$2,\ n=0,1,2,3$	$\frac{8}{3},\ n=0,1,2$	$4,\ n=0,1$
SNR_0	$\frac{2\times2.69}{.4}=13.45$	$\frac{8\cdot2.69}{3\cdot.4}=17.93$	$\frac{4\times2.69}{.4}=26.9$
SNR_1	$\frac{2\times4.09}{.4}=20.45$	$\frac{8\cdot4.09}{3\cdot.4}=27.27$	$\frac{4\times4.09}{.4}=40.9$
SNR_2	$\frac{2\times1.49}{.4}=7.45$	$\frac{8\cdot1.49}{3\cdot.4}=9.93$	0
SNR_3	$\frac{2\times0.09}{.4}=0.45$	0	0

From the first iteration where all subchannels are used, the aggregate bit rate is

$$\begin{aligned} b_{temp}(4) &= \log_2(1+13.45)+\log_2(1+20.45)+\log_2(1+7.45) \\ &\quad +\log_2(1+0.45)=11.891 \ . \end{aligned}$$

After the second iteration where the last subchannel (smallest g_n) is excluded,

$$b_{temp}(3)=\log_2(18.93)+\log_2(28.27)+\log_2(10.93)=12.51 \ .$$

After the third iteration where only the first 2 subchannels are used,

$$b_{temp}(2)=\log_2(27.9)+\log_2(41.9)=10.19<b_{temp}(3) \ .$$

Since $b_{temp}(2) < b_{temp}(3)$, the iteration process stops with $b = 12.51$ or $\bar{b} = 1.56$ bits/dimension, which is the same as that achieved by the water-filling solution. This confirms, in this example, that an on/off energy distribution is essentially as good as a water-filling energy shaping if the bandwidth used is optimized. The key OFDM-optimization aspect in this example in both Chow's and the Water-filling methods is that the last subchannel is not used. Its use corresponds to about a 5% loss in data rate, indicating that not all subchannels should always be used in OFDM. Rounding the bit distribution with $\beta = 1$ bit leads to the values tabulated in the first column of the table below. The maximum data rate is then 12 bits per symbol or 1.5 bits/dimension. The new subchannel energy distribution is tabulated in the second column. The total energy is then

$\mathcal{E}_{\boldsymbol{x}} = 7.15$, which means that the distribution of energies may all be increased by the factor 8/7.15 so that the total energy budget is maintained.

	bit distribution		subchannel energies
b_0	$\log_2(18.93) = 4.24 \xrightarrow{\beta=1} 4$	$\mathcal{E}_0$	$\frac{8}{3}\frac{2^4-1}{2^{4.24}-1} = 2.24$
b_1	$\log_2(28.27) = 4.82 \xrightarrow{\beta=1} 5$	$\mathcal{E}_1$	$\frac{8}{3}\frac{2^5-1}{2^{4.82}-1} = 3.03$
b_2	$\log_2(10.93) = 3.45 \xrightarrow{\beta=1} 3$	$\mathcal{E}_2$	$\frac{8}{3}\frac{2^3-1}{2^{3.45}-1} = 1.88$

Chow's MA loading algorithm first computes a margin, from the data rate given by Chow's Primer, with respect to the desired fixed rate. It then recomputes the bit distribution including that margin, which will make the number of bits exactly equal to the desired number. However, the number of bits will still not be integer multiples of β in general. The resultant bit distribution is then rounded as in Chow's RA loading algorithm, but the sum of the number of bits is monitored and rounding is adjusted to make the sum exactly equal to the desired number of bits per symbol.

The steps that follow Chow's Primer to complete the MA solution are:

Chow's MA Loading Algorithm

6. Compute

$$\text{SNR}_{m,u} = \Gamma \left\{ \left(\prod_{n=1}^{i} \left[1 + \frac{\text{SNR}_n}{\Gamma} \right] \right)^{\frac{1}{i}} - 1 \right\} = 2^{2\bar{b}_{max}} - 1$$

7. Compute the tentative margin

$$\gamma_{temp,m} = \frac{\frac{\text{SNR}_{m,u}}{\Gamma}}{2^{2\bar{b}} - 1}, \quad n = 1, \ldots, N^*$$

8. Compute the updated bit distribution

$$B_{temp,n} = \frac{1}{2\beta} \log_2 \left(1 + \frac{\text{SNR}_n}{\Gamma \gamma_{temp,n}} \right), \quad n = 1, ..., N^* .$$

9. Round $B_{temp,n}$ to the nearest integer (if more than one β is being used for different subchannels, then use the correct value for each subchannel) and recompute energy as

$$\mathcal{E}_{temp,n} = \frac{\Gamma\left(2^{2B_n\beta} - 1\right)}{g_n}, \quad n = 1, ..., i \ .$$

10. Compute $B_{temp} = \sum_{n=1}^{N^*} B_{temp,n}$ and $b_{temp} = B_{temp}\beta$ and check to see if $b_{temp} = b$, the desired fixed number of bits per symbol. If not, then select subchannels that were rounded (perhaps by choosing those close to one-half information granularity unit) and round the other way until the data rate is correct, recomputing energies as in step 9.

11. Compute the energy scaling factor $f_e = N\bar{\mathcal{E}}_{\boldsymbol{x}} / \sum_{n=1}^{N^*} \mathcal{E}_n$ and scale all subchannel energies by f_e, i.e., $\mathcal{E}_n \leftarrow f_e\mathcal{E}_n$ and the maximum margin becomes $\gamma_{max} = \gamma_{temp,max} f_e$.

Other variations of Chow's MA loading algorithm can be found in [54], [55].

3.3.9 Sorting and Run-time Issues

Water-filling and Chow's methods often begin with sorting of the N subchannel SNR quantities g_n. A straightforward implementation for sorting includes successively finding the maximum or minimum in a set of numbers that is reducing in size. It takes $N-1$ comparisons to find the first maximum or minimum, and $N-2$ comparisons to find the second maximum or minimum and so forth. Therefore, the total number of comparisons needed is $(N-1)N/2$. On the other hand, there are sorting algorithms based on the divide-and-conquer approach that execute in expected $O(N \log_2 N)$ time[8] [56]. The divide-and-conquer approach first selects an element at random from a set as the pivot, and then partitions the set into two subsets whose elements are respectively greater and smaller than the pivot. This process is then repeated recursively on each of the subsets until the original set of elements are sorted.

For water-filling and Chow's algorithms, absolute ordering is not essential. The important quantity is the position of the transition from "on" to

[8]there are versions of these algorithms that run in worst-case $O(N \log_2 N)$ time, but they are difficult to implement and usually not used in practice.

"off' in energy. Rather than sort g_n initially, a guess at the transition point is made. Then, with $N-1$ additional comparisons, the quantities g_n that are greater than, or less than or equal to the transition point, can be placed separately into two subsets. Next, the subchannel corresponding to the the transition point is tested for negative energy. If it is negative, the set of larger g_n's contains the correct transition point. If it is positive, then the point is in fact the transition point or it is in the set of lesser g_n's. This partitioning procedure has a worst-case (very low-probability event) complexity of $O(N^2)$ and an expected complexity of $O(N)$. Optimum discrete loading algorithms with expected $O(N)$ complexities due to the implementation of this partitioning method are presented in [52]. Also presented in [52] and [57] are slightly suboptimal but very efficient loading algorithms of worst-case complexity of $O(N)$.

3.3.10 Dynamic Loading

For typical wireless systems, the subchannel SNRs, g_n, may change with time. Then, the bit and energy distribution may need corresponding change. Such variation in the bit and energy distribution is known as *dynamic loading*, or more specifically *Dynamic Rate Adaptation (DRA)* or *Dynamic Margin Adaptation (DMA)*, where the latter is widely and colloquially called *"bit-swapping"* by this author.

With worst-case complexities as low as $O(N)$, it is conceivable that an entire loading algorithm could be run every time there is a change in the bit and/or energy distribution for a channel. However, if the channel variations were small, then "reloading" might produce a bit distribution with only minor variations from the original distribution. Generally, it is more efficient to dynamically continue loading than to reload. Usually, dynamic loading occurs through the transmission of bit or energy-distribution-change information over a reverse-direction, reliable, low-bit-rate channel. The entire bit distribution needs not be re-communicated. Rather, compressed incremental-variation information is desirable to minimize the relative data rate of the reverse channel. Dynamic loading is not possible without a reverse or feedback channel.

The LC algorithms lend themselves most readily to incremental variations. Specifically, to continuously solve the MA problem, the previous bit distribution $\boldsymbol{b}$ can be input to the EF algorithm as an initial condition for the new table of incremental energies, $e_n(b_n)$, which is simply scaled by $g_{n,old}/g_{n,new}$ for any reasonable constellation. Steps 3(a) and 3(b) of the EF algorithm execute a "swap" whenever it is more efficient energy-wise to move a β information unit from one subchannel to another. As long as a sufficient number of swaps can occur before further variation in the channel, then the EF algorithm will continue to achieve the optimum finite-granularity solution. The incremental information is the location of the two subchannels for the "swap" and which is to have its number of bits incremented or decremented. Such information is very low-rate, even for large numbers of subchannels. The energy gain or loss on each of the two subchannels must also be conveyed (from which the new incremental energy added or deleted from swapping subchannels can be computed) with use of the EF algorithm. The quantization and maintenance of the tables of g_n at both the transmitter and the receiver should be implemented exactly when this algorithm is used. In other words, the receiver should quantize a new g_n before execution of the algorithm and this exact value must be transmitted through the reverse channel and maintained identically at the transmitter to avoid anomalies. Alternatively, the incremental energy calculation for both loading and dynamic loading can be based on a suboptimum on-off distribution or any other presumed/maintained energy distribution over a given number of subchannels. Then, no gain information need be transmitted over the reverse channel.

True DRA can occur through the execution of EF and ET algorithms. The EF algorithm first produces an efficient bit distribution according to the new subchannel SNRs g_n using the current bit distribution as the initial condition. Next, the ET algorithm changes the total number of bits transmitted to conform to the best use of the allotted transmit energy. The incremental information passed through the reliable reverse channel is the index of the currently to-be-altered subchannel and whether a bit is to be added or deleted.

The instance of continuous bit-rate variation to correspond to channel change is rare (usually some higher-level communications authority prefers to arbitrate when bit rate changes are allowed to occur). Thus, bit-swapping or DMA is used even after a RA initial loading has occurred (with some extra margin in the form of an increased gap to guard against drop in capacity of the channel with time). Periodically, a higher-level entity can allow reloading.

Synchronization of dynamic loading

For maintenance of the channel, both the transmitter and the receiver need to know on which symbol a bit/energy distribution change has occurred. One possible implementation is to maintain two synchronized symbol counters, one in the transmitter and one in the receiver. They can be circular counters as long as the period of each is the same and longer than the worst-case time that it takes for a bit-swap, bit-add, or bit-delete to occur. Upon receipt of a request for a change, a transmitter then acknowledges the change and returns a counter symbol number on which it intends to execute the change. Then both the transmitter and the receiver execute the change simultaneously on the symbol corresponding to the agreed counter value.

3.3.11 Multiuser Loading

Using the Karhunen-Loève expansion, a single-user Gaussian channel with ISI can be decomposed into a set of independent, parallel, ISI-free Gaussian (sub)channels. Capacity is then achieved by allocating energy to these (sub)channels according to the water-filling distribution. In a multiuser system, where the users generally experience *different* channels with ISI, the Karhunen-Loève expansion is no longer applicable. This is because there is no kernel that simultaneously diagonalizes the channel responses of all the users. The OFDM modulation scheme is one way to circumvent this problem. OFDM systems employ cyclic extension which results in the DFT basis functions as eigenfunctions of any FIR channel whose length is less than that of the cyclic prefix. This form of orthogonal decomposition of the channel is *independent* of the channel ISI coefficients and enables a multiuser channel to be equivalently viewed as parallel multiuser channels, one for each frequency. In general, superposition coding together with successive interference cancellation is necessary for achieving multiuser capacity for ISI channels. This implies that some frequencies/subchannels need to be shared among different users which makes decoding a highly complex problem. Recently, practical ways of superimposing information from different users using embedded modulation have been proposed [58], [59]. However, for an OFDM system which has an inherent multicarrier structure in place, implementation is greatly simplified by adopting the Frequency Division Multiple-Access (FDMA) scheme where the users occupy different sets of subchannels. Given fixed subchannel assignments, the maximum multiuser rate is achieved by performing separate water-filling for each user. Effectively, the non-overlapping nature of the resource assignments decouples the multiuser system into a system of independent users. However, this multiuser rate can be far below the multiuser capacity if the resource assignment is chosen arbitrarily, resulting in many inefficiently used subchannels. This is a waste of precious bandwidth as these subchannels could have been allocated to other users. The reason for this is that the users often experience mutually independent fading, and it is highly unlikely for a subchannel or subchannels to be in deep fade for all users. Therefore, there is a need for joint optimization of subchannel *and* energy allocation among the users according to their respective channel responses. *Multiuser transmit optimization* or *Multiuser loading* addresses this need.

Multiuser transmit optimization refers to the optimization of energy allocation over frequency (subchannels) among several users that have different transmission requirements and that are experiencing different channel gains in a multicarrier system. The transmission requirements include different coding schemes, target P_e and noise margins, which are collectively characterized by the SNR gap Γ_l (Section 3.3.1). Moreover, each user may desire to transmit at a Constant Bit Rate (CBR) or at a Variable Bit Rate (VBR).

Multiuser transmit optimization requires knowledge of the channel transfer functions of all users. Hence, a feedback channel is necessary for communicating back to the transmitter the channel information estimated at the receiver of each user. Multiuser loading is practical for use in fixed wireless systems or in low-mobility environments where the channels vary slowly. If transmit re-optimization is done at a slower rate than the OFDM symbol rate, the overhead in the feedback channel can be low.

Two typical multiuser OFDM scenarios are considered. For the uplink scenario, L users, with individual power constraints, wish to communicate with a common receiver at the base station. This is also known as the *Multiple Access Channel (MAC)*. For the downlink scenario, the reverse is true whereby a base station, with a total power constraint, wishes to communicate with L users. This is also known as the *broadcast channel*. In both scenarios, the users are either fixed or mobile, and are assigned priorities. The problem of interest is then to find the capacity region, which is characterized by the set of simultaneously achievable rates $(R_1, R_2, \ldots, R_L)$, for communication in a multiple access or broadcast channel. In the next two paragraphs, an attempt, though not a comprehensive one, is made to give a brief overview of some of the work done in these two areas.

The memoryless multiple access channel capacity region was found by

Ahlswede [60] and Liao [61][9] For multiple access channels with memory, the capacity region was studied by Verdu [63], [64] and by Cheng and Verdu [65]. In [65], explicit expressions for the optimal transmit Power Spectral Densities (PSD) and a geometrical multiuser waterfilling method to characterize the capacity region were given. In [66], Diggavi proposed a multiuser water-filling algorithm to calculate the optimal PSDs for a MAC scenario where all users have the same priority. Improvement on its complexity was made by Zeng [67], [68] who also proposed optimal and suboptimal multiuser water-filling algorithms for users having different priorities. In [69], Yu presented a numerical method for characterizing the rate region achievable with FDMA for a Gaussian multiple-acess channel with ISI.

Broadcast channels were first studied by Cover [70]. The achievability of the capacity region, conjectured in [70] for the degraded broadcast channel, was proved by Bergmans [71], and the converse was established by Bergmans [72] and Gallager [73]. In [74], Goldsmith derived the capacity region for a broadcast channel with ISI and colored Gaussian noise under an input power constraint. The approach used, which was used to obtain the single-user capacity of a discrete-time Gaussian channel with ISI [75] and the multiuser capacity of a Gaussian multiple access channel with ISI [65], calls for redefining the channel of interest as a circular channel that can be decomposed into a set of parallel degraded channels. In [76], [77], [78], Hoo solves the multiuser transmit optimization problem under different performance criteria for multicarrier broadcast channels with an FDMA restriction. The optimal solution, which is achieved by multi-level water-filling, requires an exhaustive search. However, by relaxing the FDMA restriction, a technique pursued as well in [69], [79], a convex reformulation is obtained which allows for efficient computation of the optimal solution and therefore a characterization of the FDMA capacity region for a broadcast channel. To accommodate applications with relatively fast time-varying user priorities and data rate requirements, further reduction in computational complexity is necessary . This is achieved by restricting the energy distribution to be constant across the used subchannels [80], [81], [82].

[9]more historical notes can be found in [62].

FDMA Formulation

The optimization formulation for multiuser loading with the FDMA restriction is:

$$
\begin{aligned}
&\max_{\mathcal{S}_l, \mathcal{E}_{l,n}} && \sum_{l=1}^{L} \alpha_l \sum_{n\in\mathcal{S}_l} b_{l,n} = \sum_{l=1}^{L} \alpha_l \sum_{n\in\mathcal{S}_l} \log_2(1 + \frac{\mathcal{E}_{l,n} g_{l,n}}{\Gamma_l}) \\
&\text{subject to:} && \begin{cases} \sum_{l=1}^{L} \sum_{n\in\mathcal{S}_l} \mathcal{E}_{l,n} \le \mathcal{E}_{tot} & \text{Broadcast channel or} \\ \sum_{n\in\mathcal{S}_l} \mathcal{E}_{l,n} = \mathcal{E}_l, \ \forall l \in \mathcal{L} & \text{Multiple access channel} \end{cases} \\
& && \mathcal{S}_l \cap \mathcal{S}_{l'} = \emptyset, \ \forall l \ne l' \qquad \text{FDMA restriction} \\
& && \mathcal{E}_{l,n} \ge 0, \ \forall l \in \mathcal{L}, \ \forall n \in \mathcal{N} \\
&\text{where:} && \sum_{l=1}^{L} \alpha_l = 1 \quad \text{and} \quad g_{l,n} = \frac{|H_{l,n}|^2}{2\sigma_{l,n}^2}
\end{aligned}
\tag{3.3.51}
$$

In the above formulation, $\mathcal{L} = \{1, \ldots, L\}$ is the user index set, $\mathcal{N} = \{1, \ldots, N\}$ is the subchannel index set and $\mathcal{S}_l \subset \mathcal{N}$ is the subchannel assignment for user l. Moreover, α_l is a measure of priority of user l, $\mathcal{E}_{l,n}$ is the energy allocated to user l for the nth subchannel, and $g_{l,n}$ is the ratio of the magnitude squared of the channel response and the noise power spectral density for user l in the nth subchannel. Associated with each user is the SNR gap.

Given the relative priorities α_l and the effective channel responses $g_{l,n}$ of each user, the objective in (3.3.51) is to find the subchannel energy and bit distribution that maximizes a weighted sum of each user's achievable rate in an FDMA OFDM system subject to *either* individual input energy constraints *or* a total input energy constraint. Maximizing the weighted ratesum for all possible α_l's traces out the boundary of the capacity region.

For any *fixed subchannel assignments* $\mathcal{S}_l$, the optimal FDMA solution to (3.3.51) is achieved by *multi-level water-filling* whose underlying equations, assuming all subchannels are used, are tabulated in Table 3.5[10]. In general, the number of used subchannels needs to be optimized such that all energies are positive. However, finding the optimal subchannel assignments among the L users requires $L^N - 1$ searches. Therefore, the computation of the optimal FDMA solution has exponential complexity. The mathematical

[10]The optimal discrete FDMA solution is obtained by applying Campello's discrete loading algorithms in each $\mathcal{S}_l$ [76].

Broadcast Channel Formulation	Multiple Access Channel Formulation
$$\mathcal{E}_{l,n} = \left(\alpha_l K - \frac{\Gamma_l}{g_{l,n}}\right)^+, \quad \begin{array}{l}\forall n \in \mathcal{S}_l,\\ \forall l \in \mathcal{L}\end{array}$$ $$K = \frac{\mathcal{E}_{tot} + \sum_{l=1}^{L} \Gamma_l \sum_{n \in \mathcal{S}_l} \frac{1}{g_{l,n}}}{\sum_{l=1}^{L} \lvert\mathcal{S}_l\rvert \alpha_l}, \quad \forall l \in \mathcal{L}$$ (3.3.52)	$$\mathcal{E}_{l,n} = \left(\alpha_l K_l - \frac{\Gamma_l}{g_{l,n}}\right)^+, \quad \begin{array}{l}\forall n \in \mathcal{S}_l,\\ \forall l \in \mathcal{L}\end{array}$$ $$K_l = \frac{\mathcal{E}_l + \Gamma_l \sum_{n \in \mathcal{S}_l} \frac{1}{g_{l,n}}}{\lvert\mathcal{S}_l\rvert \alpha_l}, \quad \forall l \in \mathcal{L}$$ (3.3.53)

Table 3.5. Optimal FDMA solution given fixed subchannel assignments

operator $(x)^+$ used in Table 3.5 is defined as $(x)^+ = \begin{cases} x & \text{if } x \geq 0, \\ 0 & \text{if } x < 0. \end{cases}$.

For the special case when all users have equal priorities, that is, $\alpha_l = 1$, or $1/L$, $\forall l$, an FDMA scheme among the users is indeed optimal. Specifically, with R_l denoting the achievable rate for user l, the rate-sum $\sum_l R_l$ for the broadcast channel is maximized by assigning each subchannel to the user with the best composite gain-to-noise-power ratio given by $g_{l,n}/\Gamma_l$. In other words, the transmitter should transmit information only to the user with the best reception for each subchannel, and the optimal energy distribution follows the water-filling solution. For multiple access channel, likewise, an FDMA scheme with optimally selected frequency bands for each user maximizes the rate-sum [65].

FDMA-TDMA Formulation: A Convex Relaxation

If time-sharing is allowed between users for each subchannel, formulation (3.3.51) can then be recast as an FDMA-TDMA optimization problem, as follows:

$$
\begin{aligned}
&\max_{\mathcal{E}_{l,n}, w_{l,n}} \quad \sum_{l=1}^{L} \alpha_l \sum_{n=1}^{N} w_{l,n} \log_2\left(1 + \frac{\mathcal{E}_{l,n} g_{l,n}}{w_{l,n}\Gamma_l}\right) \\
&\text{subject to:} \quad \begin{cases} \sum_{l=1}^{L}\sum_{n=1}^{N} \mathcal{E}_{l,n} \leq \mathcal{E}_{tot} & \text{Broadcast channel or} \\ \sum_{n=1}^{N} \mathcal{E}_{l,n} = \mathcal{E}_l, \quad \forall l \in \mathcal{L} & \text{Multiple access channel} \end{cases} \\
&\qquad\qquad\quad \sum_{l=1}^{L} w_{l,n} \leq \omega_n, \quad \forall n \in \mathcal{N} \\
&\qquad\qquad\quad \mathcal{E}_{l,n},\ w_{l,n} \geq 0, \quad \forall l \in \mathcal{L},\ \forall n \in \mathcal{N}
\end{aligned} \tag{3.3.54}
$$

The additional variable $w_{l,n}$ is the time-sharing factor for user l of the nth subchannel. In other words, user l occupies the nth subchannel for $w_{l,n}$ fraction of the time. In this case, $\omega_n = 1$, $\forall n \in \mathcal{N}$. Another equivalent interpretation of $w_{l,n}$ for multicarrier transmission is that $w_{l,n}$ is the amount of bandwidth allocated to user l in the nth subchannel. This is equivalent to further partitioning of bandwidth in each subchannel, effectively creating an OFDM system with a total number of subchannels greater than N. As N increases, the subchannel width decreases, which implies that the increase in the aggregate rate contributed by time-sharing among the users decreases as well. In the limit, as $N \to \infty$, no time-sharing occurs and the optimum FDMA-TDMA solution is the optimum FDMA solution. Therefore, for sufficiently large N, an FDMA strategy that assigns every subchannel to the user with the largest time-sharing factor will result in negligible performance loss relative to the optimum FDMA solution.

Observation 3.3.1: *The FDMA-TDMA formulation of (3.3.54) is convex since it is the maximization of a concave cost function over a convex constraint set. First, the cost function is a linear combination of functions of the form* $f(\mathcal{E}_{l,n}, w_{l,n}) = C_1 w_{l,n} \log_2(1+C_2\mathcal{E}_{l,n}/w_{l,n})$ *where* C_1 *and* C_2 *are some constants. Since any linear combination of concave functions is concave, it suffices to prove that* $f(\mathcal{E}_{l,n}, w_{l,n})$ *is concave in* $(\mathcal{E}_{l,n}, w_{l,n})$*. This is achieved by proving that the Hessian for* $f(\mathcal{E}_{l,n}, w_{l,n})$ *is semi-positive definite over the positive quadrant of* $\mathcal{E}_{l,n}, w_{l,n}$*. Next, all the constraint functions are linear in* $\mathcal{E}_{l,n}, w_{l,n}$*, and thus give con-*

vex sets. Since the intersection of convex sets is convex [83], the constraint set in in (3.3.54) is convex.

Due to the convexity of the FDMA-TDMA optimization problem, a local maximum or minimum is also the global maximum or minimum. This optimal solution can be computed very efficiently (with polynomial complexity) using interior-point methods [83], thereby allowing us to characterize the asymptotic FDMA capacity region for the Gaussian broadcast channel with ISI [76], [77], [78] and for the Gaussian MAC channel with ISI [69].

Table 3.6 lists two other common optimization problems of interest for the downlink transmission in an OFDM system where a base station has to communicate with a set of mobile or fixed receivers. In (3.3.55), the optimization criterion is to minimize the total energy used at the base station while satisfying the constraints that each receiver achieves a certain fixed rate. It can be shown that given the constant bit rate requirements B_l, there exist weighting factors α_l and a total energy $\mathcal{E}_{tot}$ such that the minimum-total-energy formulation of (3.3.55) is equivalent to the maximum-weighted-ratesum formulation of (3.3.54). In (3.3.56), the optimization criterion is to maximize a weighted sum of rates for a subset of users while satisfying the fixed rate constraints for the rest. Assuming formulation (3.3.56) is feasible, it can be shown as well that given a total energy $\mathcal{E}_{tot}$ and the constant bit rate requirements $B_l, \forall l \in \mathcal{U}_2$, there exist weighting factors α_l, $\forall l \in \mathcal{U}_2$ such that the mixed-CBR-VBR formulation of (3.3.56) and the maximum-weighted-ratesum formulation of (3.3.54) give the same rate. Similar arguments can be made to establish the equivalence of formulations (3.3.55) and (3.3.56).

Ordered Frequency Partitioning

If the channel-to-noise responses for all the users are the same, i.e., $g_{l,n} = g_n, \forall l \in \mathcal{L}$, then, the optimal frequency partitioning or subchannel assignment among the users follows a specific ordering. *Multiservice optimization* problems are practical examples where such conditions on the channel-to-noise responses are true. Multiservice optimization refers to the allocation of resources (energy and frequency) for services that simultaneously occupy the same transmission medium. These services are either CBR (e.g. voice) or VBR (e.g. internet data, multimedia), and they require different noise immunities as characterized by their respective gaps Γ_l. The multiservice problem for multicarrier systems was first studied in [84], [85], [86]. The fast loading algorithms proposed therein are based on assigning the subchannels to the services in order of their respective gaps. Specifically, the service

Minimum Total Energy	Mixed CBR-VBR
$\begin{aligned} &\min \sum_{l=1}^{L}\sum_{n=1}^{N} \frac{w_{l,n}\Gamma_l}{g_{l,n}}(2^{\frac{b_{l,n}}{w_{l,n}}}-1) \\ &\text{s.t.} \sum_{n=1}^{N} b_{1,n} \geq B_l,\ \forall l \in \mathcal{L} \\ &\sum_{l=1}^{L} w_{l,n} \leq \omega_n,\ \forall n \in \mathcal{N} \\ &b_{l,n} \geq 0,\ w_{l,n} \geq 0,\ \forall l \in \mathcal{L}, \forall n \in \mathcal{N} \end{aligned}$ (3.3.55)	$\begin{aligned} &\max \sum_{m\in\mathcal{U}_1} \alpha_m \sum_{n=1}^{N} w_{m,n}\log_2(1+\frac{\mathcal{E}_{m,n}g_{m,n}}{w_{m,n}\Gamma_l}) \\ &\text{s.t.} \sum_{n=1}^{N} w_{l,n}\log_2(1+\frac{\mathcal{E}_{l,n}g_{l,n}}{w_{l,n}\Gamma_l}) \geq B_l,\ \forall l \in \mathcal{U}_2 \\ &\sum_{l=1}^{L}\sum_{n=1}^{N} \mathcal{E}_{l,n} \leq \mathcal{E}_{tot} \\ &\sum_{l=1}^{L} w_{l,n} \leq \omega_n,\ \forall n \in \mathcal{N} \\ &\mathcal{E}_{l,n} \geq 0,\ w_{l,n} \geq 0,\ \forall l \in \mathcal{L}, \forall n \in \mathcal{N} \end{aligned}$ (3.3.56)

Table 3.6. Possible formulations for the downlink scenario in an OFDM system

with the highest gap is assigned subchannels with the highest SNR's, and the service with the smallest gap is allocated subchannels with the worst SNR's. Therefore, the search for the optimum subchannel allocation has been reduced from an exhaustive search to a polynomial search. In [84], [85], assumptions of high SNR and equal-energy distribution are made to justify such an ordered subchannel assignment. In [76], [77], [78], a rigorous proof for the optimality of such an ordered frequency partitioning without making these assumptions is presented.

Theorem 3.3.1: *Let the channel-gain-to-noise-power function of all users in (3.3.54), (3.3.55) and (3.3.56) be the same, i.e.,* $g_{l,n} = g_n, \forall l \in \mathcal{L}$. *Furthermore, let's assume that* g_n, $\forall n \in \mathcal{N}$ *are sorted in a descending order. Then, the optimal subchannel allocation for formulations (3.3.54), (3.3.55) and (3.3.56) has an L-band structure with sharing of the boundary subchannels only, if at all. In addition, without loss of generality, let's assume that* $\Gamma_1 > \Gamma_2 > \cdots > \Gamma_L$. *Then, user* 1 *is assigned the best subchannels first, followed by user* 2, *and so forth. Specifically, there exist*

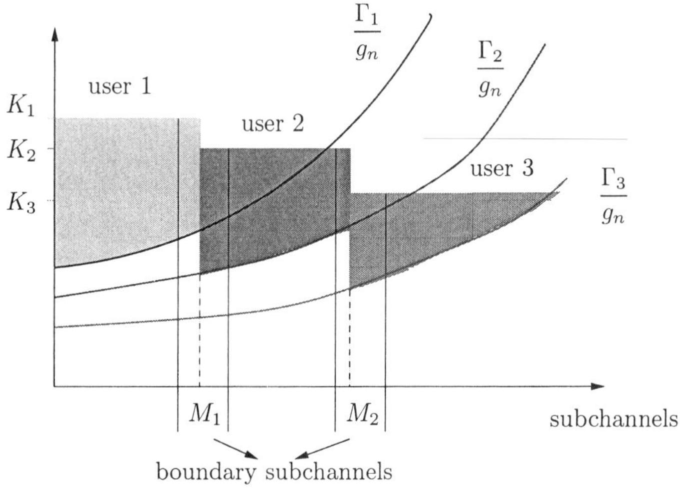

Figure 3.6. Optimal frequency partition for the 3-user multiservice FDMA-TDMA problem with $\Gamma_1 > \Gamma_2 > \Gamma_3$ and g_n sorted in a descending order.

$M_1, M_2, \ldots, M_{L-1}$, where $1 \leq M_1 \leq M_2, \cdots \leq M_{L-1} \leq N$ such that user 1 occupies subchannels 1 to M_1, user L occupies subchannels $M_{L-1}+1$ to N and user l occupies subchannels $M_{l-1}+1$ to M_l, $l \neq 1, N$. Moreover, sharing occurs, if at all, at the boundary subchannel M_l between user l and user $l+1$.

If K_l denotes the optimal water-filling level for service l, then, the following are also true.

1. $K_1 > K_2 > \cdots > K_L$
2. $\frac{K_1}{\Gamma_1} < \frac{K_2}{\Gamma_2} < \cdots < \frac{K_L}{\Gamma_L}$

Proof: The proof involves the application of the Karush-Kuhn-Tucker optimality conditions. The complete proof can be found in [76], [77], [78]. **QED.**

Figure 3.6 illustrates the optimal solution for the 3-user multiservice problem. In [69], Yu proved that the optimal frequency partitioning for a MAC channel (with individual power constraints) also has an L-band structure under the assumptions of high SNR or identical composite channel gains, i.e., $g_{l,n}/\Gamma_l = g_n, \forall l \in \mathcal{L}$.

Constant-Energy Optimization

Although optimal multiuser loading algorithms have been proposed [69], [77], [79] to solve various forms of the FDMA-TDMA optimization problem, in general, the computed time-sharing factors are non-rational and thus lead to large latencies in practice. Moreover, the computational complexity of these algorithms are not suited for fast-fading environments. This, coupled with the fact that a constant transmit energy distribution is almost as good as the water-filling energy distribution at high SNR's as long as only *useful* subchannels are used [54], [87] justify the elimination of energy optimization in problem formulation (3.3.51). Therefore, the FDMA weighted rate-sum maximization problem with constant energy distribution becomes

$$\begin{aligned}
\max_{\mathcal{S}, I_{l,n}} \quad & \sum_{l=1}^{L} \alpha_l \sum_{n \in \mathcal{S}} \log_2 \left(1 + I_{l,n} \frac{\mathcal{E}_{tot}}{|\mathcal{S}|} \frac{g_{l,n}}{\Gamma_l} \right) \\
\text{subject to :} \quad & \sum_{l=1}^{L} I_{l,n} = 1, \ \forall n \in \mathcal{N} \\
& I_{l,n} = 1 \text{ or } 0, \ \forall l \in \mathcal{L}, \ \forall n \in \mathcal{N} \\
& \mathcal{S} \subset \mathcal{N} \quad \text{– set of useful subchannels} \qquad (3.3.57)
\end{aligned}$$

where the set $\mathcal{S}$ contains the indices of the useful subchannels, and $|\mathcal{S}|$ denotes the cardinality of $\mathcal{S}$.

Given any set of used subchannels $\mathcal{S}$, the cost function in (3.3.57) can be rewritten as:

$$\sum_{n \in \mathcal{S}} \max_l \left\{ \alpha_l \log_2 \left(1 + \frac{\mathcal{E}_{tot}}{|\mathcal{S}|} \frac{g_{l,n}}{\Gamma_l} \right) \right\} \qquad (3.3.58)$$

which implies that the optimal strategy is to assign each subchannel in $\mathcal{S}$ to the user that achieves the maximum weighted rate with the same energy. To avoid wasting energy on subchannels that are experiencing deep fades, it is important to optimize the set of *useful* subchannels. The following outlines a novel and low-complexity algorithm, first proposed in [82], that achieves the optimum solution for the FDMA-based multiuser OFDM downlink problem with constant energy distribution.

For $k = 1 : N$

1. $\mathcal{E}_n = \mathcal{E}_{tot}/k = \mathcal{E}$.
2. $\forall n = \{1, \ldots, N\}$, compute $b_n = \max_{\forall l \in \mathcal{L}} \{\alpha_l \log_2(1 + \mathcal{E} g_{l,n}/\Gamma_l)\}$
3. The weighted aggregate rate is: R_k = sum of k biggest b_n

The maximum weighted aggregate rate is: $\max_{1 \leq k \leq N} R_k$.

For each iteration, step 2 requires $N(L-1)$ comparisons and NL log operations while step 3 requires $O(N \log_2 N)$ operations if using the divide-and-conquer based sorting algorithms [56]. Therefore, the worst-case complexity of this optimal constant-energy FDMA multiuser loading algorithm is $O(N^2 \max(L, \log_2 N))$. Also proposed in [82] is a slightly suboptimal but $O(NL)$ algorithm that starts off with using all subchannels, and then removing the subchannel from $\mathcal{S}$ that contributes the least weighted rate until no further increase in the weighted rate-sum occurs.

The constant-energy formulation simplifies the multiuser transmit optimization problem tremendously but with negligible performance loss even at low SNR's [76], [82]. Although slightly better performance can be achieved by relaxing the restriction that each user be allocated the same amount of energy, this approach will involve set assignments making the problem combinatorial. Previously known constant-energy multiuser loading algorithms can be found in [80], [81]. In [80], optimal and suboptimal methods were proposed for minimizing the total energy used while satisfying the constraint that each user is allotted a fixed number of subchannels[11]. In [81], a heuristic method was proposed to maximize the minimum achievable rate among all users. However, none of these methods optimize over the set of useful subchannels.

3.4 Optimization through Coding

OFDM transmission can also be optimized through the use of codes. Often known codes designed for an intersymbol-interference-free AWGN channel are very good when used in combination with OFDM, even if the channel has severe ISI. This is because the OFDM system creates a set of parallel AWGN channels from an ISI channel, and these AWGN codes thus directly apply to each of these subchannels. Considered mostly for wireless applications, OFDM has the advantage of spreading a fade over subchannels so

[11] fixed-rate constraints are more practical.

that all symbols in an OFDM block are slightly distorted but can still be correctly decoded, while a few adjacent symbols without OFDM are completely destroyed.

As is often the case in general with coding, there are two objectives of code use in OFDM transmission:

1. **coding gain** - to gain an effective SNR improvement in the channel because of the code.

2. **robustness** - to prevent intermittent disturbances from causing channel outage

The first coding-gain objective receives great attention in most designs, while the second robustness objective may be yet more important. Codes with good coding gain typically have good separation or distance between adjacent messages/codewords, and the optimized OFDM system should preserve this good coding gain. Robust codes typically have good product distance, distributing the distance between messages/codewords over many OFDM tones in optimized OFDM. The distribution of coding benefit prevents an outage on any small number of subchannels from causing unacceptable performance, by exploiting the frequency diversity in a fading channel. Usually Robust codes are augmented by interleaving methods, possibly at several levels with OFDM, to disperse impulsive or intermittent-fade effects over a number of independent codewords. The word "optimization" is used more loosely here in this section than with loading or partitioning because the codes used with OFDM may just be good codes and not necessary the best – the best codes for use with OFDM for the second robustness objective may not presently be known.

Optimized Coded transmission systems go under 3 categories of increasingly improved performance that are discussed in the 3 subsections of this section: Coded OFDM or COFDM, Coded DMT, and Turbo-DMT.

3.4.1 COFDM

Coded OFDM systems are presently used heavily in wireless transmission, both in broadcast and portable or mobile communications. Examples include the Terrestrial Television and Radio broadcast standards [88], DAB [89] and Digital Video Broadcast (DVB) [90] [91] in Europe. Herein, the key to achieving an error-free transmission as near as possible to the Shannon limit is how to design codewords and perform the interleaving. There are typically

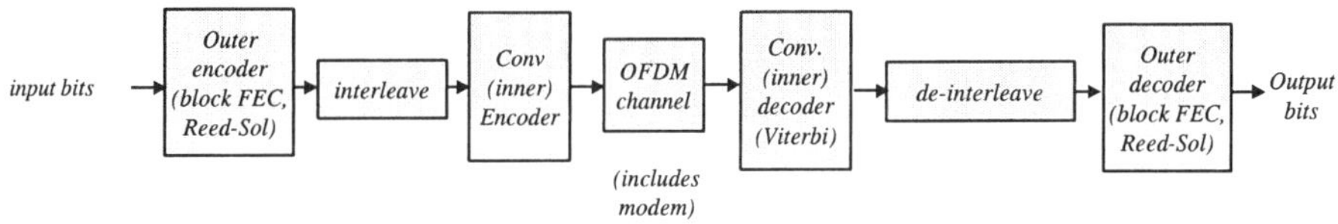

Figure 3.7. Coded OFDM showing inner and outer codes.

two levels of coding, called concatenated coding [92], in such transmission systems as illustrated in Figure 3.7.

In a wireless channel, diversity can be introduced through coding and interleaving. With perfect interleaving, symbols that are uncorrelated are fed to the decoder so that the decoder achieves the full error correction capability. In a single-carrier system, interleaving is done only in time, which can result in huge delays for slow fading channel. In an OFDM system, codewords can be arranged, and interleaving is done in frequency as well as in time [93] to fully exploit time and frequency diversities.

The *inner code* is a *convolutional code*, often rate-1/2, which means one bit of input is mapped to two coded bits of output depending on the current and previous input bit values[12]. For a rate-1/2 code, as an example, each tone carries 2 bits of information that represents one bit of input data in a QPSK constellation. Other convolutional codes may have other rates, but generally 1 to a few input bits into a slightly larger set of output bits, see [94]. Decoding of the code proceeds through the tones in the receiver through the use of a maximum-likelihood Viterbi (or other approximately max likelihood receiver) that nominally should gain the coding-gain improvement. Decoding proceeds through the OFDM tones on each subchannel/tone as if they had all been transmitted over the same AWGN channel. Because the individual subchannels may have different SNR_n at the channel output, the maximum-likelihood receiver should weight each tone in proportion to its SNR in determining branch metrics in the receiver. This inverse weighting can be derived from the multichannel normalizer (FEQ) discussed in Section 3.3. The decisions in that FEQ are not used for bit decoding, but instead only are used for adaptive updating of the FEQ and determination of the SNR as in Section 3.3. Convolutional inner codes for OFDM are typically

[12]The 64-state convolutional optimum encoder (designed originally for the AWGN channel) listed with free distance 10 and coding gain 7 dB is common in its use, see [94] but other convolutional codes can be used also.

useful only for improving SNR performance. Thus, these codes achieve only objective 1 above. The justification for use of rate-1/2 codes is somewhat heuristic and may have more to do with the availability of hardware for such codes plus the present heavy use of QPSK in OFDM systems. QPSK transmission maps to rate 1/2 easily by letting the two output bits of a convolutional encoder map directly to one of the 4 points in QPSK. Rate 3/4 codes are typically found in those few applications, *e.g.*, DAB and DVB, where 16 QAM is used on each of the tones because those codes map easily. Care must be taken in the design of these codes to ensure good SNR properties and not just any good convolutional code maps into a constellation like 16 QAM (trellis codes, discussed later, are often better for this situation).

This section attempts to provide some "rules-of-thumb" guidelines for the design of OFDM systems with coding for wireless transmission. Largely, this area to date has been one of individual designs for specific transmission systems without a general approach being developed. In wireless transmission, Rayleigh fading can be "flat," meaning that it affects essentially all OFDM tones, or "frequency selective," meaning it affects only a few adjacent tones (perhaps in multiple positions within the set of OFDM tones). The convolutional code does little to mitigate this type of fading, largely because the SNR loss during a fade is much greater than the gain of the code. The convolutional code however, through the FEQ SNR measurements and metrics of the Viterbi decoder, can provide a strong indication of fading often known as an *erasure* to the *outer code.* For this reason, most wireless COFDM systems, which in practice must consider Rayleigh fading, also use block forward error correcting codes. Usually, these are Reed-Solomon byte-error correcting codes, which are well-known to be able to use erasure information. An RS code has parameters (n, k) and t where $n < 255$ is the total number of bytes in a code word, $k < n$ is the corresponding number of data bytes and $2t = n - k \leq 32$ is the number of parity or "check" bytes that are found as the remainder after dividing the message byte polynomial by a generator byte polynomial [95] (thus the numbers n and k are chosen so that their difference is an even number). The RS decoder can then correct up to t bytes in error within a single codeword with a corresponding ML decoder in the receiver, and up to $2t$ bytes in error if the location of the error bytes is flagged by the erasures of the inner code.

The multipath channel has memory so that it exhibits correlated signal transmission impairments, both in the time and frequency domains. This correlation reduces the effectiveness of coding and typically *interleaving* is introduced to reduce the size of error bursts, spreading bad bits over the code space, and thereby making errors more correctable. This interleaving is done in two dimensions: in time or in frequency. Widely-used *block interleaving* accepts the coded symbols in blocks from the encoder, permutes the symbols, and then feeds the rearranged symbols to the modulator. The usual permutation of the block is accomplished by filling the columns of an $M \times N$ array, and these symbols are then fed to the modulator in each row. At the receiver, the de-interleaver preforms the inverse operation. On the other hand, *convolutional interleaving* ([96], Chapter 8) takes a group of bytes that may correspond to the bytes of adjacent tones in several successive symbols of a frequency selective fade or to all the bytes in a flat fade and disperses these likely error bytes into many codewords. The number of codewords over which a burst is dispersed is known as the interleaver depth, L. With convolutional interleaving, one can show that any burst of M consecutive bytes will thus result in at most $\lceil M/L \rceil$ bytes in error in each of L consecutive codewords. As long as the error bytes from multiple fades are not mixed in the interleaver, the power of the code-correction capability (t or $2t$ bytes with erasures) is multiplied by the interleaver depth L, resulting in a very powerful and robust structure.

The designer of a COFDM system then also needs to decide the interleave depth and amount of error-correction overhead (the receiver complexity grows exponentially with t for a RS code). Thus, the crucial parameters for COFDM code design are:

1. the mean time between fades τ_m.

2. the coherence bandwidth of each fade f_{cb}, which is approximately equal to $2 \cdot (1/\tau_{rms})$ where τ_{rms} is the rms delay spread of the channel [97]

3. the number of tones N and associated symbol rate $1/T$

4. the interleave depth L

5. the correction capability t or $2t$ with erasures and FEC parameters n and k

6. the number of bits per tone $2\bar{b}$ and the total length of the error burst in bytes M.

For OFDM, the possibility of several frequency selective fades occurring simultaneously increases with the number of tones (or actually the used bandwidth). To compute the number of bytes per fade

$$M = \underbrace{\frac{2\bar{b}}{8}}_{bytes/tone} \cdot \underbrace{\frac{N/T}{f_{cb}}}_{tones/fade} = \frac{N\bar{b}\tau_{rms}}{8T} \text{ bytes/fade.} \tag{3.4.1}$$

This formula can be construed as approximately correct even if $\tau_{rms} > T$, which essentially means the fade spans several successive symbols and is flat as long as the mean-time between fades exceeds $4\tau_{rms}$. However, if more than one fade per N tones, then M in (3.4.1) needs to be increased by that same factor.

The mean-time between fades allows bounding of the interleave depth and codeword length according to

$$\tau_m > 3 \cdot \underbrace{L}_{codewords} \cdot \underbrace{n}_{bytes/codeword} \cdot \underbrace{\frac{8}{b}}_{tones/byte} \cdot \underbrace{\frac{N}{T}}_{seconds/tone} , \tag{3.4.2}$$

essentially meaning that the "interleave depth" is less than 1/3 the average time between fades. An extra factor can be applied for the situation where 1/3 mean-time between fades is not sufficient to ensure adequately low probability of successive fades occurring too closely in time. After choosing an L and n above in (3.4.2) and computing M in (3.4.1), the value for t must be such that

$$t \geq \frac{M}{L} \ . \tag{3.4.3}$$

or $2t$ should exceed this value with erasures. Some flexibility is allowed in that L can be increased and large values of t used to improve robustness and reduce necessary interleave depth. However, use of too large t or overhead percentage can result in a significant loss in coding gain of the resultant concatenated coding scheme. Usually, Rayleigh fading dominates channel SNR effects and so code rates for FEC of 1/2 or higher are often found, especially when the convolutional coding rate is greater than 1/2. However, in systems where error-correction objectives can be met with lower overhead, they should. Best coding gain occurs when FEC overhead is less than 10%.

Alard [89] considered the problem of the optimum decoding of a code associated with OFDM modulation in a Rayleigh selective channel. He showed that a concatenated coding for OFDM with 4 PSK achieves between 0.8 and 1.6 bits/Hz, which is much higher than that of any other techniques. This implies that the OFDM technique permits an optimum weighting of the samples at the decoding level. For the Rician fading channel, Nicolas [98] found the error bound of the probability of error by assuming independent and identically distributed channel frequency responses.

For wireless applications with *multiple antennas*, there is an additional dimension, *space*. Thus, there is more freedom to assign codewords and to perform the interleaving over 3 dimensions, namely, time, frequency and space. Kim [93] presented a concatenated coding scheme that combines the attributes of an RS code to handle bursty errors and a convolutional code to handle random bit errors. In the proposed coding scheme, successive bits in each subchannel form an RS symbol in time, which is protected by a low constraint-length convolutional code arranged and interleaved in frequency and space. The demodulator outputs soft quantized code symbols to the inner convolutional decoder, which in turn outputs hard quantized code symbols, with bursty errors, to the RS decoder. This combined scheme is shown to achieve a higher error correction capability with lower complexity. It also shows that if successive bits are properly assigned in a codeword and transmitted on different antennas, randomization of bits in a codeword can be achieved without using conventional time interleaving. New coding techniques such as Space-Time coding [99] can also be a good alternative. It achieves superior performance in a flat fading channel by fully utilizing antenna diversity. Since each subchannel sees a flat channel, this space-time coding technique can be successfully employed in OFDM [100].

The design of codes can be improved to encompass Rayleigh fading and other non-stationary channels as long as the worst degree of fading is known. This leads away from traditional coding and requires a special code search procedure that tests for code distance and periodic product distance (the product of the distance of each of the tones within an error event for a code, a measure of the distribution of code strength over affected symbols), where periodic corresponds to the known interleaving pattern. When this is done, dramatic improvement is possible and was first noted in the rather remarkable codes of Wesel [101].

3.4.2 Coded DMT

COFDM essentially ignores the potential benefit of Section 3.3's loading in transmission. This benefit can be large, but depends on the channel and the disparity between the optimum number of tones that should be used and the actual number used. When this difference is large, the benefit is also large. Essentially, the robust coding is replaced by the more channel-intelligent aversion of the fading tones. The problem for wireless transmission is the rate of variation of the channel versus the ability of the transmission channel to monitor such variation and react in an optimized-OFDM (DMT) system. When such tracking is possible, gains are moderate to large [102]. The transmitted signal can be coordinated to maximize the transmission efficiency over the space dimension. In this way, a two-dimensional water-filling, both in frequency and space can be performed [103].

Coded DMT replaces the convolutional inner code of COFDM by a trellis code [43]. The key observation is that trellis codes are based on a principle of constellation partitioning that leaves a single code capable of implementing essentially any number of bits in the constellation. The only difference between different constellations in the use of a trellis code is the number of points within each of the partitioned subsets of the constellation. Thus, the same encoder and decoder can be used for all the subchannels of a DMT system even if the number of bits varies, unlike the convolutional encoder, which was first observed by Ruiz [104].

For outer-code design, the coded DMT system design proceeds the same as the COFDM system with the trellis code replacing the convolutional code. The computation of the number of bytes in error needs to be slightly altered, but use of the average number of bits per tone is adequate for the approximate analysis. Zogakis [105] has investigated the concatenated used of Reed-Solomon codes, trellis codes, interleaving, and shaping codes, reporting very high coding gain with significant robustness.

3.4.3 Turbo DMT

Turbo codes were introduced by Berrou [106] for the AWGN channel. They combine interleaving and soft-convolutional codes into a single design of a code with enormous number of states, for which a recursive suboptimal decoding algorithm is used to attain coding gains that are near capacity limits for the AWGN channel. Turbo-equalizers were later introduced [107] to extend the resultant gains to channels with intersymbol interference, unfortunately at huge complexity increase of the decoding algorithm

Recently Lauer [108] has found a way, through puncturing, to apply turbo codes to a DMT system with variable optimized number of bits per tone. Remarkably, puncturing of some of the original codes of Berrou in this fashion, known as *Turbo DMT*, leads to near capacity-limit performance on any bandlimited channel. It appears also that the interleaving inherent in the codes offers some immunity to multipath fading as long as distribution of burst errors occurs. FEC and outer codes can be used for final correction of errors, but may be less necessary than in COFDM or Coded DMT. The straightforward use of turbo codes on the inner code of COFDM also shows strong promise [109], but of course without the gain afforded by adaptive loading.

3.5 Conclusion and Projections

Optimization of OFDM leads to significant performance improvement when channel bandlimitation is severe. The optimization was divided in this chapter into the problems of channel partitioning, loading, and coding. Using any of the three optimizations will lead to improvement with the combination of all three leading to the highest possible performance levels on a bandlimited channel.

The greatest difficulty for wireless transmission and optimized OFDM is the time variation of the channel . The optimized system requires handshaking between transmitter and receiver, which may not be feasible when the time variation is fast. Efforts are ongoing to try to speed this process as well as to reduce the amount of handshaking required, such that future wireless transmission systems may benefit from optimization of OFDM. Presently, however, optimization through loading is not yet used in wireless transmission.

Perhaps the most promising area for optimization is that of wireless spatial-temporal techniques [7] [103], where enormously efficient wireless transmission has been projected. Herein, it is proved that the system capacity can dramatically increase with multiple antennas when the space dimension is fully utilized. In "fixed" wireless transmission (local area networks or wireless local loops without mobile movement during use), time variation (especially in space) is significantly less. These areas thus represent the most fruitful areas of future research in optimized OFDM transmission, allowing potential for the OFDM system adequate time to optimize to best performance.

Chapter 4

SYNCHRONIZATION

Sarah Kate Wilson

OFDM, like any other digital communication system, requires synchronization. However, OFDM as a multicarrier system has a different structure than a single-carrier system and so has different requirements and different resources. For example in OFDM, one can tolerate larger errors in estimating the start of a symbol than in a single-carrier system. This is due to OFDM's longer symbol period and its cyclic prefix. On the other hand, frequency synchronization in OFDM must be tighter than that in single-carrier systems, due to the narrowness of the OFDM subcarriers. In terms of resources, OFDM has a structure that is not available in single-carrier systems that is useful for synchronization. For example, most OFDM systems have a cyclic prefix that, as we will see, can be used for synchronization. The cyclic prefix can act as pilot data. Often, an OFDM symbol itself is used as pilot data. In this case, the structure of the OFDM symbol can be exploited for time and frequency offset estimation. The choice of pilots versus no-pilots depends on many parameters: the operating SNR, the size of the cyclic prefix, coherent versus differential modulation. Whether to insert pilot data or use OFDM symbols as pilot data often depends on how much overhead the system can tolerate.

After briefly introducing synchronization schemes in Section 4.1, we will present methods for timing and frequency offset estimation in Sections 4.2 and 4.3, respectively, and joint timing and frequency offset estimation in Section 4.4. In addition, we will briefly address sampling clock offset estimation and tracking in Section 4.5 and discuss further issues and reading in Section 4.6.

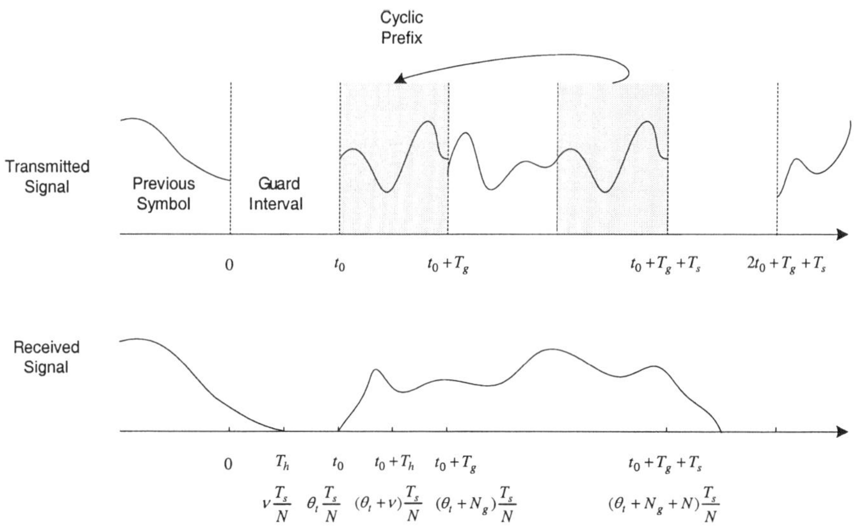

Figure 4.1. Transmitted and received OFDM signal with guard interval and cyclic extension.

4.1 Overview of Synchronization Schemes

Consider the transmitted and received baseband OFDM signals shown in Figure 4.1. In this figure t_o indicates the true starting position of the OFDM signal while T_g and T_s are the length of the cyclic extension and the symbol duration, respectively. By 'true' starting position, we mean the boundary between the current OFDM symbol and the one before it.

For a wireless channel with length T_h, the received signal between $t_o + T_h$ and $t_o + T_g + T_s$ can be expressed as

$$y(t) = \sum_{k=0}^{N-1} H_k s_k e^{\jmath 2\pi (k\Delta f + \delta f)(t - T_g - t_o)} + n(t), \tag{4.1.1}$$

where $n(t)$ is the AWGN; N, Δf, and δf are the number of subchannels in an OFDM symbol, the subchannel space and frequency offset, respectively; H_k is the frequency response of the channel at the k-th subchannel. The sampled baseband OFDM signal can be written as

$$y_n = y\left(\frac{n}{N}T_s\right) = x_n + n_n, \quad n \in [\theta_t + \nu, \theta_t + N + N_g],$$

where

$$x_n = \sum_{k=0}^{N-1} H_k s_k e^{\jmath 2\pi \frac{(k+k_o+\epsilon)(n-N_g-\theta_t)}{N}},$$

$$n_n = n\left(\frac{n}{N}T_s\right),$$

$\theta_t = \frac{t_o}{T_s}N$ and $N_g = \frac{T_g}{T_s}N$, θ_t represents the integer offset in the OFDM frame, while N_g can be regarded as the number of samples in the cyclic prefix. In addition, the variable $\nu = \frac{T_h}{T_s}N$ is the length of the channel in samples, k_o is an integer carrier frequency offset and ϵ with $|\epsilon| \leq 1/2$ is a fractional carrier frequency offset. The fractional carrier offset ϵ is a function of the frequency offset, δf. From Section 2.2.1, we have $\frac{\delta f}{\Delta f} = k_o + \epsilon$. This chapter focuses on ways to estimate the fractional time offset, θ_t, the tone index, k_0 and the fraction offset, ϵ.

4.1.1 Timing Offset Estimation

The goal in timing offset estimation is to find a place to start the N-point FFT for demodulating an OFDM symbol. Assuming that the length of the channel is less than the length of the cyclic prefix ($T_h < T_g$ or $\nu < N_g$) and ignoring frequency offset,

$$y_n = \begin{cases} \sum_{k=0}^{N-1} H_k s_k e^{\jmath\frac{2\pi k(n-N_g-\theta_t)}{N}} & + n_n + \text{ISI} + \text{ICI}, \\ & \text{for } n \in [\theta_t, \theta_t + \nu], \\ \sum_{k=0}^{N-1} H_k s_k e^{\jmath\frac{2\pi k(n-N_g-\theta_t)}{N}} & + n_n, \\ & \text{for } n \in [\theta_t + \nu, \theta_t + N + N_g], \\ \text{different symbol}, & n > \theta_t + N + N_g \text{ or } n < \theta_t. \end{cases}$$

The ISI comes from the previous OFDM symbol while the ICI results from missing samples in the current OFDM symbol. Suppose that we have an estimate of the true starting point, $\widehat{\theta}$, and start the FFT at some point after our estimate, say $\widehat{\theta} + M$, for some positive M. To avoid the ISI and ICI, it is required

$$\theta_t + \nu < M + \widehat{\theta} \leq \theta_t + N_g\,.$$

If $\widehat{\theta}$ lies within this region, the only consequence of timing offset estimation error is a phase rotation. The best region for starting the DFT is illustrated in Figure 4.2. This means that the larger the cyclic prefix, the more error our system can tolerate, as we can see from the following example.

EXAMPLE 4.1.1: *If the length of the channel* $\nu = 10$ *samples, then there is a 6 symbol margin. The receiver starts the DFT at time* $\widehat{\theta} + M$*, where* $M = 13$*. Suppose that* $\theta_t = 0$*, while* $\widehat{\theta} = -1$*, then we start the DFT at the 12-th sample, so our 128-point sample is*

$$\left(\sum_{k=0}^{127} s_k H_k e^{\jmath \frac{2\pi k(12-16)}{128}}, \ldots, \sum_{k=0}^{127} s_k H_k e^{\jmath \frac{2\pi k(139-16)}{128}} \right) + noise.$$

After the DFT, we have

$$Y_k = s_k H_k e^{-\jmath \frac{2\pi 4k}{128}} + N_k,$$

where N_k *is the DFT of AWGN. The ISI and ICI are avoided because the FFT starts within the safe region* $10 < M + \widehat{\theta} \leq 16$*. However, there is still a phase rotation, which can be corrected by channel estimation [110]. As long as* $-3 < \widehat{\theta} - \theta_t < 3$*, the DFT starts in the safe region, requiring only additional channel estimation to clean up the phase rotation. If* $\widehat{\theta} - \theta_t < -3$*, the system will experience ISI and ICI. But, if* $\widehat{\theta} - \theta_t > 3$*, then our FFT will include samples from the next symbol.*

This example illustrates how some errors in time-offset estimation can be tolerated given a channel estimator. It also illustrates that in a differentially-detected system, we may have residual phase error due to time-offset estimation [111]. The goal is to have an efficient time-offset estimator with reasonable accuracy and a cyclic prefix that is long enough to accommodate mistakes in the estimator, but not so long that the system is inefficient.

4.1.2 Frequency Offset Estimation

As mentioned earlier, the normalized frequency offset can be written as $\frac{\delta f}{\Delta f} = k_o + \epsilon$, where k_o is an integer and $|\epsilon| < \frac{1}{2}$. Accordingly, we can divide the problem into two parts [112]:

- *fine frequency offset estimation:* estimating the center frequencies of each subchannel, ϵ;
- *coarse frequency offset estimation:* estimating the tone numbering index, k_o.

From Chapter 2.2, we know that the fractional frequency offset ($k_o = 0$ and $|\epsilon| < 1/2$) results in ICI between tones. An incorrect tone numbering with

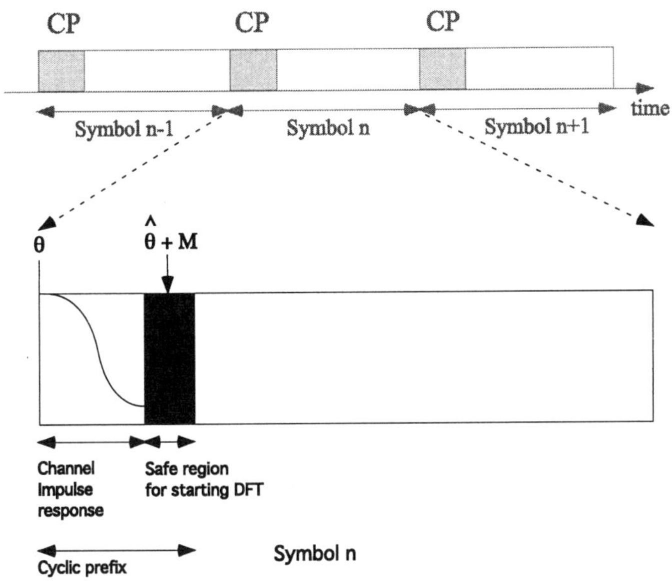

Figure 4.2. Regions of timing synchronization.

no fractional offset ($k_o \neq 0$ and $|\epsilon| = 0$) shuffles the tones for a receiver with a sufficiently large bandwidth. For example, if $k_o > 0$, then

$$y_n = \sum_{k=0}^{N-1} s_k H_k e^{\jmath \frac{2\pi n(k-k_o)}{N}} + n_n, \tag{4.1.2}$$

for $n = 0, \ldots, N-1$ and the demodulated output is

$$Y_k = \begin{cases} s_{N-k_o-k} H_{N-k_o-k} + N_k, & k < k_o, \\ s_{k-k_o} H_{k-k_o} + N_k, & k_o \geq k < N. \end{cases} \tag{4.1.3}$$

However, a non-zero k_o does not generally affect fine frequency synchronization algorithms, so the frequency synchronization can first correct the fractional offset, then correct the integer offset.

4.1.3 Acquisition Versus Tracking

Synchronization schemes generally have two parts: acquisition and tracking. Acquisition obtains an initial rough or coarse estimate of timing and/or frequency parameters. Tracking is an on-going process where this rough estimate is refined to get a better estimation. Often, the paradigm is fast acquisition followed by tracking [113]. This is very similar to how an artist paints a picture. First the painting is sketched in black and white (acquisition). Then after the artist has drawn a satisfactory sketch, the details and colors are added (tracking). Acquisition parameter estimation schemes generally have a wide range, but low accuracy. Tracking algorithms have a narrower range and finer accuracy.

It was demonstrated in Section 4.1.1 that given a sufficiently long cyclic prefix and reasonably accurate estimation of the symbol start, $\widehat{\theta}$, the only penalty is a phase rotation of the channel frequency response. This phase rotation can be adjusted by subsequent channel estimation and equalization. In this sense, we can think of symbol start estimation as acquisition and channel estimation as tracking. Channel estimation essentially cleans up any small mistakes made by the synchronization algorithm. However, in a synchronization scheme, the initial estimate of the symbol start may need to cover a wide range of possibilities. Depending on the estimation method, the receiver may need to search over $N + N_g$ samples to determine the approximate starting time. Once the initial $\widehat{\theta}$ has been found, phase tracking may involve searching over a narrower range. For differential demodulation, some form of fine timing estimation may be necessary to cancel the phase rotation induced by the non-ideal starting point.

However, as we shall see later in this chapter, one can have better results by first finding the fine frequency offset, followed by finding the correct tone numbering.

Frequency synchronization in OFDM can be divided into three stages: finding the correct integer offset, finding a coarse estimate of the fractional frequency offset, then refining this fractional frequency offset. However, often the last two steps are combined into one. These lead to the following OFDM synchronization model. Coarse frequency acquisition finds the correct tone numbering and fine frequency acquisition finds the fractional offset from the received subcarriers. This principle is illustrated in Figure 4.3. However, as we shall see later in this chapter, one can also first estimate the fractional frequency offset and then integer frequency offset.

4.2 Timing Offset Estimation

The goal of timing offset estimation is to determine where to start the FFT at the receiver to avoid ISI and ICI.

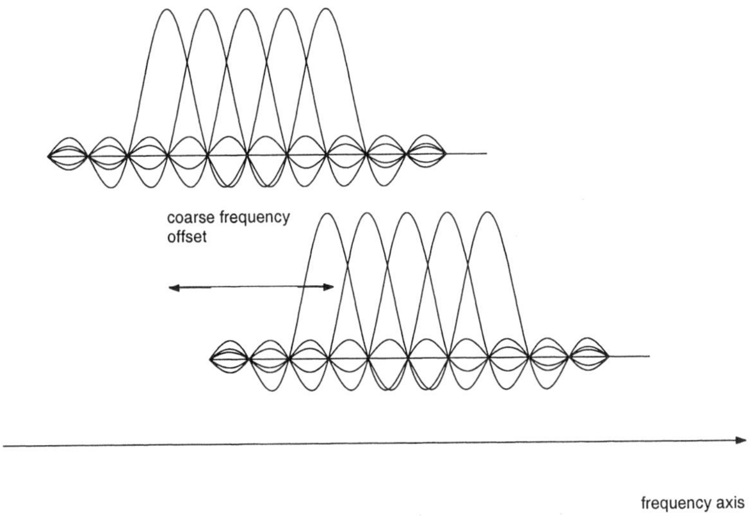

Figure 4.3. Coarse frequency synchronization.

4.2.1 Pilot-based Methods

Pilots, such as pseudo-random sequences or null symbols, can be used to determine the start of an OFDM symbol. They are particularly useful for systems with low SNR where synchronization might be otherwise difficult. Some timing synchronization methods for single-carrier can be also extended to OFDM systems [110].

The pilot symbols can be OFDM-based or non-OFDM-based. If OFDM-based pilot symbols are used, the frame size may need to be increased to lower the overhead. Otherwise, for shorter frames, non-OFDM-based pilot synchronization bursts may be more appropriate.

Another issue to consider when designing pilots for a system is, *is the data continuous or bursty?* If the data is continuous, a null signal can be used to signal the start of a symbol. This is the method employed by the European DAB standard [114]. If the data is bursty, a null signal is inappropriate.

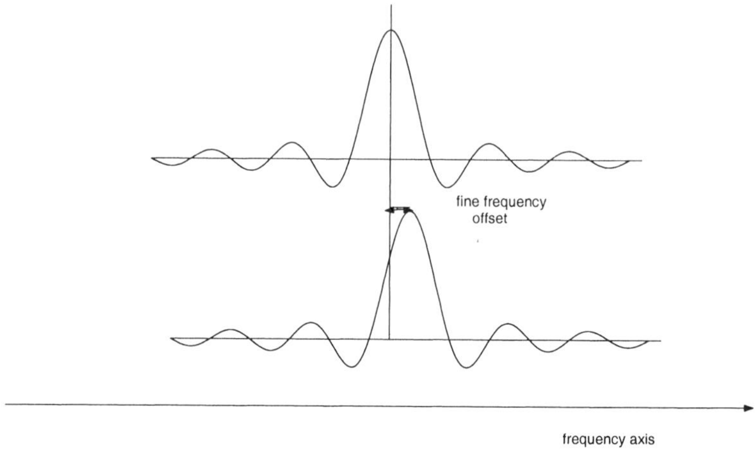

Figure 4.4. Fine frequency synchronization.

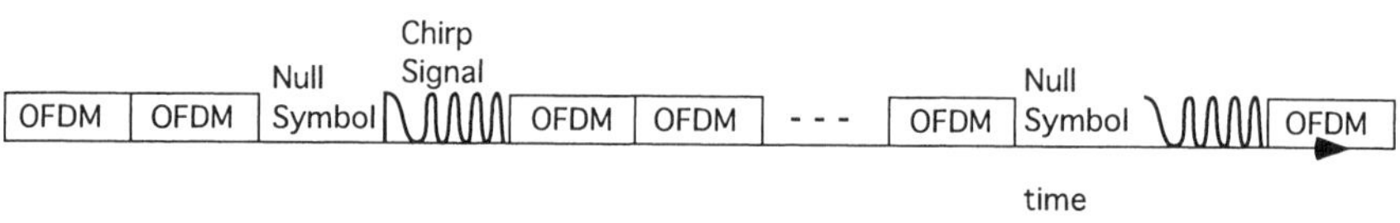

Figure 4.5. Pilot scheme for the European DAB systems

Non-OFDM-based Pilot Symbols

As shown in Figure 4.5, the DAB system inserts a null signal at the beginning of a group of OFDM symbols [114], [115]. The number of OFDM symbols following the null symbol depends on the operating mode but is on the order of 80 to 150. In addition there is an OFDM symbol used as a reference for the differential modulation scheme. This reference symbol is a chirp and can be also used for synchronization. The overhead of this scheme is relatively low – about 1%. The receiver first estimates the power of the incoming signals. When the estimated power is below a certain threshold, the receiver determines that a null signal is present and that the reference OFDM symbol is about to start. It then uses the chirp signal to estimate channel parameters [116] and to refine timing synchronization. This DAB synchronization scheme follows the traditional model of acquisition using the null symbol followed by fine estimation using the chirp signal.

OFDM-based Pilot Symbols

There are several ways to use pilots in OFDM symbols. In particular, we will focus on the pilot symbol proposed in [117], [118]. Figure 4.6 shows time and frequency characteristics of the pilot symbol. In this pilot symbol, the second half is equal to the first half, excluding the cyclic prefix. This is equivalent to only using every other tone in the OFDM symbol. To ensure that the time-domain pilot signal has the same average energy as the data-bearing symbols, the energy on each used subchannel is doubled. We will present two timing offset estimators, an *ML estimator* [119] and a *normalized correlation estimator* [117], based on this pilot symbol.

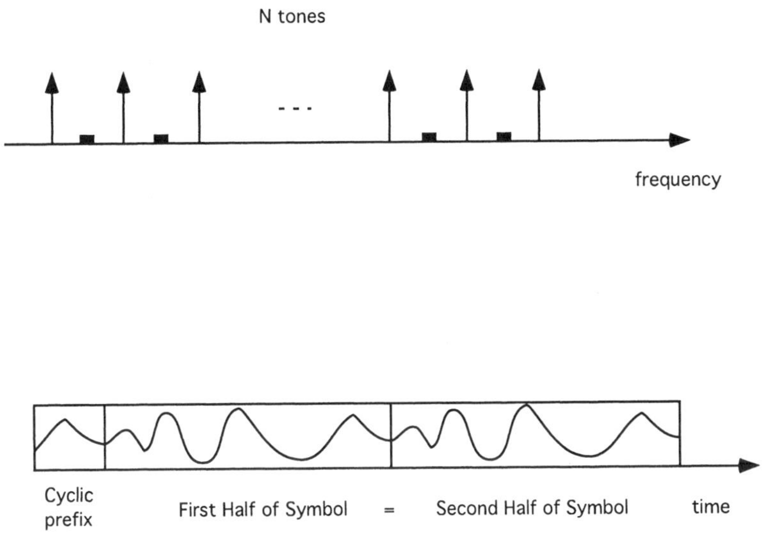

Figure 4.6. Time and frequency characteristics of Moose's pilot symbol

To derive the ML estimator, we make the following two assumptions:

1. The received OFDM signal is zero-mean Gaussian.
2. The only correlation in the time-domain data is due to the replication of the first half of the signal.

Assumption 1 makes the derivation much easier. Assumption 2 ignores the correlation due to the delay spread of wireless channel or the cyclic prefix. In this derivation we do not use explicit knowledge of the pilot values, but focus on the repetitive nature of the pilots. This proves to be relatively robust in the presence of an unknown frequency-selective channel. A method that incorporate both the pattern of the pilots as well as the value of the pilots can be found in [134].

Following the derivation for the cyclic-prefix based ML estimator presented in [120], [121], we can obtain the ML estimator,

$$\widehat{\theta}_{ML} = \arg\max_{\theta}(\Lambda_{ML}(\theta)), \tag{4.2.1}$$

where

$$\Lambda_{ML}(\theta) = \sum_{n=N_g+\theta}^{N_g+\theta+N/2-1} \left(\mathcal{R}\{y_n^* y_{n+N/2}\} - \frac{\zeta}{2}(|y_n|^2 + |y_{n+N/2}|^2) \right).$$

The term ζ is defined by

$$\zeta = \frac{E_x}{E_x + N_o} = \frac{\text{SNR}}{\text{SNR}+1},$$

$$E_x = E\{|x_n|^2\}, \text{ and } N_o = E\{|n_n|^2\}.$$

This estimator exploits the correlation between the first and the second halfs of the pilot symbol with an additional normalization factor based on the SNR and the relative power of the signal terms.

For the normalized correlation estimator,

$$\widehat{\theta}_{norm} = \arg\max_{\theta} \Lambda_{norm}(\theta), \tag{4.2.2}$$

where the normalized metric is

$$\Lambda_{norm}(\theta) = \left| \frac{\sum_{n=N_g+\theta}^{N_g+\theta+N/2-1} y_n^* y_{n+N/2}}{\sum_{n=N_g+\theta}^{N_g+\theta+N/2-1} |y_{n+N/2}|^2} \right|^2. \tag{4.2.3}$$

This estimator differs from the ML estimator in (4.2.1) in two ways. Frequency offset will cause a phase change in the correlation function. We will examine this effect further in Section 4.4 when we study joint frequency and timing offset estimators.

Both the ML timing metric and the normalized correlation timing metric are random variables, but the mean of both timing metrics is constant when θ is inside the cyclic prefix. Since $\Lambda_{norm}(\theta)$ is a nonlinear function, finding its accurate mean is difficult. However, under some approximations, it is found in [117] that

$$E\{\Lambda_{norm}(\theta)\} = \frac{E_x^2}{(E_x + N_o)^2} = \zeta^2,$$

when θ is inside the cyclic prefix.

Simply choosing the maximum of the timing metric could cause a false start outside the plateau due to noise. Since the center of the plateau is the best place to put the start of the symbol, a method to determine the center is proposed in [117]. The peak (the maximum value) of the timing metric is first found. Then the closest point to the left that is 90 % of the maximum value is chosen as the left boundary. This value will appear somewhere on the positive slope of the timing metric. After determining a similar boundary point to the right of the maximum value, the estimated symbol starting point is the average of the left and right boundary points. It is reported that searching for the center of the plateau region ensures fewer errors in the start of the timing signal. Depending on the estimated length of the channel response and the length of the cyclic prefix, the designer may want to add some positive offset to ensure that the DFT starts within the safe region.

EXAMPLE 4.2.1: *Consider an OFDM system with* 128 *subchannels and a cyclic prefix of* 16 *samples. We apply both the ML estimator and the normalized correlation estimator in an AWGN channel. In this case, as long as the DFT starts anywhere between the start of the symbol,* θ_t*, and the end of the cyclic prefix,* $\theta_t + N_g$*, channel estimation can mitigate any phase offset. Figure 4.7 shows the standard deviation and the probability of being outside the safe region of the pilot symbol. We also apply the* 90 % *rule mentioned above to find the center in the plateau for the normalized estimator. For the ML estimator, we use* 95 % *rather than* 90 %*. Both estimators perform very well for SNR's greater than* 10 *dB. This estimator not only finds the start of any OFDM symbol, but the start of the pilot symbol. The standard deviation takes into account how close the timing offset estimator is to the start of the pilot symbol.*

Though the performance of the estimators in an AWGN channel is a good starting benchmark, the real test is how they perform in a dispersive wireless channel.

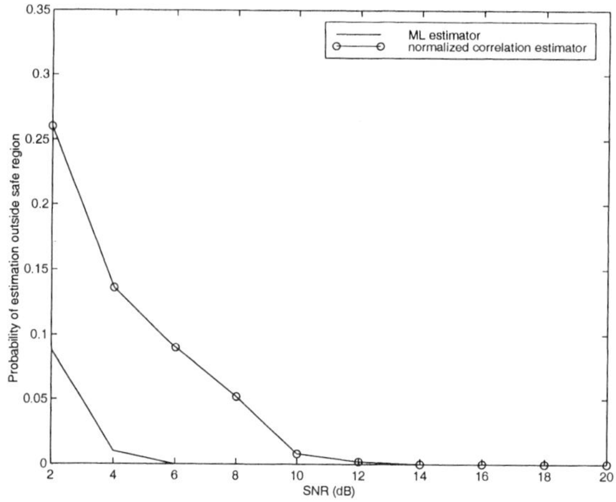

(a)

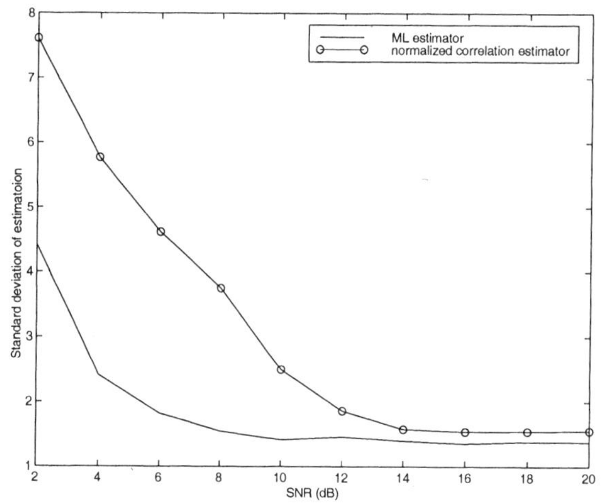

(b)

Figure 4.7. (a) The probability of the estimated symbol start outside the safe region and (b) the standard deviations of the estimators for pilot-based time-offset estimators in an AWGN channel.

EXAMPLE 4.2.2: *In this example, we use the same parameters as in the previous example, but assume a two-tap multipath channel. The two taps vary randomly and are with Gaussian distribution, but the overall average*

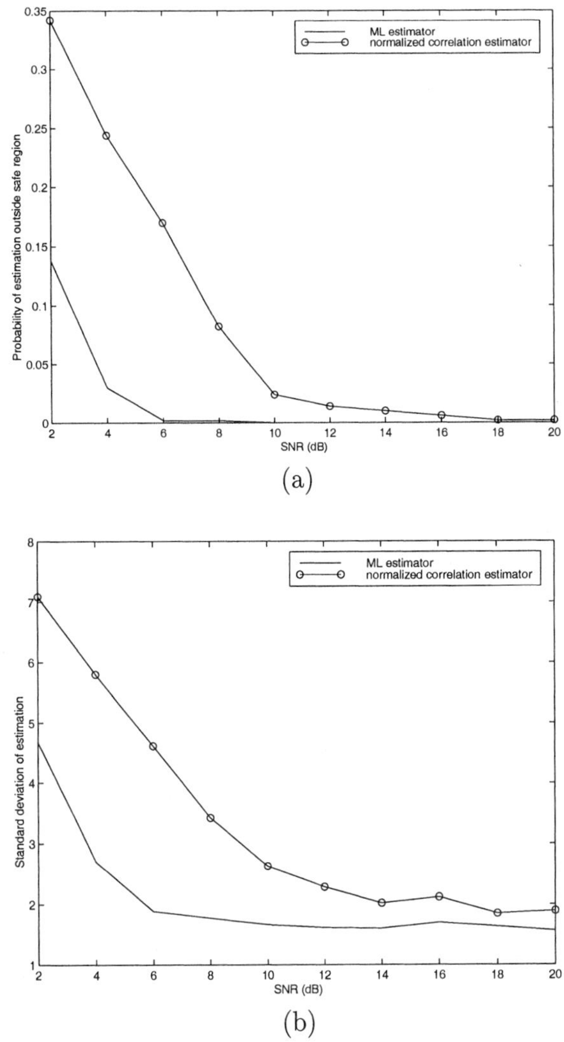

Figure 4.8. (a) The probability of the estimated symbol start outside the safe region and (b) the standard deviations of the estimators for pilot-based time-offset estimators in two-tap channels.

energy of the two taps is normalized. The first tap is at sample 0 *while the second one is at sample* 3. *This means the safe region for starting the DFT has* $16-4=12$ *samples. The channel is varied over 5000 trials but the SNR is fixed. This means that sometimes the first tap is larger and sometimes the second tap is larger. The probability that the estimated DFT start is outside the* safe *region, along with the standard deviation of the estimator, is shown in Figure 4.8. From the figure, there is a slightly larger irreducible error floor in the two-tap channel than that in the AWGN channel. With a multipath channel, it is less clear what the true starting point of the symbol is, especially if the second tap of the channel is larger than the first. With this in mind, the larger error floor is not surprising.*

In general there are two opposing forces in both of the timing metrics described above: the height of the peak versus the noise floor of the signal. The noise floor is not only related to the noise in the signal, but also to the energy of the signal. The other point to consider is the slope of the peak. The slope of the ML estimator is the same regardless of the number of samples in the pilot symbol. This is the main reason there is an error floor in the timing offset estimator. This will be explained in more detail in the next section.

4.2.2 Non-Pilot-based Methods

Most non-pilot-based methods for timing offset estimation are based on the redundancy of the cyclic prefix. Even through many of these non-pilot-based methods are derived for an AWGN channel, with minor modification, these methods can be also used for dispersive channels.

Many algorithms use the periodicity of the correlation function of the time-domain OFDM symbol [121], [122], [123]. This periodicity can be exploited to find the start of an OFDM signal. How to use this periodicity varies from method to method. Some methods look only at the correlation function while the others include the relative power of the samples as well. As in Section 4.2.1, we will begin with an ML estimator in an AWGN channel and then compare it with others.

Under the assumption that the time-domain data is Gaussian and the samples outside the cyclic prefix are uncorrelated, it has been derived in [121]

$$\widehat{\theta}_{ML} = \arg\max_{\theta}(\Lambda_{ML}(\theta)),$$

where

$$\Lambda_{ML}(\theta) = \sum_{n=\theta}^{\theta+N_g-1} \mathcal{R}\{y_n^* y_{n+N}\} - \frac{\zeta}{2}(|y_n|^2 + |y_{n+N}|^2),$$

and as before

$$\zeta = \frac{E_x}{E_x + N_o}.$$

Intuitively, the correlation between points within the cyclic prefix and those N samples away should reach the peak when $\widehat{\theta} = \theta_t$. Subtracting the power accommodates any fluctuations in the signal so that a peak in the function is less likely to be confused with a rise in the signal power.

At a high SNR, the weighting factor, $\zeta \approx 1$. This means for high SNR, the ML estimator is approximately equivalent to a *square-difference estimator* in [123], which finds the $\widehat{\theta}_d$ to minimize

$$\Lambda_d(\theta) \sum_{n=\theta}^{\theta+N_g-1} |y_n - y_{n+N}|^2.$$

At a low SNR, the weighting factor, ζ, is small. In particular, at an extremely low SNR, $\zeta \approx 0$. In this case, the ML estimator is equivalent to a *correlation estimator*, which finds the $\widehat{\theta}_c$ to maximize

$$\Lambda_c(\theta) = \sum_{n=\theta}^{\theta+N_g-1} \mathcal{R}\{y_n^* y_{n+N}\}.$$

In the following example, we compare the performance of the above three estimators.

EXAMPLE 4.2.3: *Consider the same OFDM system as in Example 4.2 for an AWGN channel. In this example, we compare three different timing offset estimators. Figure 4.9 shows the standard deviation and the probability of starting the DFT in the safe region. We assume that the DFT starts* 8 *samples after our estimated start of the timing period. From the figure, the ML estimator has the best performance over all SNR's. As the SNR increases, the square-difference estimator approaches the performance of the ML estimator. This is because for large SNR's the two estimators are same. The correlation estimator performs relatively poorly, showing that the extra power terms are necessary in the ML metric.*

It may be surprising that the estimators without pilots in this example and with pilots in Example 4.2.1 have comparable performance. This is because the pilot-based timing offset estimators have a *plateau* due to the redundancy in the cyclic prefix. This makes finding the exact start of an

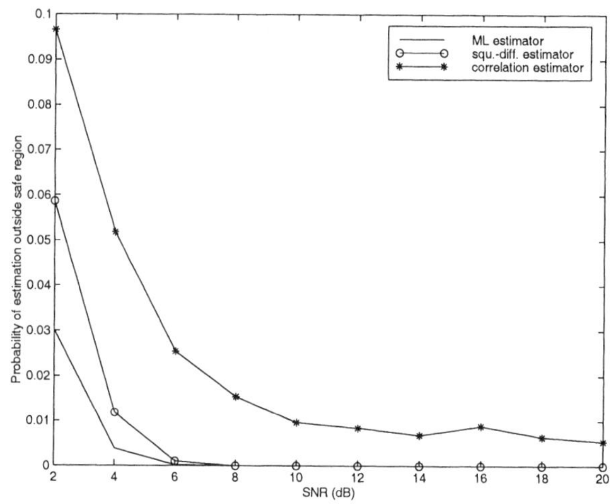

(a)

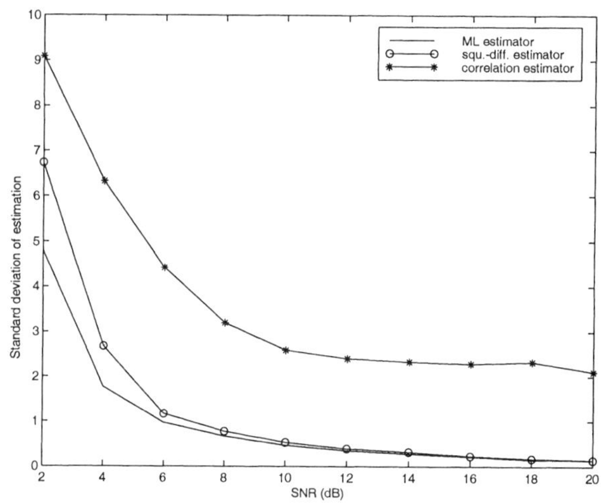

(b)

Figure 4.9. (a) The probability of the estimated symbol start outside the safe region and (b) the standard deviations for the three different estimators in an AWGN channel.

OFDM symbol difficult. The second reason is that extending the size of the redundant period does not necessarily increase the accuracy of the estimator.

To see how the length of the cyclic prefix affects the performance of the timing offsets, consider the squared-difference metric, $\Lambda_d(\theta)$. The expected value of this metric is,

$$\begin{aligned} E\{\Lambda_d(\theta)\} &= E\left\{\sum_{n=\theta}^{\theta+N_g-1} |y_n - y_{n+N}|^2\right\} \qquad (4.2.4) \\ &= \begin{cases} 2N_g(E_x+N_o) - 2E_xN_g\left(1-\frac{|\theta-\theta_t|}{N_g}\right) & \text{if } |\theta-\theta_t| \le N_g, \\ 2N_g(E_x+N_o) & \text{otherwise.} \end{cases} \end{aligned}$$

From the above equation, we can see that the slope of the mean of the estimator depends only on the difference between the true starting point, θ_t, and the estimated one, $\widehat{\theta}$. Increasing the size of the cyclic prefix does not guarantee a sharper peak in the metric.

Furthermore, the error floor of the timing offset estimator does not depend on the size of the cyclic prefix, rather it depends on the SNR of the signal. To see this, we use the following set of inequalities,

$$\begin{aligned} Var(\widehat{\theta}) &= \sum_{n=1}^{\infty} n^2(Pr\{\widehat{\theta}=\theta_t+n\} + Pr\{\widehat{\theta}=\theta_t-n\}) \\ &\ge \sum_{n=1}^{\infty}(Pr\{\widehat{\theta}=\theta_t+n\} + Pr\{\widehat{\theta}=\theta_t-n\}) \\ &= Pr\{\widehat{\theta}\neq\theta_t\} \\ &\ge Pr\left\{\sum_{n=\theta_t+1}^{N_g+\theta_t}|y_n-y_{n+N}|^2 < \sum_{n=\theta_t}^{N_g+\theta_t-1}|y_n-y_{n+N}|^2\right\} \\ &= Pr\left\{|y_{N_g+\theta_t}-y_{N_g+\theta_t+N}|^2 < |n_{\theta_t}-n_{\theta_t+N}|^2\right\}. \end{aligned}$$

Because the square of the absolute value of complex Gaussian random variables is with exponential distribution, we have

$$\begin{aligned} Var(\widehat{\theta}) &\ge Pr\{|y_{N_g+\theta_t}-y_{N_g+\theta_t+N}|^2 < |n_{\theta_t}-n_{\theta_t+N}|^2\} \\ &= \frac{N_o}{2N_o+E_x} = \frac{1}{SNR+2}. \qquad (4.2.5) \end{aligned}$$

This means that regardless of the size of the cyclic prefix, the variance of the estimator is lower bounded by a positive number that depends on the

SNR and has nothing to do with the size of the cyclic prefix. Therefore, increasing the size of the redundant information has limited benefits.

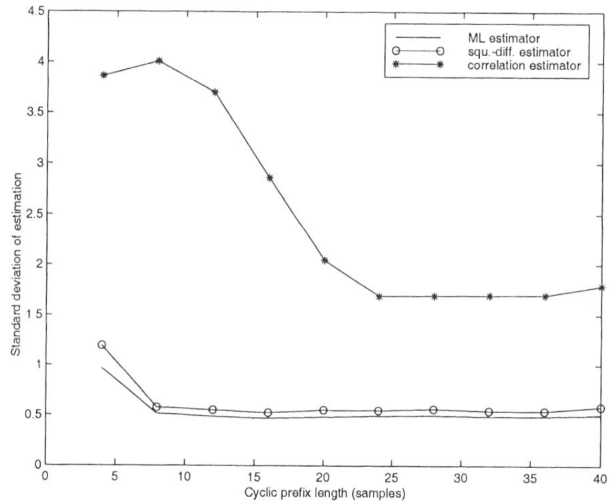

Figure 4.10. Performance of the three estimators using different lengths of cyclic prefix in an AWGN channel with an SNR of 10 dB.

EXAMPLE 4.2.4: *The three estimators have been applied to an OFDM system with 128 subchannels and different sizes of cyclic prefix to an AWGN channel. Both the square-difference estimator and the ML estimator converge to a standard deviation of about* 0.5, *while the lower bound on the variance, from Equation (4.2.5) is* 0.08. *Therefore, the lower bound on the variance is not tight and is only a way to demonstrate that the variance is lower bounded by some positive number.*

This leads to our previous discussion on the pilot-based timing offset estimators. Increasing the amount of redundant information beyond a certain point does not increase the accuracy of the estimators, which implies that non-pilot based timing offset estimators can work as well as pilot-based estimators in certain circumstances.

Up to this point, we have only considered an AWGN channel for these non-pilot-based estimators. As with the pilot-based estimators, the question is, *how do these metrics work in a multipath channel?* The ISI and ICI at the beginning of an OFDM symbol will reduce the redundancy in the cyclic prefix. If the length of the channel is less than the size of the cyclic prefix, there will be a window at the end of the cyclic prefix without ISI from the previous symbol [123]. In [123], the size of the cyclic prefix is extended to make the non-ISI portion of the cyclic prefix larger. In their squared-difference metric, the number of points in the summation is decreased to $N_g - \nu$.

EXAMPLE 4.2.5: *In this example, we compare the performance of the above three estimators plus the shortened square-difference estimator in a dispersive fading channel. Both the OFDM system and the two-tap channel are the same as in Example 4.3. This means that the ISI-free portion of the OFDM symbol is* $16 - 4 = 12$ *samples long. The results are shown in Figure 4.11. The shortened square-difference estimator has a slightly smaller standard deviation than the full-length estimators. However, we note that the shortened square-difference metric estimates the timing offset 2 samples later than the non-shortened metric. This is not a drawback of the estimator, but something that should be kept in mind when estimating where to start the DFT. As previously noted in the AWGN case, the detector based on correlation alone has relatively poor performance. This is the same channel that has been used in Example 4.4. Note that the performance of the estimators here are comparable to the performance of the OFDM-based timing offset estimators.*

4.3 Frequency Offset Estimation

Frequency offset estimation refines the initial analog estimate of the carrier frequency. In this section, we will assume that the analog frequency offset is not so large nor the bandwidth of the receiver so broad that no tones in the OFDM symbol have been lost through the receiver filter.

From Chapter 2 and Section 4.1, when there is no timing offset and channel noise and channel gain is unit, the received signal due to frequency offset can be expressed as

$$\begin{aligned} y_n &= \sum_{k=0}^{N-1} H_k s_k e^{\jmath \frac{2\pi(k+k_o+\epsilon)n}{N}} \\ &= x_n e^{\jmath \frac{2\pi(k_o+\epsilon)n}{N}}, \end{aligned}$$

where k_o is an integer and represents the coarse frequency offset, $|\epsilon| < 1/2$

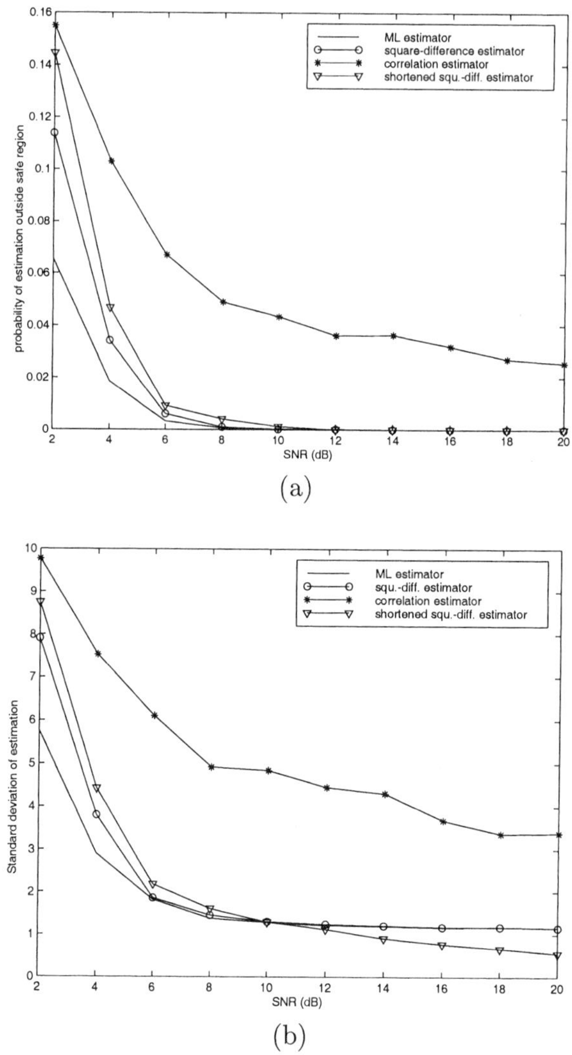

Figure 4.11. (a) The probability of the estimated symbol start outside the safe region and (b) the standard deviations for the three different estimators in a two-tap channel.

represents the fine frequency offset, and

$$x_n = \sum_{k=0}^{N-1} H_k s_k e^{j\frac{2\pi kn}{N}}.$$

Many frequency synchronization algorithms work by taking the argument of some noisy version of $\exp(j2\pi(k_o+\epsilon)K/N)$ for some constant K that depends on the chosen frequency estimator. Because the argument of a complex exponential is restricted to the region $(-\pi, \pi)$, the value of K determines the range of the estimated frequency offset. If $|K(k_o + \epsilon)/N| > 1/2$, there will be a wrap-around effect and some useful information will get lost. For fine frequency estimation, $K \approx N$ so that the argument of $\exp(j2\pi(k_o + \epsilon)K/N)$ is on the order of $j2\pi\epsilon$. For coarse frequency estimation, K is usually smaller. For example, if $K = N/4$, then $|2\pi(k_o+\epsilon)K/N| < \pi$ implies $-2 < k_o+\epsilon < 2$.

Coarse frequency synchronization usually uses pilot-based approaches. They basically rely on pattern matching. Non-pilot-based methods can be used for fine frequency offset estimation, ϵ. We will first discuss some pilot-based methods for both coarse and fine frequency offset estimation and then focus on non-pilot-based methods.

4.3.1 Pilot-based Methods

Pilot-based methods can be used for both coarse and fine frequency synchronization. Pilots can be an extra sequence outside an OFDM symbol or known data interspersed within the OFDM symbol. One way is to use short bursts of single-carrier training data interspersed with OFDM symbols [124] and directly use the methods developed for single-carrier synchronization. However, for fine synchronization, more accurate methods are required. Both OFDM-based and non-OFDM-based pilot symbols can be used for frequency offset estimation and synchronization.

Non-OFDM-based Pilot Symbols

Methods that use non-OFDM-based pilot symbols rely on single-carrier synchronization methods [124], [125]. Because OFDM requires finer frequency synchronization than an equivalent single-carrier system, these methods have been adapted for better accuracy. All of these methods find phase changes within the pilot sequences and estimate $k_o + \epsilon$ or ϵ.

In [124], a scheme based on inserting repeated short *constant-amplitude zero-autocorrelation* (CAZAC) sequences between OFDM symbols has been proposed. This CAZAC scheme performs both coarse and fine frequency offset estimation. The basic idea is to measure the phase change due to the frequency offset. If the same symbol, x, is inserted in time domain at n and $n+K$, then the received symbols at n and $n+K$ will be

$$y_n = xe^{\jmath\frac{2\pi n\epsilon}{N}} + n_n,$$

and

$$y_{n+K} = xe^{\jmath\frac{2\pi(n+K)\epsilon}{N}} + n_{n+K}.$$

Consequently,

$$y_n^* y_{n+K} \approx |x|^2 e^{\jmath\frac{2\pi K\epsilon}{N}} + \text{noise}. \tag{4.3.1}$$

As we saw above, the larger K, the finer the frequency estimation. Thus coarse estimation involves comparing the phases of two symbols close together while fine frequency synchronization relies on comparing symbols further apart. The requirement that symbols must be far apart for fine frequency synchronization puts some limitations on the system, especially if the channel is with fast fading. Any significant change in the channel phase will affect the fine frequency offset estimation. This could be mitigated by channel estimation. In the subsequent section of this chapter, we will also introduce a fine frequency offset estimation method based on one OFDM symbol to avoid the problem.

OFDM-based Pilot Symbols

Many methods [117], [118], [126] use pilot symbols that are part of OFDM symbols. In both [117] and [118], data are repeated within a time-domain OFDM symbol. In [126], known symbols are evenly spaced among OFDM subchannels. Here we will introduce some methods for fine and coarse frequency synchronization, respectively.

Fine frequency synchronization

Consider fine frequency synchronization using Moose's double OFDM symbol [118] similar to Figure 4.6. This is a super OFDM symbol with $2N + N_g$ points, where the first and the last N points are the same. A simple and elegant frequency offset estimator based on this pilot-symbol has been derived in [118]. This frequency offset can be estimated by

$$\widehat{\epsilon} = \frac{1}{2\pi}\arg\left(\sum_{k\in\mathcal{K}} Y_{1,k}^* Y_{2,k}\right), \tag{4.3.2}$$

where $Y_{1,k}$ and $Y_{2,k}$ are the DFT of the first and the second N samples, respectively, and $\mathcal{K}_p$ is the index set of the used tones. It can be shown that Equation (4.3.2) is equivalent to

$$\widehat{\epsilon} = \frac{1}{2\pi} \arg \left(\sum_{n=0}^{N-1} y_n^* y_{n+N} \right). \tag{4.3.3}$$

Equation (4.3.3) is very similar to the fine frequency offset estimator proposed by [117]. The difference is that two repeated OFDM symbols are used in [118], while one OFDM symbol where the first and the second half of the time-domain signal are equal is used in [117]. When there is no timing offset,

$$\begin{aligned} y_n^* y_{n+N} &= \left(x_n^* e^{-\jmath \frac{2\pi\epsilon n}{N}} + n_n^* \right) \left(x_{n+N} e^{\jmath \frac{2\pi\epsilon(n+N)}{N}} + n_{n+N} \right) \\ &= |x_n|^2 e^{\jmath 2\pi\epsilon} + \widehat{n}_n, \end{aligned} \tag{4.3.4}$$

where

$$\widehat{n}_n = x_n^* n_{n+N} e^{-\jmath \frac{2\pi\epsilon n}{N}} + x_{n+N} n_n^* e^{\jmath \frac{2\pi\epsilon(n+N)}{N}} + n_n^* n_{n+N}.$$

It is shown in [118] that both the time-domain estimator (4.3.3) and the frequency domain estimator (4.3.2) are ML for an AWGN channel. Using a trick for noise analysis in frequency modulation [127, pp. 414-418], it is also demonstrated in [118] that the above estimator is zero-mean and and with variance

$$E \left| \widehat{\epsilon} - \epsilon \right|^2 \frac{1}{(2\pi)^2 SNR \sum_{k \in \mathcal{K}_p} |H_k|^2}. \tag{4.3.5}$$

EXAMPLE 4.3.1: *In Figure 4.12, the MSE of fine frequency estimation is displayed for the time-domain and the frequency domain approaches. It is based on 500 trials. The OFDM symbol has* 128 *subchannels, but only* 98 *of them are used. The same two-tap dispersive channel as in Example 4.2.5 is used. The true fractional frequency offset is* $\epsilon = 0.3$. *The time-domain estimator uses* 128 *terms in the summation, while the frequency domain estimator uses* 98 *subchannels. The frequency domain estimator has a slightly lower MSE as it ignores unused subchannels that have only noise.*

Coarse frequency synchronization

Some schemes can be used for both fine and coarse frequency synchronization. For example, with minor modification, the fine frequency offset

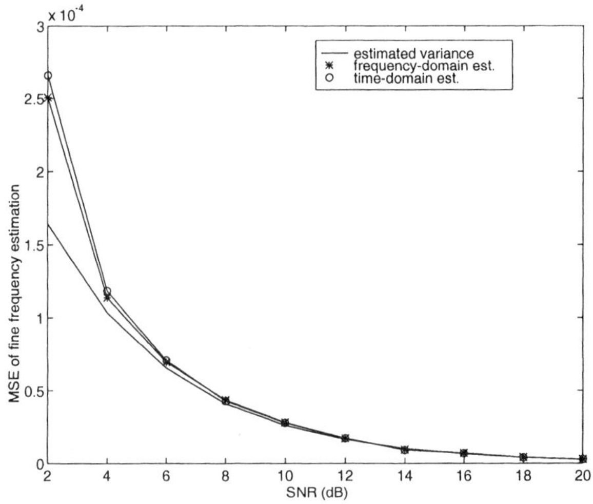

Figure 4.12. MSE of frequency offset estimators

estimator introduced above can estimate the subchannel numbering offset [118]. Here, we will present a frequency synchronization scheme developed by Classen and Meyr in [126]. We focus only on the coarse frequency synchronization, although the scheme can be also used for fine frequency synchronization. This scheme is illustrated in Figure 4.13. Pilot channels are

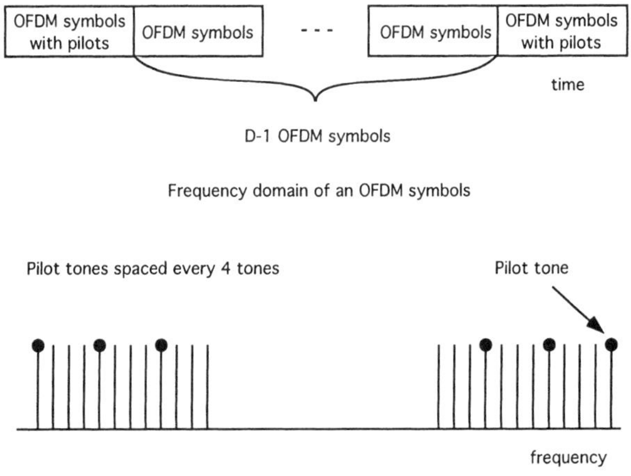

Figure 4.13. OFDM-based pilot scheme for coarse and fine frequency synchronization

placed to every D^{th} OFDM symbol. Within an OFDM symbol with pilots, there are L_f evenly spaced pilot tones. For an OFDM symbol with pilots, if the tone numbering is correct, $(k_o = 0)$, then the demodulated symbols at pilot position will have a high correlation with the known pilots. Otherwise, the correlation is very low. Based on the above observation, the tone numbering index can be estimated by

$$\widehat{k}_0 = \arg\max_l \left| \sum_{k \in \mathcal{K}_p} Y_k(l) s_k^* \right|,$$

where $\mathcal{K}_p$ is the pilot tone index set, s_k for $k \in \mathcal{K}_p$ is known at the receiver, and $\{Y_k(l)\}_{k=0}^{N-1}$ is the DFT of $\{y_n, y_{n+1}e^{-\jmath 2\pi l/N}, \cdots, y_{n+N-1}e^{-\jmath 2\pi l(N-1)/N}\}$. Let $s_k(l)$ and $N_k(l)$ be signal and noise components, respectively, in $Y_k(l)$. Then

$$Y_k(l) = H s_k(l) + N_k(l),$$

where H is the gain of the flat fading channel. If $l = k_o$, then s_k will be equal to $s_k(l)$ for $k \in \mathcal{K}_p$ and the correlation between $\{s_k : k \in \mathcal{K}_p\}$ and $\{s_k(l) : k \in \mathcal{K}_p\}$ will reach the peak.

The above algorithm is not applicable to frequency selective channels because the phase of the frequency response will affect the correlation. To compensate for this, a principle similar to differential decoding is used in [126]. Let $Y_{1,k}(l) = H_k(l)s_{1,k}(l) + N_{1,k}(l)$ and $Y_{2,k}(l) = H_k(l)s_{2,k}(l) + N_{2,k}(l)$ correspond to two OFDM symbols with pilots, $s_{1,k}$ and $s_{2,k}$ for $k \in \mathcal{K}_p$, respectively, and $\{H_k(l)\}_{k=0}^{N-1}$ be the N-point DFT of $\{h_0, h_1 e^{-\jmath 2\pi l/N}, \cdots\}$. The tone numbering offset can be estimated by

$$\widehat{k}_o = \arg\max_l \left| \sum_{k \in \mathcal{K}_p} Y_{1,k}(l) Y_{2,k}^*(l) \left(s_{1,k} s_{2,k}^*\right)^* \right|. \tag{4.3.6}$$

If the channel noise is ignored, then Equation (4.3.6) will be equivalent to

$$\widehat{k}_o \approx \arg\max_l \left| \sum_{k \in \mathcal{K}_p} |H_k(l)|^2 s_{1,k}(l) s_{2,k}^*(l) s_{1,k}^* s_{2,k} \right|.$$

It should be indicated that this algorithm has been developed for a coarse frequency estimation where ϵ is usually non-zero; therefore, it requires frequency correction in the time domain and repeated DFT's. However, if ϵ is zero, we can circularly shift the demodulated samples in the frequency domain by $-l$ points to avoid repeating DFT's [112]. When the tone numbering index, k_o, is estimated, we can either directly correct the coarse frequency offset in the time domain or feed it back to a numerically controlled oscillator.

EXAMPLE 4.3.2: *We apply the above coarse frequency offset estimation algorithm to the same OFDM system in the same two-tap channel as in Example 4.3. For an OFDM symbol with pilot tones, there are* 32 *evenly spaced pilot tones. The pilot symbols,* $s_{1,k}$ *and* $s_{2,k}$ *for* $k \in \mathcal{K}_p$*, are chosen such that* $s_{1,k}^* s_{2,k}$ *is a binary pseudo-random sequence. Figure 4.14 shows the probability that* $\widehat{k}_o \neq k_o$ *for different* ϵ*'s. The result is based on* 500 *trials. From the figure, when* $\epsilon \leq 0.2$ *the algorithm works well. However, the performance is not so good when* $\epsilon > 0.2$*. This means that good fractional frequency offset estimation and correction can significantly help tone numbering (coarse) estimation.*

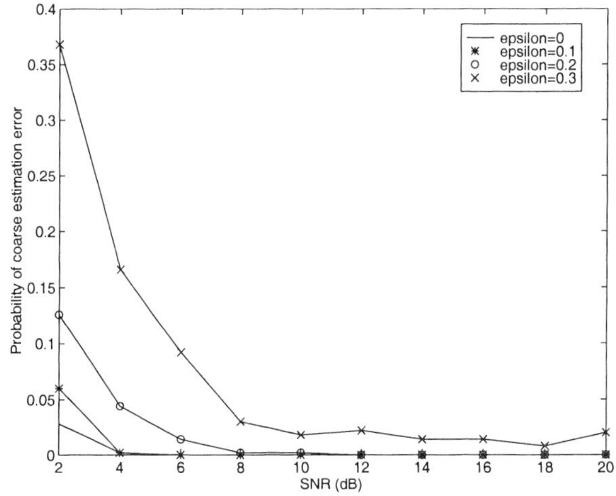

Figure 4.14. Probability that $\widehat{k}_o \neq k_o$ versus SNR for different fractional offsets, ϵ's.

The OFDM-based pilot symbols proposed in [117] can be also used for coarse frequency acquisition. In that case, there are two OFDM symbols with pilots. The first one is a shortened Moose pilot symbol and is used for timing and fine-frequency acquisition. The second one is combined with the first one for coarse frequency acquisition.

4.3.2 Non-Pilot-based Methods

Redundancy in the cyclic prefix can be also used for frequency offset estimation [121], [128], [129]. These estimators are very similar to the pilot-based ones discussed in the previous section except that they rely on the correlation within the OFDM symbol rather than adding known pilot symbols. For example, the ML estimator developed in [121] exploits the cross-correlation between samples inside the cyclic prefix and samples N points away. These non-pilot-based schemes are generally used for fine-frequency synchronization only.

The derivation of the ML frequency offset estimator based on the redundancy in the cyclic prefix is similar to the derivation of estimator in [118]. Assuming the timing offset, θ_t, is known, then the frequency offset can be estimated by

$$\widehat{\epsilon} = \arg\left(\sum_{n=\theta_t}^{\theta_t+N_g-1} y_n y_{n+N}^*\right). \tag{4.3.7}$$

Although this estimator has been originally derived for an AWGN channel, it can be also applied to a dispersive channel. However, for a dispersive channel, there will be additional noise in the estimator due to the ICI and the ISI in y_n for $n \in [\theta, \theta+\nu]$.

The variance of the above fine frequency offset estimator can be derived using the method in [117] and is very similar to (4.3.5). It is shown in [117] that increasing the size of the cyclic prefix will improve the performance of the estimator. This is different from the cyclic-prefix based timing offset estimator, where increasing the size of the cyclic prefix has only limited benefit.

EXAMPLE 4.3.3: *In this example, the fractional frequency offset, ϵ, is estimated using (4.3.7) for the same OFDM system in the same two-tap channel as in Example 4.3. The true fractional offset is $\epsilon = 0.25$. Figure 4.15 compares the MSE from simulation and the theoretical MSE (using the formula similar to (4.3.5)) that ignores the ICI and ISI. Note that there is an error floor in the MSE from simulation due to the ICI and the ISI.*

4.4 Joint Time- and Frequency Offset Estimation

In the previous sections, we derived the time-offset estimator assuming the frequency offset was known and the frequency offset assuming the time-offset

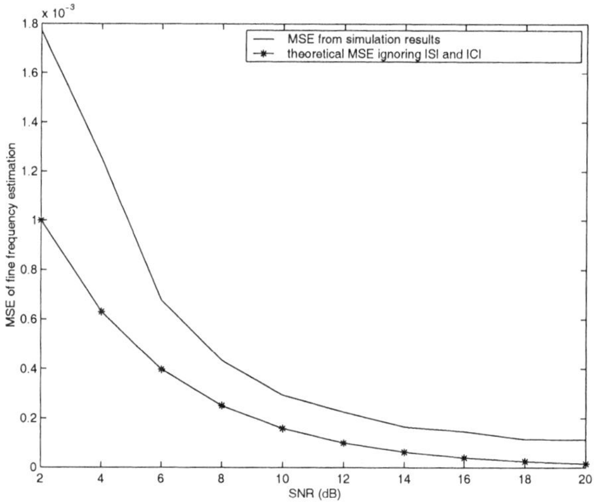

Figure 4.15. Theoretical and simulated MSE's of the fine frequency offset estimator based a 16 point cyclic prefix.

was known. In practice this will not be true and inaccuracies in our estimates of θ and ϵ will degrade the performance of the estimators. In this section, we discuss some of these impairments and how to jointly estimate time and frequency offsets.

4.4.1 Pilot-based

A non-zero frequency-offset, ϵ causes ICI which makes the timing metric noisier. Similarly, if $\widehat{\theta} - \theta$ is too large, our data will have ICI and ISI making the frequency metric noisier. In [117], the authors avoided this problem by taking the absolute value of the sum of the cross-terms. An ML derivation of the joint ML estimator based on Moose's pilot symbols as in [121] shows that this can be the right thing to do.

If a shortened Moose's pilot symbol, as in Figure 4.6, is transmitted, samples at the receiver can be expressed as

$$y_n = \sum_{k \text{ is even}} H_k s_k e^{\jmath \frac{2\pi(k+k_o+\epsilon)(n-N_g-\theta)}{N}},$$

for $n = \theta_t + N_g, \cdots, \theta_t + N_g + N - 1$. Therefore,

$$\begin{aligned} y_{n+N/2} &= \sum_{k \text{ is even}} H_k s_k e^{\jmath\frac{2\pi(k+k_o+\epsilon)(n+N/2-N_g-\theta_t)}{N}} \\ &= \sum_{k \text{ is even}} H_k s_k e^{\jmath\frac{2\pi(k+k_o+\epsilon)(n+-N_g-\theta_t)}{N}} e^{\jmath\pi(k+k_o+\epsilon)} \\ &= \sum_{k \text{ is even}} H_k s_k e^{\jmath\frac{2\pi(k+k_o+\epsilon)(n-N_g-\theta_t)}{N}} e^{\jmath\pi(k_o+\epsilon)} \\ &= y_n e^{\jmath\pi(k_o+\epsilon)}, \end{aligned} \tag{4.4.1}$$

for $n = \theta + N_g, \cdots, \theta + N_g + N/2 - 1$. Consequently,

$$y_n y^*_{n+N/2} = |y_n|^2 e^{-\jmath\pi(k_o+\epsilon)}. \tag{4.4.2}$$

From Equation (4.4.2), we first estimate the timing offset by finding the value $\widehat{\theta}$ that maximizes the following modified cost function

$$\Lambda(\theta) = \left| \sum_{n=\theta+N_g}^{\theta+N/2+N_g-1} y^*_n y_{n+N/2} \right| - \frac{\zeta}{2} \sum_{n=\theta+N_g}^{\theta+N+N_g-1} \left(|y_n|^2 + |y_{n+N/2}|^2\right).$$

Then, from the estimated θ_t, the fractional frequency offset can be estimated by the following two steps:

- calculating ϵ' by

$$\widehat{\epsilon'} = \frac{1}{\pi} \arg \left(\sum_{n=\widehat{\theta}+N_g}^{\widehat{\theta}+N/2+N_g-1} y^*_n y_{n+N/2} \right), \tag{4.4.3}$$

- estimating the fractional frequency offset by

$$\widehat{\epsilon} = \begin{cases} \widehat{\epsilon'} & \text{if } |\widehat{\epsilon'}| < \frac{1}{2}, \\ \widehat{\epsilon'} - 1 & \text{if } \frac{1}{2} \le \widehat{\epsilon'} < 1, \\ \widehat{\epsilon'} + 1 & \text{if } -1 < \widehat{\epsilon'} \le -\frac{1}{2}. \end{cases}$$

In fact, $\widehat{\epsilon'}$ carries more information than fractional frequency offset. We can not only estimate fractional frequency offset but also determine whether the tone numbering index, k_o, is even or odd.

Alternatively, we can jointly estimate both ϵ and θ by maximizing the following cost function:

$$\Lambda(\theta, \epsilon') = \mathcal{R}\left(e^{-\jmath\pi\epsilon'} \sum_{n=\theta+N_g}^{\theta+N/2+N_g-1} y_n^* y_{n+N/2} \right) - \frac{\zeta}{2} \sum_{n=\theta+N_g}^{\theta+N/2+N_g-1} \left(|y_n|^2 + |y_{n+N/2}|^2\right).$$

Many other timing offset estimators introduced in the previous sections, such as the normalized correlation estimator and the square-difference estimator, can be similarly extended into joint timing and frequency offset estimators.

The ML timing offset estimator based on the cyclic prefix can also be extended to a joint ML estimator. This is addressed in detail in [121]. However, due to limited redundancy, the cyclic-prefix based joint ML estimators are much more sensitive to timing offset than the pilot-based ML estimator.

4.5 Sampling Clock Offset Estimation and Correction

In any communication system, the sampling clocks of the *digital-to-analog converter* (DAC) at the transmitter and the *analog-to-digital converter* (ADC) of the receiver will be different. As indicated in Chapter 2 and [39], the sampling clock offset will cause two effects: a subchannel-dependent phase rotation and an ICI. In this section, we will briefly introduce a *digital phase-lock loop* (DPLL) [39] to correct the sampling clock offset.

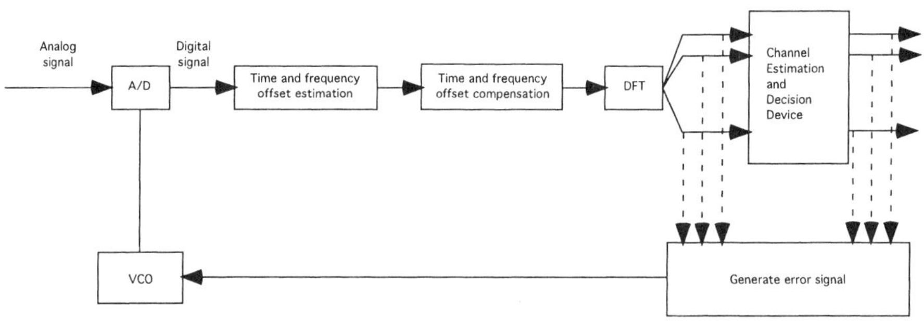

Figure 4.16. Synchronization scheme with digital time-domain timing and frequency offset estimation and compensation

Figure 4.16 shows sampling clock correction through a DPLL. An error signal, which contains information on sampling clock offset, is required to adjust the sampling clock. It can be obtained by examining the difference between the decided symbols and the received noisy signal [39]. Alternatively, it can be calculated by

$$\begin{aligned} \frac{Y_{k+1}}{\widehat{s}_{k+1}} \frac{Y_k^*}{\widehat{s}_k^*} &= H_k H_{k+1}^* e^{j2\pi\beta} + \text{ noise} \\ &\approx |H_k|^2 e^{j2\pi\beta} + \text{ noise}. \end{aligned} \tag{4.5.1}$$

The normalized angle in (4.5.1) gives a noisy estimate of the sampling clock offset. Averaging across all subchannels in an OFDM symbol can significantly reduce the effect. The details on DPLL can be found in a standard textbook [113].

Figure 4.16 also includes timing and frequency correction. After the ADC, the receiver can estimate and adjust the timing and frequency offsets digitally. Timing offset adjustment means starting the DFT at an appropriate place. For frequency offset correction, we can multiply the sampled signal by $\exp(j2\pi\widehat{\epsilon}n/N)$.

It should be noted that the frequency offset can be corrected digitally as above, or it can be corrected before the ADC in a frequency recovery loop [125]. In this case, coarse frequency estimation is a wider estimate of the frequency offset instead of just tone numbering. Furthermore, the carrier frequency must be corrected in the analog domain if the frequency offset is so large that part of an OFDM symbol gets lost due to filtering.

4.6 Summary and Further Reading

In this chapter, we have presented techniques to synchronize OFDM systems with the help of pilots with redundant information or by exploiting redundancy in the OFDM symbol itself. We have concentrated on timing and frequency offset estimation and synchronization using the special structure of OFDM symbols that is unavailable in single-carrier systems. We have introduced some basic techniques, however, more recently developed algorithms are not included. There are several other approaches on timing offset estimation based on pilot symbols [130], [131], [132], based on the cyclic prefix [133], and based on both [134]. A frequency domain approach for timing offset estimation can be found in [135]. Frequency offset estimation based on pilot symbols and based on the cyclic prefix has been also investigated in

[136], [137], [138] and in [139], respectively. Frequency synchronization using single-carrier based methods can be found in [125], [140] and the references therein. A closed-loop frequency synchronizer is addressed in [141], [142]. Joint timing and frequency offset estimation can be also found in [143], [144], [145], [146], [147], [148]. Recently, blind techniques have been also applied in timing and frequency synchronization [149], [150], [151], [152], [153]. There are synchronization methods developed for some special applications. Timing offset estimation for high-bit rate *digital subscriber line* (DSL) systems has been proposed in [154]. Joint timing and frequency offset estimation for the DAB OFDM systems can be found in [115], [116], [155], [156], [157].

Chapter 5

CHANNEL ESTIMATION

Ye (Geoffrey) Li

5.1 Introduction

Channel parameters are required by many applications. With channel estimation and tracking, OFDM systems can use *coherent PSK* to obtain a 3 dB signal-to-noise ratio (SNR) gain over *differential PSK* (DPSK). Adaptive loading algorithms described in Chapter 3 also depend on the accurate channel information. For OFDM systems with multiple transmit and/or receive antennas for system capacity or performance improvement, channel information is essential to diversity combining, interference suppression, and signal detection. In summary, the accuracy of channel state information greatly influences the overall system performance. Therefore, in this chapter, we present channel parameter estimation in OFDM systems.

Before introducing various channel estimation algorithms, we first compare the performance of differential and coherent detection to demonstrate the importance of this topic.

5.1.1 Differential and Coherent Detection

PSK is the most popular modulation for fading channels since its performance is not sensitive to channel amplitude variation. There are two ways to demodulate the PSK signal: *differential* and *coherent.* Coherent PSK carries the information to be transmitted by the phase of a tone in OFDM. Therefore, channel knowledge is required for demodulation. DPSK uses the phase difference to carry the information to be transmitted. For OFDM, the phase difference may be between adjacent tones of the same OFDM block or the tones at the same position of adjacent OFDM blocks. In either case, channel knowledge is not required for demodulation. However, as we see next, coherent PSK has much better performance than DPSK.

From [34], for a channel with only AWGN, the *bit error probability* (BEP) for binary differential and coherent PSK is

$$P_D(\rho) = \frac{1}{2}e^{-\rho}, \tag{5.1.1}$$

and

$$P_C(\rho) = Q(\sqrt{2\rho}), \tag{5.1.2}$$

respectively, where ρ is the SNR per bit and $Q(x)$ is defined as

$$Q(x) = \frac{1}{2\pi}\int_x^\infty e^{-t^2}dt. \tag{5.1.3}$$

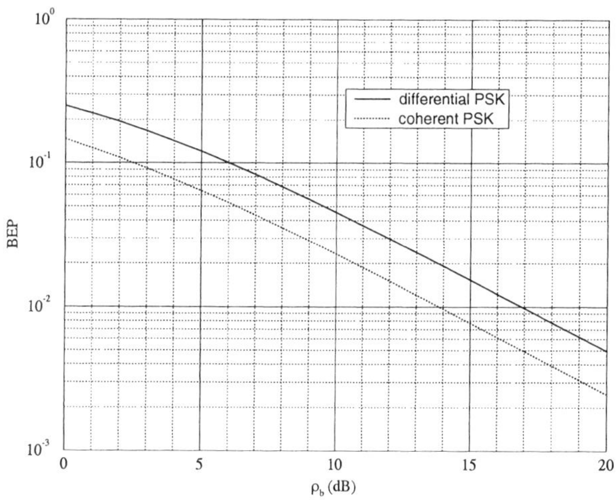

Figure 5.1. BEP of differential and coherent PSK for Rayleigh fading channel

In a fading environment, the received SNR is a random variable. For a Rayleigh fading channel, its probability density function can be proved to be,

$$p(\rho) = \frac{1}{\rho_b}e^{-\frac{\rho}{\rho_b}}, \quad (\rho \geq 0), \tag{5.1.4}$$

where ρ_b is the average SNR per bit. In this case, the BEP for differential and coherent PSK can be directly calculated from (5.1.1), (5.1.2), and (5.1.4) as

$$\bar{P}_D(\rho_b) = \frac{1}{2(1+\rho_b)}, \tag{5.1.5}$$

and

$$\bar{P}_C(\rho_b) = \frac{1}{2}(1 - \frac{1}{\sqrt{1+1/\rho_b}}), \tag{5.1.6}$$

respectively. For larger average SNR's, we have

$$\bar{P}_D(\rho_b) \approx \frac{1}{2\rho_b}, \quad \bar{P}_C(\rho_b) \approx \frac{1}{4\rho_b}, \tag{5.1.7}$$

which implies that coherent detection provides a 3-dB SNR gain compared with differential detection (see Figure 5.1).

However, to achieve this gain, channel state information is required. In this chapter, we discuss channel estimation for OFDM systems in a rapid dispersive fading environment.

5.1.2 OFDM Systems with Channel Estimator

The OFDM system with channel estimation considered in this chapter is shown in Figure 5.2. An error correction code across tones is utilized in the system to correct the errors resulting from frequency selective fading. Since the phase of each tone can be obtained by the channel estimator, coherent PSK modulation is used here to enhance the system performance.

For a diversity receiver, the signal from the m-th antenna at the k-th tone of the n-th block can be expressed as

$$X_{n,k}^{(m)} = H_{n,k}^{(m)} s_{n,k} + N_{n,k}^{(m)}. \tag{5.1.8}$$

In (5.1.8), $N_{n,k}^{(m)}$ is *additive white Gaussian noise* (AWGN) from the m-th antenna at the k-th subcarrier of the n-th block, which is assumed to be with zero-mean and variance ρ. We also assume that the additive channel noise, $N_{n,k}^{(m)}$, is independent for different OFDM blocks, tones, and antennas. $H_{n,k}^{(m)}$, the frequency response at the k-th tone of the n-th block corresponding to the m-th antenna, is assumed independent for different m's, but with the same statistics. For simplicity, we assume that $E\left|H_{n,k}^{(m)}\right|^2 = 1$. $s_{n,k}$ is the signal modulating the k-th tone during the n-th block, and is assumed to have unit-variance and be independent for different k's and n's.

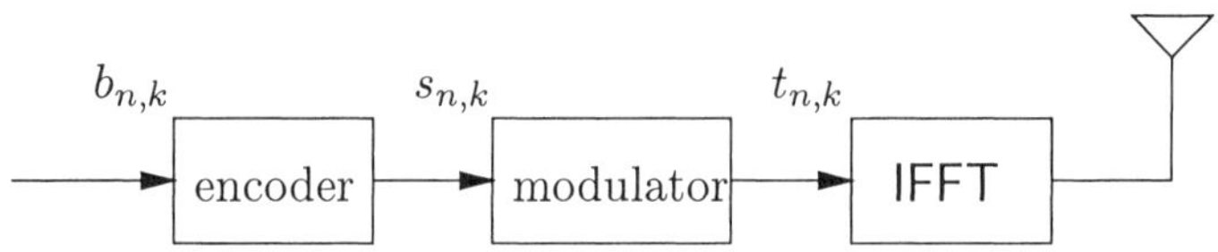

(a) transmitter

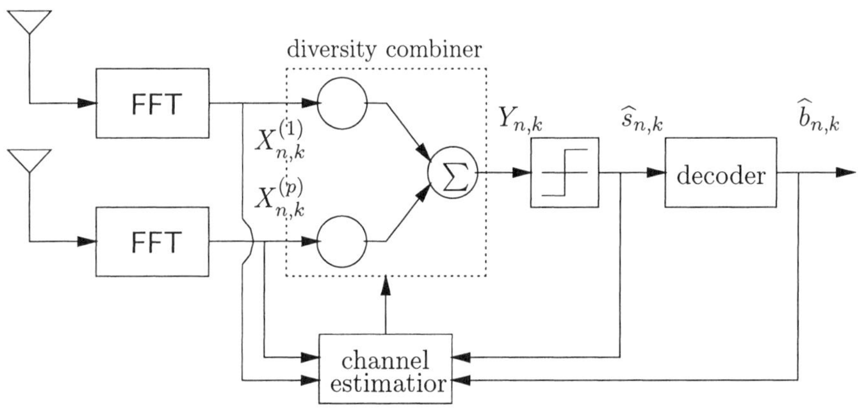

(b) receiver

Figure 5.2. OFDM system with channel estimator

With knowledge of the channel parameters, $s_{n,k}$ can be estimated as $Y_{n,k}$, by a *maximal ratio combiner* (MRC) ,

$$Y_{n,k} = \frac{\sum_{m=1}^{p}(H_{n,k}^{(m)})^* X_{n,k}^{(m)}}{\sum_{m=1}^{p}|H_{n,k}^{(m)}|^2}. \tag{5.1.9}$$

However, since the multipath channel parameters are time-varying and are usually unknown, a channel estimator must be used to obtain accurate estimation of the channel parameters.

5.1.3 Organization of the Chapter

In this chapter, we focus on channel parameter estimation for OFDM systems in dispersive fading environments. Both *decision-directed* and *pilot-symbol-aided* channel estimation approaches will be presented. In Section 5.2, we introduce decision-directed channel estimation. We first derive the basic MMSE channel estimators by exploiting frequency domain correlation of the channel frequency response and exploiting both time- and frequency domain correlations, respectively. Then, we present a robust channel estimator design approach and demonstrate the effectiveness of channel estimator by computer simulation examples. In Section 5.3, we describe the pilot-symbol-based channel estimation that includes filtering and FFT based approaches. In Section 5.4, we will summarize the results on channel estimation and indicate further reading on this topic.

5.2 Decision-Directed Channel Estimation

Channel estimation that exploits the decided (sliced) symbols as reference is called the *decision-directed channel estimation.* If an ideal reference, $s_{n,k}$, can be generated in Figure 5.2, then a temporal estimation of $H_{n,k}$ can be obtained as

$$\begin{aligned}\tilde{H}_{n,k} &= \frac{X_{n,k}}{s_{n,k}} \\ &= H_{n,k} + \frac{N_{n,k}}{s_{n,k}} \\ &= H_{n,k} + \tilde{N}_{n,k},\end{aligned} \tag{5.2.1}$$

where $\tilde{N}_{n,k}$ is the *temporal estimation error* due to AWGN and it can be expressed as

$$\tilde{N}_{n,k} = \frac{N_{n,k}}{s_{n,k}}.$$

It can be easily checked that $\tilde{N}_{n,k}$'s for different n's and k's are uncorrelated if $N_{n,k}$'s are.

In this section, we first introduce MMSE channel estimation using only frequency domain correlation and using both the time- and frequency domain correlations, respectively. And then, we present robust channel estimation that provides good performance even if the exact channel statistics are unknown. Finally, the performance of these estimators is demonstrated through a particular example.

5.2.1 MMSE Estimation using Frequency Domain Correlation

The channel frequency responses at different tones are correlated. Exploiting this correlation, the performance of channel parameter estimation can be significantly improved.

Denote the $N \times 1$ *temporally estimated parameter vector* as

$$\tilde{\mathbf{H}}_n = (\tilde{H}_{n,1}, \cdots, \tilde{H}_{n,N})^T, \tag{5.2.2}$$

where the superscript T denotes the transpose of a vector or a matrix. Then

$$\tilde{\mathbf{H}}_n = \mathbf{H}_n + \tilde{\mathbf{N}}_n,$$

where $\mathbf{H}_n$ and $\tilde{\mathbf{N}}_n$ are the $N \times 1$ *channel frequency response vector* and the $N \times 1$ *temporal estimation error vector*, respectively. They are defined as

$$\mathbf{H}_n = (H_{n,1}, \cdots, H_{n,N})^T,$$

and

$$\tilde{\mathbf{N}}_n = (\tilde{N}_{n,1}, \cdots, \tilde{N}_{n,N})^T.$$

Exploiting the frequency domain correlation of channel parameters, a better estimate can be obtained

$$\hat{\mathbf{H}}_n = \mathbf{C}\tilde{\mathbf{H}}_n, \tag{5.2.3}$$

where $\mathbf{C}$ is a $N \times N$ *estimator coefficient matrix.* For MMSE estimation [158], [159], the matrix $\mathbf{C}$ is determined by minimizing the mean-square error,

$$\text{MSE}(\mathbf{C}) = \text{E}\left\|\hat{\mathbf{H}}_n - \mathbf{H}_n\right\|^2,$$

which, using the *orthogonality principle* [160], reduces to solving

$$\text{E}\{(\hat{\mathbf{H}}_n - \mathbf{H}_n)\tilde{\mathbf{H}}_n^H\} = 0,$$

or equivalently,

$$\mathbf{C}\text{E}\{\tilde{\mathbf{H}}_n\tilde{\mathbf{H}}_n^H\} - \text{E}\{\mathbf{H}_n\tilde{\mathbf{H}}_n^H\} = 0, \tag{5.2.4}$$

where the superscript H denotes the Hermitian of a vector or a matrix. The expectations in (5.2.4) can be expressed as

$$\text{E}\{\tilde{\mathbf{H}}_n\tilde{\mathbf{H}}_n^H\} = \mathbf{R}_f + \rho\mathbf{I}, \text{ and } \text{E}\{\mathbf{H}_n\tilde{\mathbf{H}}_n^H\} = \mathbf{R}_f,$$

where ρ is the variance of temporal channel estimation error that can be expressed as

$$\rho = E\left|\frac{N_{n,k}}{s_{n,k}}\right|^2,$$

and $\mathbf{R}_f$ is the *frequency domain correlation matrix of channel parameters*, defined as

$$\mathbf{R}_f = \begin{pmatrix} r_f[0] & r_f[1] & \cdots & r_f[N-1] \\ r_f[-1] & r_f[0] & \ddots & r_f[N-2] \\ \vdots & \ddots & \ddots & \vdots \\ r_f[-N+1] & r_f[-N+2] & \cdots & r_f[0] \end{pmatrix}.$$

The elements of $\mathbf{R}_f$ represent the frequency domain correlation of the channel defined as

$$r_f[k] = E\{H_{n,k+k_o} H^*_{n,k_o}\}.$$

It is obvious that

$$r_f[0] = E\left|H_{n,k}\right|^2 = 1.$$

From (5.2.4), the coefficient matrix for the MMSE estimator is

$$\mathbf{C} = \mathbf{R}_f(\mathbf{R}_f + \rho\mathbf{I})^{-1},$$

and the MMSE estimate of the channel is

$$\widehat{\mathbf{H}}_n = \mathbf{C}\tilde{\mathbf{H}}_n = \mathbf{R}_f(\mathbf{R}_f + \rho\mathbf{I})^{-1}\tilde{\mathbf{H}}_n.$$

From the above discussion, we see that an MMSE frequency domain channel estimator can be obtained once the correlation of channel frequency response at different frequencies is known.

Let the eigen-decomposition of $\mathbf{R}_f$ be

$$\mathbf{R}_f = \mathbf{U}\mathbf{D}\mathbf{U}^H, \tag{5.2.5}$$

where $\mathbf{U}$ is a unitary matrix and $\mathbf{D}$ is a diagonal matrix with ordered diagonal elements d_k ($d_1 \geq d_2 \geq \cdots \geq d_N$). It is clear that

$$\sum_{k=1}^{N} d_k = \mathrm{Tr}\{\mathbf{R}_f\} = K r_f[0] = N.$$

Then,

$$\widehat{\mathbf{H}}_n = \mathbf{U}\mathbf{\Phi}\mathbf{U}^H\tilde{\mathbf{H}}_n, \tag{5.2.6}$$

where $\mathbf{\Phi}$ is a diagonal matrix defined as

$$\mathbf{\Phi} = \mathrm{diag}\{\phi_1, \phi_2, \cdots, \phi_N\}, \tag{5.2.7}$$

with

$$\phi_k = \frac{d_k}{d_k + \rho}, \tag{5.2.8}$$

for $k = 1, \cdots, N$.

Even though the MMSE estimator has the best performance when the statistics for estimator match the real channel, the performance degrades significantly when the estimator does not match the channel. To overcome this drawback, optimum K_o-rank estimator [159] has been introduced, which has less mismatch error than the MMSE estimator.

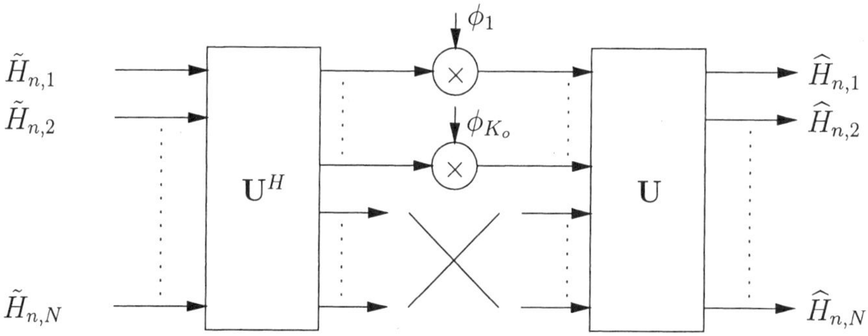

Figure 5.3. Optimal rank-K_o estimator

Since the delay spread in OFDM is usually much less than the symbol duration to mitigate ISI, the channel frequency responses at different frequencies are highly correlated. As a result, $\mathbf{R}_f$ has only a few eigenvalues that are significantly larger than zero. Without loss of generality, let the $d_1,\ d_2, \cdots, d_{K_o}$ be the K_o largest eigenvalues of the frequency domain correlation matrix of the channel. If

$$\phi_k = \begin{cases} \frac{d_k}{d_k+\rho}, & 1 \geq K_o, \\ 0, & K_o + 1 \geq k \geq N, \end{cases} \tag{5.2.9}$$

in (5.2.6), then an optimal rank-K_o estimator is obtained, which is shown as in Figure 5.3.

It is shown in [158], with tolerable leakage, the unitary matrix $\mathbf{U}$ can be approximated by the DFT matrix defined as

$$\mathbf{W} = \frac{1}{\sqrt{K}} \begin{pmatrix} 1 & 1 & \cdots & 1 \\ 1 & e^{-\jmath\frac{2\pi}{N}} & \cdots & e^{-\jmath\frac{2\pi(N-1)}{N}} \\ \vdots & \cdots & \cdots & \vdots \\ 1 & e^{-\jmath\frac{2\pi(N-1)}{N}} & \cdots & e^{-\jmath\frac{2\pi(N-1)(N-1)}{N}} \end{pmatrix}, \tag{5.2.10}$$

to further reduce the complexity of channel estimation. The leakage of the above approximation depends on the guard interval and the channel multipath delay profile. If the delay τ of a path is an integer multiple of the sampling interval T_s, that is, $\tau = lT_s$, then all the energy from the path will be mapped to d_l. Otherwise, if the delay is a non-integer multiple of the sampling interval, that is, $(l-1)T_s < \tau < lT_s$, then most of its energy will be contained in d_{l-1} and d_l, although the energy will leak to all d_k's. Hence, if the maximum delay spread is t_d, then for all $l \leq K_o$ ($K_o = \lceil Kt_d/T_f \rceil$), $d_l \approx 0$.

For an OFDM system with a large number of tones, a *partitioning approach* can be used along with the optimal rank estimator to reduce the complexity. This approach is described in detail in [159].

5.2.2 MMSE Estimation using both Time- and Frequency Domain Correlations

Since the channel parameters, $H_{n,k}$'s, are correlated for different OFDM blocks and frequencies, instead of using only the frequency domain correlation, an improved MMSE channel estimator can be constructed exploiting both time and frequency corrections. In this case, the estimated channel parameter vector can be expressed as,

$$\widehat{\mathbf{H}}_n = \sum_{m=0}^{+\infty} \mathbf{C}_m \tilde{\mathbf{H}}_{n-m},$$

where $\{\mathbf{C}_m\}$ is a $K \times K$ matrix *sequence* that is chosen to minimize

$$\text{MSE}(\{\mathbf{C}_m\}) = E\left\|\widehat{H}_{n,k} - H_{n,k}\right\|^2. \tag{5.2.11}$$

From Appendix A, the coefficients of the estimator are determined by

$$\mathbf{C}(\omega) = \mathbf{U}\mathbf{\Phi}(\omega)\mathbf{U}^H, \tag{5.2.12}$$

where

$$\mathbf{C}(\omega) = \sum_{m=0}^{+\infty} \mathbf{C}_m e^{-jm\omega},$$

$$\mathbf{\Phi}(\omega) = \text{diag}\{\phi_1(\omega), \cdots, \phi_N(\omega)\}.$$

In the above expression,

$$\phi_k(\omega) = \begin{cases} 0 & d_k = 0, \\ 1 - \frac{1}{M_k(\omega)\gamma_k[0]} & d_k \neq 0, \end{cases} \tag{5.2.13}$$

where $M_k(\omega)$ is a stable one-sided Fourier transform

$$M_k(\omega) = \sum_{n=0}^{\infty} \gamma_n^{(k)} e^{-jn\omega}, \tag{5.2.14}$$

which is uniquely determined by

$$M_k(\omega)M_k(-\omega) = \frac{d_k}{\rho} p_t(\omega) + 1, \tag{5.2.15}$$

with

$$p_t(\omega) = \sum_{n=-\infty}^{\infty} r_t[n] e^{-jn\omega},$$

and

$$r_t[n] = E\{H_{m+n,k} H_{m,k}^*\}.$$

The dc component $\gamma_k[0]$ in $M_k(\omega)$ can be found by

$$\gamma_k^2[0] = \exp\left\{\frac{1}{2\pi}\int_{-\pi}^{\pi} \ln[\frac{d_k}{\rho} p_t(\omega) + 1] d\omega\right\}. \tag{5.2.16}$$

The MMSE channel estimator in (5.2.12) is shown in Figure 5.4. Similar to the MMSE estimator using only frequency domain correlation, the unitary transform, $\mathbf{U}$, in the figure is determined by the delay profile. It exploits the correlation of channel at different frequencies. For those significant d_k's, linear filters between $\mathbf{U}^H$ and $\mathbf{U}$ in Fig. 5.4 are used, instead of the constants in Fig. 5.3, to take advantage of the time-domain correlation.

The average MSE over different tones is defined as

$$\overline{\text{MSE}} = \frac{1}{N} \text{E} \left\| \widehat{\mathbf{H}}_n - \mathbf{H}_n \right\|^2.$$

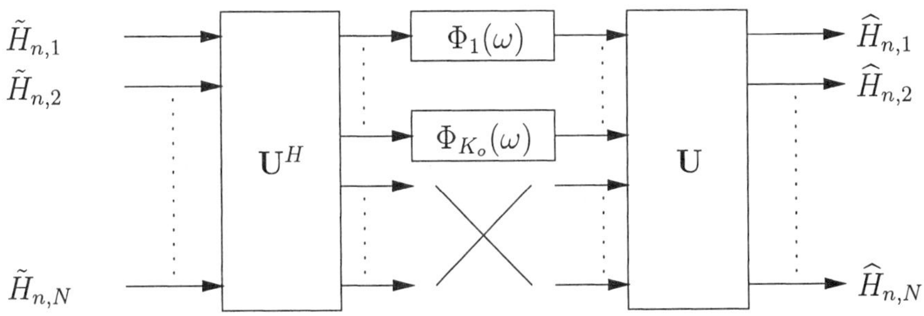

Figure 5.4. Channel estimator for OFDM systems

From Appendix B, the MSE of the MMSE estimator is

$$\overline{\text{MMSE}} = \frac{\rho}{N}\sum_{k=1}^{N}\left(1 - \exp\left\{-\frac{1}{2\pi}\int_{-\pi}^{\pi}\ln[\frac{d_k}{\rho}p_t(\omega) + 1]d\omega\right\}\right), \quad (5.2.17)$$

which is determined by the channel's delay profile and Doppler spectrum.

In [161], we have also derived the $\overline{\text{MMSE}}$ when channel's Doppler spectrum is Jake's model and ω_d-bandliminited. If the time-domain correlation is ideal ω_d-bandlimited, *i.e.*,

$$p_B(\omega) = \begin{cases} \frac{\pi}{\omega_d} & |\omega| \le \omega_d, \\ 0 & \text{otherwise,} \end{cases}$$

then, from (5.2.17)

$$\overline{\text{MMSE}}_B(\omega_d) = \frac{\rho}{N}\sum_{k=1}^{N} 1 - \frac{1}{\left(1 + \frac{\pi d_k}{\omega_d \rho}\right)^{\frac{\omega_d}{\pi}}}. \quad (5.2.18)$$

5.2.3 Robust Estimation

Once the channel statistics, such as the time- and frequency domain correlations, are known, the optimum channel estimator can be designed. However, in mobile wireless links, the channel statistics depend on the particular environment, for example, indoor or outdoor, urban or suburban, and change with time. The MMSE estimator, which *tightly* matches the channel statistics, is sensitive to channel statistics; hence, it is not robust. In this section, we will develop a design approach to robust estimator that is insensitive to channel statistics.

If an MMSE channel estimator is designed according to time- and frequency domain correlations $\bar{r}_t[m]$ and $\bar{r}_f[l]$, respectively, then its coefficients, $\bar{c}_{m,l,k}$'s, are determined from (5.2.12) by

$$\bar{\mathbf{C}}(\omega) = \bar{\mathbf{U}}\bar{\Phi}_{(}\omega)\bar{\mathbf{U}}^H, \tag{5.2.19}$$

where the definitions of $\bar{\mathbf{U}}$ and $\bar{\Phi}(\omega)$ are similar to those of $\mathbf{U}$ and $\Phi(\omega)$ except that $r_t[m]$ and $r_f[l]$ are substituted by the designed correlations $\bar{r}_t[m]$ and $\bar{r}_f[l]$, respectively. For a channel with time- and frequency domain correlations $r_t[m]$ and $r_f[l]$, rather than $\bar{r}_t[m]$ and $\bar{r}_f[l]$, it can be derived in Appendix C that the MSE of the estimator will be

$$\begin{aligned}
&\text{MSE}(\{\bar{\mathbf{C}}_m\}) \\
= &\frac{1}{2\pi}\int_{-\pi}^{\pi} p_t(\omega)\text{T}r\{\bar{\mathbf{U}}\mathbf{R}_f\bar{\mathbf{U}}^H(\bar{\Phi}(\omega)-\mathbf{I})^H(\bar{\Phi}(\omega)-\mathbf{I})\}d\omega \\
&+\frac{\rho}{2\pi}\int_{-\pi}^{\pi} \text{T}r\{\bar{\Phi}(\omega)\bar{\Phi}(\omega)^H\}d\omega.
\end{aligned} \tag{5.2.20}$$

To understand the effect of mismatch in time- and frequency domains, we first consider an estimator matching the correlation in time-domain and mismatching in frequency domain, that is,

$$\bar{r}_f[k] = r_f[k],$$

for $k = 1, \cdots, N$, and the time-domain correlation $r_t[m]$ is mismatched. Then from Appendix C, the MSE of the estimator will be

$$\text{MSE}(\{\bar{\mathbf{C}}_m\}) = \rho\sum_{k=1}^{N}\frac{1}{\bar{\gamma}_k^2[0]}\left(\frac{1}{2\pi}\int_{-\pi}^{\pi}\frac{\bar{d}_k p_t(\omega)+\rho}{\bar{d}_k\bar{p}_t(\omega)+\rho}d\omega - 1\right) + \overline{\text{MMSE}},$$

where $\overline{\text{MMSE}}$ is the MSE of the optimal estimator that is determined by (5.2.17) and

$$\bar{\gamma}_k^2[0] = \exp\left\{\frac{1}{2\pi}\int_{-\pi}^{\pi}\ln[\frac{\bar{d}_k}{\rho}\bar{p}_t(\omega)+1]d\omega\right\}.$$

The first term in the above expression represents the MSE variation due to the mismatch.

If

$$\bar{p}_t(\omega) = p_B(\omega) = \begin{cases} \frac{\pi}{\omega_d}, & |\omega| \le \omega_d, \\ 0, & \text{otherwise.} \end{cases}$$

That is, the time-domain correlation of the designed estimator is ideal ω_d-bandlimited, then, it can be shown in Appendix C that the first term in (5.2.20) is zero for all $p_t(\omega)$; therefore, the MSE of the estimator is the same as that of the optimal MMSE estimator for channels with ideal ω_d-bandlimited time-domain correlations, which can be expressed by (5.2.18). Hence, if an OFDM channel estimator is designed using $p_B(\omega)$ as the time-domain correlation, then the time-domain correlation mismatch of the estimator will not degrade its performance. This suggests that a robust channel estimator should use $p_B(\omega)$ as the time-domain correlation.

To analyze the frequency domain correlation mismatch, we assume that (i) the time-domain correlation of the designed estimator is the same as that of the channel, that is, $\bar{p}_t(\omega) = p_t(\omega)$, and (ii) the frequency correlation matrix of the designed estimator has the same eigenvectors as that of the channel. That is, $\mathbf{R}_f$ can be eigen-decomposed into

$$\mathbf{R}_f = \bar{\mathbf{U}}^H \mathbf{D} \bar{\mathbf{U}}, \text{ or } \bar{\mathbf{U}} \mathbf{R}_f \bar{\mathbf{U}}^H = \mathbf{D},$$

where $\mathbf{D} = \text{diag}\{d_1, \cdots, d_N\}$ and $\sum_k d_k = N$. d_k and $\bar{d}_k$ for $k = 1, \cdots, N$ are generally different. Although Assumption (ii) seems strange, it is, in fact reasonable. As indicated in Section 5.2.1, with tolerable leakage, both matrices $\bar{\mathbf{U}}$ and $\mathbf{U}$ can be approximated by the DFT matrix $\mathbf{W}$.

Applying the above two assumptions to (5.2.20), from Appendix C, we get

$$\text{MSE}(\{\bar{\mathbf{C}}_m\}) = \sum_{k=1}^{N} (d_k - \bar{d}_k)\Gamma(\bar{d}_k) + \overline{\text{MMSE}}, \tag{5.2.21}$$

where

$$\Gamma(\bar{d}_k) = \frac{1}{2\pi\bar{\gamma}_k^2[0]} \int_{-\pi}^{\pi} \frac{\bar{p}_t(\omega)}{\frac{\bar{d}_k}{\rho}\bar{p}_t(\omega) + 1} d\omega.$$

Therefore, if the channel estimator is designed such that

$$\bar{d}_k = \begin{cases} N/K_o & \text{for } 1 \le k \le K_o, \\ 0 & \text{for } K_o + 1 \le k \le N, \end{cases}$$

then, for any channel with $d_k = 0$ for $K_o + 1 \le k \le N$ and $\sum_{k=1}^{K_o} d_k = N$, we have

$$\text{MSE}(\{\bar{\mathbf{C}}_m\}) = \overline{\text{MMSE}}.$$

From the above discussion, a robust estimator, which is insensitive to the channel statistics (both time- and frequency domain correlations), should have an ideal bandlimited time-domain correlation $\bar{r}_t[n] = \mathcal{F}^{-1}\{p_B(\omega)\} = \frac{\sin(n\omega_d)}{n\omega_d}$ and a frequency domain correlation matrix

$$\begin{pmatrix} \bar{r}_f[0] & \cdots & \bar{r}_f[N-1] \\ \vdots & \ddots & \vdots \\ \bar{r}_f[-N+1] & \cdots & \bar{r}_f[0] \end{pmatrix} = \mathbf{W}^H \bar{\mathbf{D}} \mathbf{W},$$

where

$$\bar{\mathbf{D}} = \text{diag}\{\underbrace{N/K_o, \cdots, N/K_o,}_{K_o \text{ elements}}\ 0, \cdots,\ 0\},$$

$$K_o = \lceil \Delta f t_{max} N \rceil = \lceil t_{max}/T_s N \rceil,$$

and $\omega_d = 2\pi T_f f_{max} = 2\pi f_{max}/R_b$. Note that $\mathbf{W}$ is the DFT matrix defined in (5.2.10). In this case, the average MSE of the robust estimator is

$$\overline{\text{MMSE}}_B = \frac{K_o \rho}{N} \left(1 - \frac{1}{\left(\frac{\pi N}{K_o \rho \omega_d} + 1 \right)^{\frac{\omega_d}{\pi}}} \right).$$

For any channel with $f_d \leq f_{max}$ and $t_d \leq t_{max}$, the average MSE should be $\overline{\text{MMSE}}_B$.

For the design example in Chapter 2, $T_s = 160$ μsec and $R_b = 5$ kbaud, the MSE of the robust estimator that matches different Doppler frequencies and delay spreads is shown in Figure 5.5. From the figure, it is almost a constant if $t_{max} f_{max}$ is fixed. In particular, let

$$\mu = \frac{2\rho t_{max} f_{max}}{R_b T_s}.$$

For the OFDM systems satisfying $\mu << 1$ and $\frac{\omega_d}{\pi} \ln \frac{1}{\mu} >> 1$, the average MSE of the robust channel estimator can be approximately expressed as [161]

$$\overline{\text{MMSE}}_B \approx \mu \ln \frac{1}{\mu}. \tag{5.2.22}$$

If the channel estimator is designed to match the Jakes' Doppler spectrum given by

$$p_J(\omega) = \begin{cases} \frac{2}{\omega_d} \frac{1}{\sqrt{1 - \left(\frac{\omega}{\omega_c} \right)^2}} & |\omega| \leq \omega_d, \\ 0 & \text{otherwise} \end{cases}$$

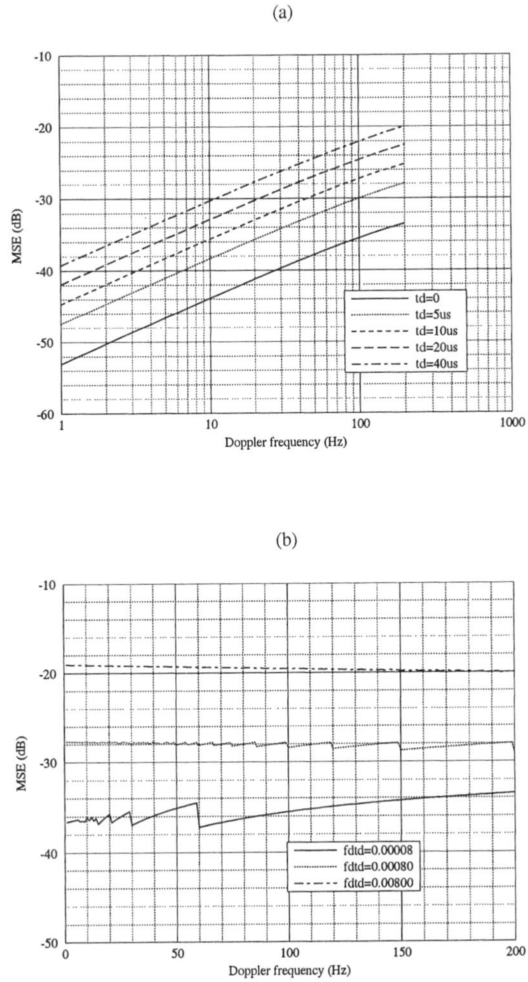

Figure 5.5. MSE of channel estimator versus Doppler frequency (a) when SNR=10 dB and t_d=0, 5, 10, 20, and 40 μsec and (b) when SNR=10 dB and $f_d t_d$=0.00008, 0.0008, and 0.008, respectively

then from (B.4), we have

$$\overline{\text{MMSE}}_J \approx \mu(\ln \frac{1}{\mu} - 0.1447). \tag{5.2.23}$$

Hence, compared with the estimator that tightly matches the Doppler spectrum, the performance degradation of the robust channel estimator is negligible.

An ideal reference is assumed in the derivation of the channel estimation and tracking before. In practical systems, a reference can be generated during a training block. In subsequent blocks, a reference is generated using the received signals. Four possible reference generating schemes have been introduced in [161]. The following are two of them that give better performance:

- *Undecoded/decoded dual mode reference*: If the decoder can successfully correct all errors in an OFDM block, the reference for the block can be generated by the decoded data; Otherwise, the sliced (decided) signal is used as reference.
- *Undecoded reference*: The sliced (decided) signal is used as reference, no matter whether the decoder can successfully correct all errors in a block, or not.

5.2.4 Performance Evaluation

In this section, we demonstrate the performance of the decision-directed channel estimator by a computer simulation example.

In our simulation, we use a two-path Rayleigh fading channel model with equal power on each path and a delay span, duration between the two impulses, from 0 to 40 μs and Doppler frequencies from 10 Hz to 200 Hz. In this example, we assume that two antennas are used for receiver diversity. The parameters for OFDM are the same as those in the design example in Section 2.1.5.

To compare the performance with and without the channel estimation, coherent PSK and differential PSK are used, respectively. In this example, we will use the same parameters as in [26]. For error correction, assume that a (40,20) Reed-Solomon (R-S) code is used, with each code symbol consisting of 3 QPSK/DQPSK symbols (6 bits) grouped in frequency. Consequently, each OFDM block forms a single R-S codeword. We assume that R-S decoder erases 10 symbols based on signal strength and corrects 5 additional random errors. Hence, the simulated system can transmit data at 1.2 Mbits/sec before decoding, or 600 kbits/sec after decoding, over an 800 kHz channel.

Figure 5.6 demonstrates the *word-error-rate* (WER) of the channel estimator using different references under different channel conditions. To get

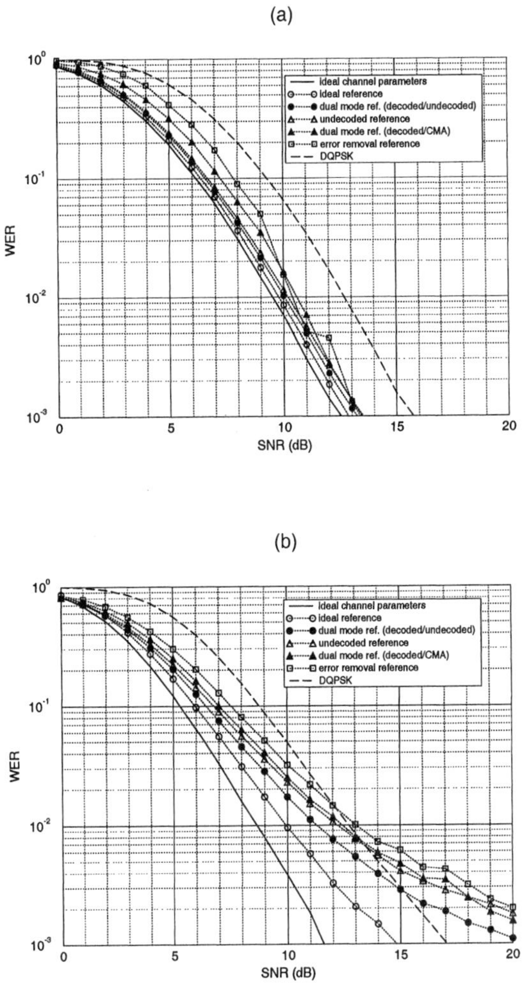

Figure 5.6. WER versus SNR for estimator with different references and the channels with (a) $f_d = 40$ Hz and (b) $f_d = 200$ Hz, respectively.

insight into the average behavior of the channel estimator, the performance has been averaged over 10,000 OFDM blocks. From the figure, the channel estimator using the decoded/undecoded dual mode reference or the unde-

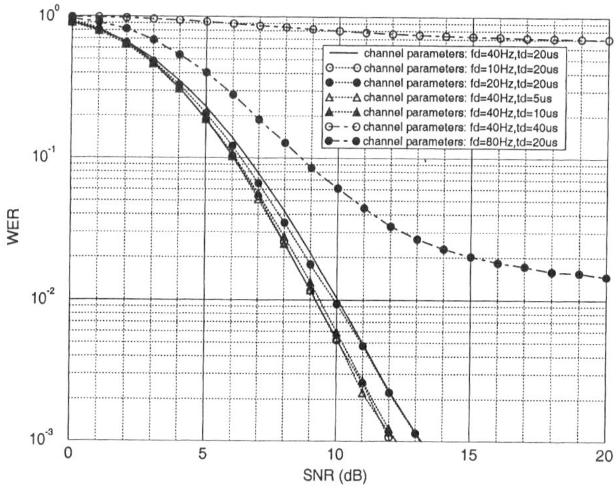

Figure 5.7. An estimator matching a 40 Hz Doppler frequency and a 20 μsec delay span for channels with different channel parameters

coded reference has the best performance. In particular, for a channel with a 40 Hz Doppler frequency, the WER of the system is almost the same as the performance assuming ideal knowledge of the channel parameters, which is about 3 dB better than using differential detection. The system performance degrades with the increasing of Doppler frequency. However, for a channel with a Doppler frequency as large as 200 Hz, then the required SNR is reduced by 2.5 dB for a 10% WER and by 1.5 dB for a 1% WER, respectively, compared with differential detection.

Figures 5.7 and 5.8 illustrate the robustness of the estimator. As indicated before, if a channel estimator is designed to match the channel with 40 Hz maximum Doppler frequency and 20 μsec maximum delay span, then for all channels with a smaller Doppler frequency and delay span, the mismatch does not incur performance degradation, as confirmed by Figure 5.7. However, for channels with a larger Doppler frequency or a larger delay span, the system performance degrades dramatically. On the other hand, as we can see from Figure 5.8, if the estimator is designed to match the Doppler frequency or delay spread larger than the actual ones, the system performance

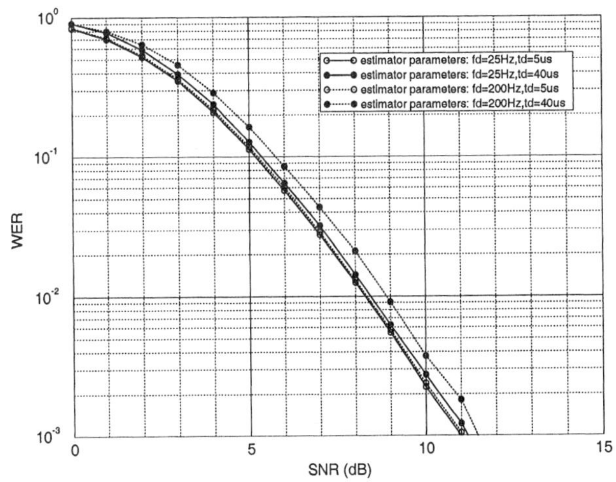

Figure 5.8. Estimators using different parameters for a channel with a 10 Hz Doppler frequency and a 5 μsec delay span.

degrades only slightly compared with the estimator that exactly matches the channel Doppler frequency and delay spread.

5.3 Pilot-Symbol-Aided Estimation

In the previous section, we have presented the decision-directed channel estimation approaches. The decision-directed estimator can be effectively used in systems with packet data transmission mode, such as mobile wireless cellular communications. However, when system is in continuous data transmission mode, the error propagation of the decision-directed estimator will cause serious performance degradation. In this case, the pilot-system-aided estimation provides better performance than the decision-directed estimation, especially for systems with large Doppler frequencies. Here, we introduce the pilot-symbol-aided channel estimation approaches.

For a system using pilot symbols to estimate channel, the pilot symbols need to be scattered at different OFDM blocks (or times) and tones. The parameters at the pilot tones are first estimated and then the parameters at the data tones are obtained by using the correlations of the channel parameters in time- and/or frequency domains. Several interpolation and filtering approaches will be described in this section.

Before presenting different pilot-symbol-aided approaches, we first introduce the pilot grid design.

5.3.1 Grid Design

As discussed in [162], [163], the grid density of the pilot symbols must satisfy the 2-D sampling theorem in order to recover channel parameters, that is

$$f_{max} T_f v_1 \leq 1/2 \quad \text{and} \quad t_{max} \Delta f v_2 \leq 1/2, \tag{5.3.1}$$

where v_1 and v_2 denote the spaces between the time- and the frequency domains respectively; and as before, f_{max} and t_{max} are the maximum Doppler frequency and delay span. It is suggested in [162], [163] oversampling by a factor of 2 for the pilot symbols to suppress channel noise and improve the estimation performance.

Rectangular pilot symbol grids are often used because of its convenience for channel estimation using the direct interpolation approaches. The pilot symbol grids may be any other regular shapes such as the one in Figure 5.9. Furthermore, for a system with a given overhead for pilot symbols, we can improve channel estimation performance by choosing a proper shape of pilot symbol grid. In [164], a regular, but non-rectangular, pilot grid is discussed. Figure 5.10 shows the 2-D spectrum contour of the non-rectangular pilot symbol grid in Fig. 5.9. From Fig. 5.10, if a rectangular pilot symbol grid is used, there will be aliasing on its 2-D spectrum, which will result in a larger estimation error.

The index set for any regular grid can be expressed as

$$\mathcal{P} = \{\mathbf{p} = \mathbf{V}\mathbf{m} : \ \mathbf{m} = (m_1, m_2)^T \ m_1, m_2 \text{ are integers}\},$$

where $\mathbf{V}$ is a 2×2 non-singular integer matrix. In particular, if

$$\mathbf{V} = \begin{pmatrix} v_1 & 0 \\ 0 & v_2 \end{pmatrix},$$

the corresponding grid is rectangular with spaces v_1 and v_2 for time and frequency indexes, respectively. For the grid shown in Figure 5.9,

$$\mathbf{V} = \begin{pmatrix} 2 & 5 \\ 2 & 0 \end{pmatrix}.$$

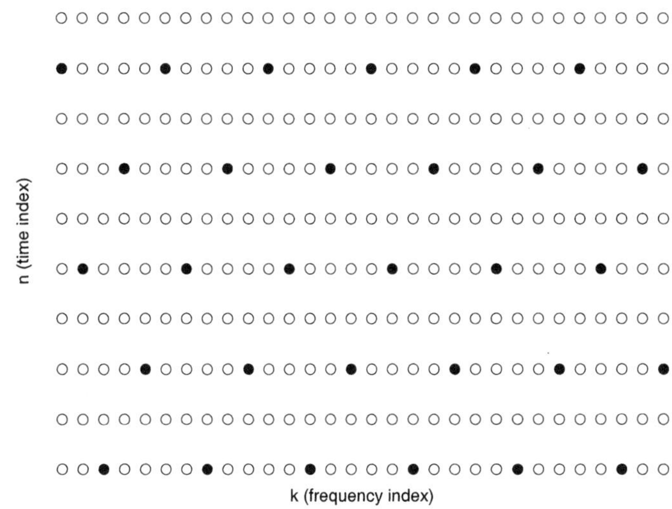

Figure 5.9. A pilot symbol grid.

According to [165, page 563], for any $\mathbf{V}$, there is an accompany 2-element integer vector set, $\mathcal{I}$, with $|\det(\mathbf{V})|$ elements, such that any 2-element integer vector $\mathbf{n} = (n, k)^T$ can be uniquely decomposed into

$$\mathbf{n} = \mathbf{r} + \mathbf{V}\mathbf{m},$$

for some $\mathbf{m} = (l_1, l_2)^T \in \mathcal{Z}^2$ and $\mathbf{r} = (r_1, r_2)^T \in \mathcal{I}$. Therefore, for an OFDM system with a pilot grid determined by $\mathbf{V}$, the overhead of pilot symbols is $1/|\det(\mathbf{V})|$ of the total symbols.

5.3.2 Direct Interpolation

Direct interpolation is a simple way to obtain channel parameters using pilot symbols. It may be based on sample (sinc) function or polynomials.

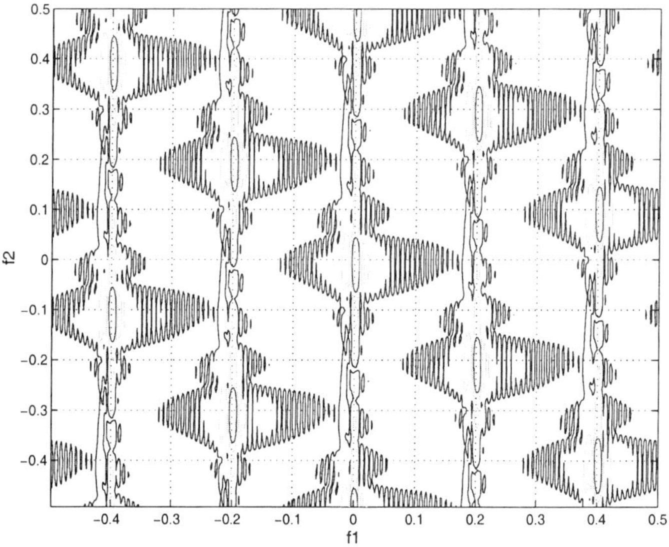

Figure 5.10. The contour of 2-D spectrum for a TU channel.

For these approaches, rectangular grids for pilot positions are preferred for the simplicity.

For a system with pilot symbols at every v_1 OFDM blocks at every v_2 tones, the channel parameters at pilot positions can be estimated by

$$\tilde{H}_{v_1 l_1, v_2 l_2} = \frac{x_{v_1 l_1, v_2 l_2}}{s_{v_1 l_1, v_2 l_2}}, \tag{5.3.2}$$

as before, where $x_{n,k}$ is the received signal at the k-th tone of the n-block, and $s_{n,k}$ is the transmitted symbol that is assumed to be known at the receiver at the pilot position.

From the estimated channel parameters at pilot positions, the channel parameters at other positions can be obtained using either of the interpolation approaches described next.

Sample function based approach

If the pilot density satisfies (5.3.1), then from the sampling theorem, the channel parameters can be estimated by

$$\widehat{H}_{n,k} = \sum_{l_1,l_2} \tilde{H}_{v_1 l_1, v_2 l_2} \mathrm{sinc}\left(2\pi(\frac{n}{v_1} - l_1)\right) \mathrm{sinc}\left(2\pi(\frac{k}{v_2} - l_2)\right), \tag{5.3.3}$$

where

$$\mathrm{sinc}(x) \triangleq \frac{\sin(x)}{x}.$$

In particular,

$$\widehat{H}_{v_1 n_o,k} = \sum_{l_2} \tilde{H}_{v_1 n_o, v_2 l_2} \mathrm{sinc}\left(2\pi(\frac{k}{v_2} - l_2)\right),$$

then, from (5.3.3)

$$\widehat{H}_{n,k} = \sum_{l_1} \widehat{H}_{v_1 l_1, k} \mathrm{sinc}\left(2\pi(\frac{n}{v_1} - l_1)\right).$$

Consequently, a 2-dimensional (2-D) interpolation can be simplified into two one-dimensional (1-D) interpolations and the complexity of channel estimation is reduced.

Any n between $v_1 n_o$ and $v_1(n_o+1)$ and k between $v_2 k_o$ and $v_2(k_o+1)$ can be expressed as

$$n = v_1 n_o + r_1, \text{ and } k = v_2 k_o + r_2,$$

respectively, where $0 \leq r_1 \leq v_1 - 1$ and $0 \leq r_2 \leq v_2 - 1$. Therefore, from (5.3.3)

$$\widehat{H}_{n,k} = \sum_{l_1,l_2} \tilde{H}_{v_1(n_o - l_1), v_2(k - l_2)} \mathrm{sinc}\left(2\pi(l_1 + \frac{r_1}{v_1})\right) \mathrm{sinc}\left(2\pi(l_2 + \frac{r_2}{v_2})\right), \tag{5.3.4}$$

From (5.3.4), when $l_1,\ l_2 = 0,\ -1$, the $\mathrm{sinc}\left(2\pi(l_1 + \frac{r_1}{v_1}\right) \mathrm{sinc}\left(2\pi(l_2 + \frac{r_2}{v_2}\right)$ is large, and it is small for the rest of l_1's or l_2's. Therefore, $\widehat{H}_{n,k}$ is *almost* determined by $\tilde{H}_{v_1 n_o, v_2 k_o}$, $\tilde{H}_{v_1(n_o+1), v_2 k_o}$, $\tilde{H}_{v_1 n_o, v_2(k_o+1)}$, and $\tilde{H}_{v_1(n_o+1), v_2(k_o+1)}$. In practical systems, the summation in (5.3.4) is taken only over a limited range. More details on the interpolation based on the sample function can be found in [166].

To analyze the performance of the estimator, denote $w_{n,k}$ the additive white Gaussian noise at the k-th tone of the n-th OFDM block. Assume that the pilot symbols are with constant modulus, that is, $|s_{v_1 n_o, v_2 k_o}| = 1$. Furthermore, we also assume that $\mathrm{E}\{|H_{n,k}|^2\} = 1$ and SNR of the OFDM system is ρ; then $E\{|w_{n,k}|^2\} = \rho$. From (5.3.2),

$$\tilde{H}_{v_1 n_o, v_2 k_o} = H_{v_1 n_o, v_2 k_o} + \bar{w}_{v_1 n_o, v_2 k_o},$$

where

$$\bar{w}_{v_1 n_o, v_2 k_o} = \frac{w_{v_1 n_o, v_2 k_o}}{s_{v_1 n_o, v_2 k_o}},$$

and

$$\mathrm{E}\left|\bar{w}_{v_1 n_o, v_2 k_o}\right|^2 = \rho.$$

From (5.3.3),

$$\begin{aligned} \widehat{H}_{n,k} &= \sum_{l_1, l_2} H_{v_1 l_1, v_2 l_2} \mathrm{sinc}\left(2\pi(\frac{n}{v_1} - l_1)\right) \mathrm{sinc}\left(2\pi(\frac{k}{v_2} - l_2)\right) \\ &+ \sum_{l_1, l_2} \bar{w}_{v_1 l_1, v_2 l_2} \mathrm{sinc}\left(2\pi(\frac{n}{v_1} - l_1)\right) \mathrm{sinc}\left(2\pi(\frac{k}{v_2} - l_2)\right). \end{aligned} \tag{5.3.5}$$

From the sampling theorem [167], when (5.3.1) is satisfied,

$$H_{n,k} = \sum_{l_1, l_2} H_{v_1 l_1, v_2 l_2} \mathrm{sinc}\left(2\pi(\frac{n}{v_1} - l_1)\right) \mathrm{sinc}\left(2\pi(\frac{k}{v_2} - l_2)\right).$$

Therefore,

$$\tilde{H}_{n,k} = H_{n,k} + \sum_{l_1, l_2} \bar{w}_{v_1 l_1, v_2 l_2} \mathrm{sinc}\left(2\pi(\frac{n}{v_1} - l_1)\right) \mathrm{sinc}\left(2\pi(\frac{k}{v_2} - l_2)\right),$$

and

$$\begin{aligned} \mathrm{MSE} &\triangleq \mathrm{E}\left|\widehat{H}_{n,k} - H_{n,k}\right|^2 \\ &= \mathrm{E}\left|\sum_{l_1, l_2} \bar{w}_{v_1 l_1, v_2 l_2} \mathrm{sinc}\left(2\pi(\frac{n}{v_1} - l_1)\right) \mathrm{sinc}\left(2\pi(\frac{k}{v_2} - l_2)\right)\right|^2 \\ &= \rho \sum_{l_1, l_2} \mathrm{sinc}^2\left(2\pi(\frac{n}{v_1} - l_1)\right) \mathrm{sinc}^2\left(2\pi(\frac{k}{v_2} - l_2)\right) \\ &= \rho \sum_{l_1} \mathrm{sinc}^2\left(2\pi(\frac{n}{v_1} - l_1)\right) \sum_{l_2} \mathrm{sinc}^2\left(2\pi(\frac{k}{v_2} - l_2)\right). \end{aligned} \tag{5.3.6}$$

Using Parseval's theorem,

$$\sum_{l} \operatorname{sinc}^2 \left(2\pi(\frac{n}{v} - l)\right) = 1,$$

for any n. Consequently, MSE $= \rho$, which implies that the performance of estimation is same for all n's and k's.

Linear interpolation approach

Channel parameters can be also obtained using polynomial fitting [168], [169]. Here we only introduce a special case of polynomial fitting approaches - *linear interpolation approach.*

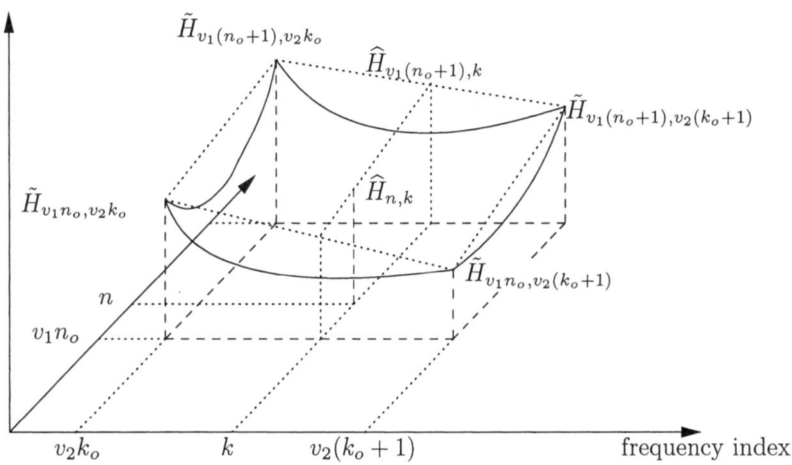

Figure 5.11. Linear interpolation approach.

Figure 5.11 illustrates the linear interpolation approach. For any k between v_2k_o and $v_2(k_o+1)$, that is, $k = v_2k_o + r_2$ with $0 < r_2 < v_2$, channel parameters at v_1n_o-th OFDM block can be estimated by

$$\widehat{H}_{v_1n_o,k} = \left(1 - \frac{r_2}{v_2}\right)\tilde{H}_{v_1n_o,v_2k_o} + \frac{r_2}{v_2}\tilde{H}_{v_1n_o,v_2(k_o+1)}.$$

From $\widehat{H}_{v_1n_o,k}$, channel parameters at any time, n, and tone, k, can be obtained by

$$\begin{aligned}
\widehat{H}_{n,k} &= \left(1-\frac{r_1}{v_1}\right)\widehat{H}_{v_1 n_o,k} + \frac{r_1}{v_1}\widehat{H}_{v_1(n_o+1),k} \\
&= \left(1-\frac{r_1}{v_1}\right)\left(1-\frac{r_2}{v_2}\right)\tilde{H}_{v_1 n_o, v_2 k_o} + \left(1-\frac{r_1}{v_1}\right)\frac{r_2}{v_2}\tilde{H}_{v_1 n_o, v_2(k_o+1)} \\
&\quad + \frac{r_1}{v_1}\left(1-\frac{r_2}{v_2}\right)\tilde{H}_{v_1(n_o+1), v_2 k_o} + \frac{r_1}{v_1}\frac{r_2}{v_2}\tilde{H}_{v_1(n_o+1), v_2(k_o+1)}
\end{aligned}$$

where $n = v_1 n_o + r_1$ with $0 < r_1 < v_1$. Compared with the sample function based estimator, direct linear interpolation approach is much simpler. However, its performance is usually not so good as that of the sample function based estimator. More details on the interpolation based algorithm can be found in [31].

5.3.3 Filtering Approaches

Interpolation can be also implemented using a filter. For filtering based interpolation, the pilot symbol grid is not necessarily rectangular; it could be any regular shape.

Direct filtering

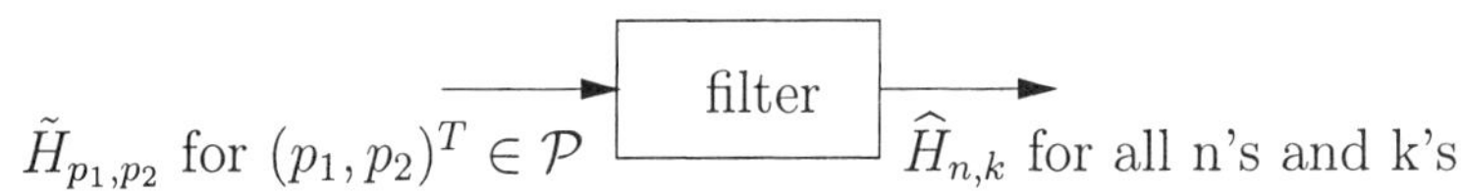

Figure 5.12. Structure of the filtering approaches for pilot-symbol-aided estimation.

For the direct filtering approach, the channel parameters are obtained by

$$\widehat{H}_{n,k} = \sum_{(m,l)^T \in \mathcal{T}_{n,k}} c_{m,l;n,k}\tilde{H}_{m,l}, \tag{5.3.7}$$

where $c_{m,l;n,k}$ is the filter coefficient and $\mathcal{T}_{n,k}$ is the index set of pilot symbols used to estimate $H_{n,k}$, which is obviously a subset of $\mathcal{P} = \{\mathbf{p} = \mathbf{Vm} : \mathbf{m} = (m_1, m_2)^T\ m_1,\ m_2 \text{ are integers}\}$. The general structure of the filtering approaches is shown in Figure 5.12. Note that the filter may be *shift-variant or invariant.* The filter coefficients may be determined by *MMSE filtering* or *robust filtering.*

Generally, the filter in Figure 5.12 is a two-dimensional (2-D) filter. However, under certain conditions, it may be just an one-dimensional (1-D) filter, or consist of two 1-D filters. $\mathcal{T}_{n,k}$ determines whether the filter is one-dimensional (1-D) or not. If $\mathcal{T}_{n,k} = \{(n,l)^T : (n,l)^T \in \mathcal{P}\}$, then it is a *1-D frequency domain filter*, and the channel parameters at the n-th OFDM block are interpolated only by the temporal estimated channel parameters in this block. Similarly if $\mathcal{T}_{n,k} = \{(m,k)^T : (m,k)^T \in \mathcal{P}\}$, then it is a *1-D time-domain filter.* Mathematically, there is no difference between the time- and frequency domain filtering. However, the performance of the 1-D time- and frequency domain filtering is generally different. It depends on the time- and frequency domain correlations of the channel frequency response and the pilot symbol grid. Sometimes, two 1-D filters are used to enhance the performance and reduce the computational complexity. In this case, $\mathcal{T}_{n,k} = \{(n,l)^T \in \mathcal{P} \text{ or } (m,k)^T \in \mathcal{P}\}$.

Once $\mathcal{T}_{n,k}$ is selected, the filter coefficients can be chosen to minimize $E\{|\widehat{H}_{n,k} - H_{n,k}|^2\}$. A detailed discussion on the performance of the filtering approaches can be found in [166], [162], [163], [170], [171], and the references therein.

Transform-domain filtering

For systems with pilot symbols, filtering for channel estimation can be in transform-domain. For the *transform-domain filtering approaches*, the *fast Fourier transform* (FFT) can be used to reduce the complexity of the estimator.

From the temporal estimation at pilot positions, $\tilde{H}_{m_1,m_2}$ for $(m_1,m_2)^T \in \mathcal{P}$, a 2-D filter is used to estimate channel parameters at other positions, that is,

$$\widehat{H}_{n,k} = \sum_{(m_1,m_2)^T \in \mathcal{P}} c_{n-m_1,k-m_2} \tilde{H}_{m_1,m_2},$$

where $c_{n,k}$ is the coefficient of the filter. The filter coefficient, $c_{n,k}$, is chosen to minimize the MSE of the estimator, that is,

$$\text{MSE} = \text{E}\left|\widehat{H}_{n,k} - H_{n,k}\right|^2.$$

Using the orthogonality principle, it is shown in Appendix D that the

2-D Fourier transform of the filter coefficients is

$$C(\omega_1, \omega_2) = \sum_{n=-\infty}^{+\infty} \sum_{k=-\infty}^{+\infty} c_{n,k} e^{-\jmath(\omega_1 n + \omega_2 k)} = \frac{R(\omega_1, \omega_2)}{\frac{R(\omega_1,\omega_2)}{|\det(\mathbf{V})|} + \rho}, \tag{5.3.8}$$

where $R(\omega_1, \omega_2)$ is the 2-D Fourier transform of channel's time and frequency correlation function, that is,

$$R_H(\omega_1, \omega_2) = \sum_{n,k} r_H[n,k] e^{-(\omega_1 n + \omega_2 k)}.$$

The MSE for the optimal transform estimator is

$$\overline{\text{MMSE}} = \frac{\rho}{(2\pi)^2} \iint_D \frac{R(\omega_1, \omega_2)}{\frac{R(\omega_1,\omega_2)}{\det(\mathbf{V})} + \rho} d\omega_1 d\omega_2.$$

From the above discussion, MMSE interpolation can be designed once channel's time- and frequency domain correlation is known. The 2-D fast Fourier transform (FFT) can be used to reduce the complexity of the estimator.

When channel's statistics are unknown, a robust interpolator has been developed in [172], which is very similar to robust estimation introduced in Section 5.2.3. For the robust interpolator, an ideal 2-D low-pass filter is used in (5.3.8) instead of channel's time- and frequency domain correlation.

An enhanced channel parameter estimation algorithm has been proposed in [173], [174] to improve the performance of the channel parameter estimation. Based on the initial parameter estimation using the pilot symbols, the transmitted symbols can be coherently detected; therefore the temporal estimation for all tones of all OFDM blocks can be obtained with the help of the detected symbols. Better parameter estimation can then be obtained from the temporal parameter estimator. Computer simulation demonstrates that the enhanced estimation can significantly improve the system performance even though symbol detection errors may occur.

5.3.4 Performance Evaluation

In this section, we present the performance of the pilot-symbol-aided parameter estimation using the transform based interpolation approach. The system parameters are the same as those in Section 5.2.5. Pilot symbols are inserted in the system at a 10% rate with the grid shown in Figure 5.9.

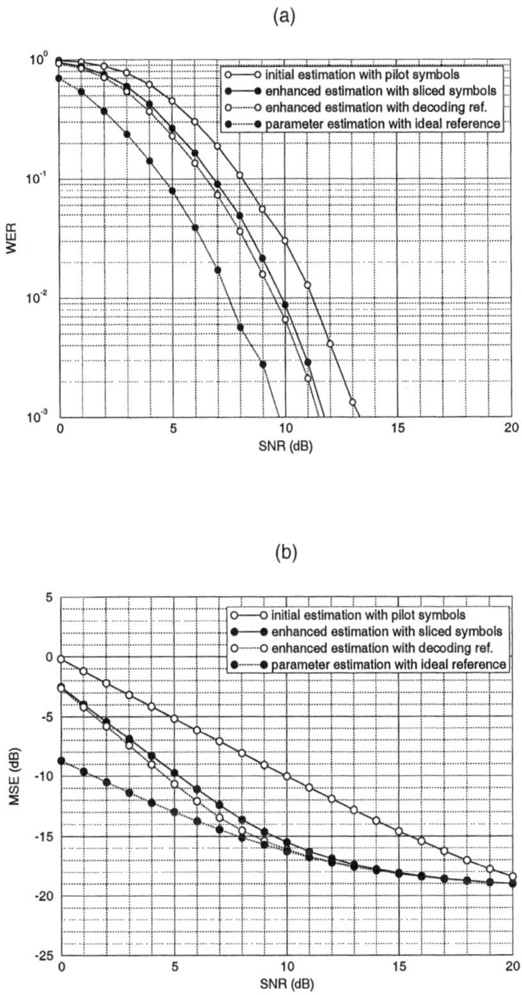

Figure 5.13. Effects of different references on (a) WER and (b) MSE over a TU channel with f_d=40 Hz.

Effects of references on system performance

Figure 5.13 shows the effect of different references on the WER over a TU channel with 40 Hz Doppler frequency. When 10% tones are assigned to

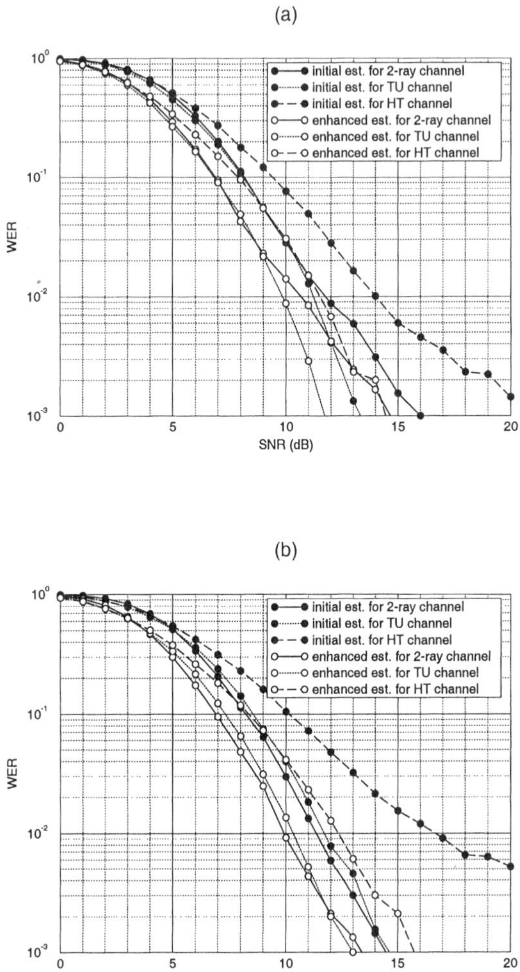

Figure 5.14. WER versus SNR of the initial and the enhanced estimators over channels with (a) a 40 Hz Doppler frequency and (b) a 200 Hz Doppler frequency

pilot symbols, the MSE of the estimated parameters is almost same as the system's SNR. The required SNR's for 10% and 1% WER's are about 8 dB and 11.2 dB, respectively. When the decided (sliced) symbols along with the pilot symbols are used to enhance channel parameter estimation, the performance can be improved significantly, and the MSE of the enhanced estimator is reduced from -10 dB to -15.5 dB when SNR=10 dB. The required SNR's for 10% and 1% WER's are improved by about 1.4 dB. If the undecoded/decoded dual mode reference introduced in Section 5.2.3 or [161] is used, another 0.3 dB SNR improvement can be obtained. However, when an ideal reference is used for channel parameter estimation, the system performance is much better than that of the initial or enhanced channel parameter estimation.

Parameter estimation performance under different environments

Figure 5.14 shows the performance of the initial and enhanced parameter estimators for channels with the two-ray (with delay span 5 μsec), TU, and HT delay profiles and 40 Hz and 200 Hz Doppler frequencies, respectively. From the figures, the system with a pilot-symbol-aided estimator is very robust to Doppler shift, and there is minimal degradation as the Doppler frequency increases from 40 Hz to 200 Hz. The required SNR for a 10% WER is 7 dB for the two-ray and TU channels and about 8 dB for the HT channel.

Comparison with decision-directed estimator

Figure 5.15 compares the pilot-symbol-aided estimator and the decision-directed estimator in Section 5.2. From Figure 5.15, for the TU channel with a 40 Hz Doppler frequency, the pilot-symbol-aided and the decision-directed estimators provide similar performance with the required SNR's for 10% and 1% WER's about 7 dB and 10 dB, respectively. However, for the TU channel with $f_d = 200$ Hz, the performance of the decision-directed estimator is degraded significantly, and the required SNR for a 10% WER is as large as 9.3 dB while it is only about 7.3 dB for the pilot-symbol-aided estimator. Figure 5.15(b) shows the WER performance for the HT channel. It shows that the decision-directed estimator is 1 dB better than the pilot-symbol-aided estimator for lower Doppler frequency ($f_d = 40$ Hz); however, the pilot-symbol-aided estimator is much better for higher Doppler frequency ($f_d = 200$ Hz).

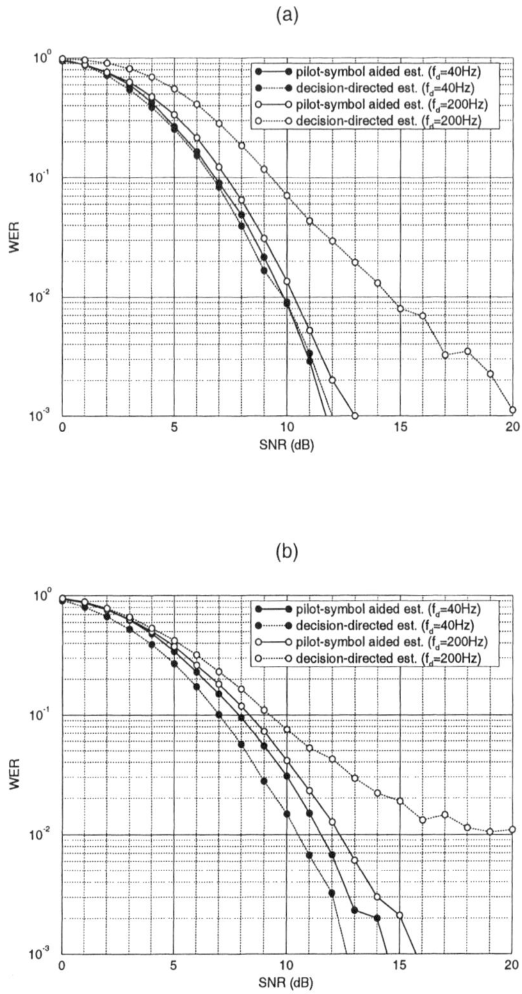

Figure 5.15. WER of the pilot-symbol aided and decision-directed estimators for (a) the TU and (b) the HT channels with different f_d's.

5.4 MIMO Channel Estimation

As indicated in Chapter 1 and in [7], [6], multiple transmit and receive antennas can be used to improve the performance and increase the capacity of wireless communications. It is proved in [7], [6] that, when multiple transmit and receive antennas are used to form a *multiple-input and multiple-output* (MIMO) system, the system capacity can be improved by a factor of the minimum of the number of transmit and receive antennas compared with a *single-input single-output* (SISO) system with flat Rayleigh fading or narrowband channels. However, for wideband channels, *orthogonal frequency division multiplexing* (OFDM) has to be used with MIMO techniques for ISI mitigation and capacity improvement. MIMO-OFDM has been investigated in [175], [176], [177]. In this section, we concentrate on the training sequence design and channel estimation for MIMO-OFDM systems.

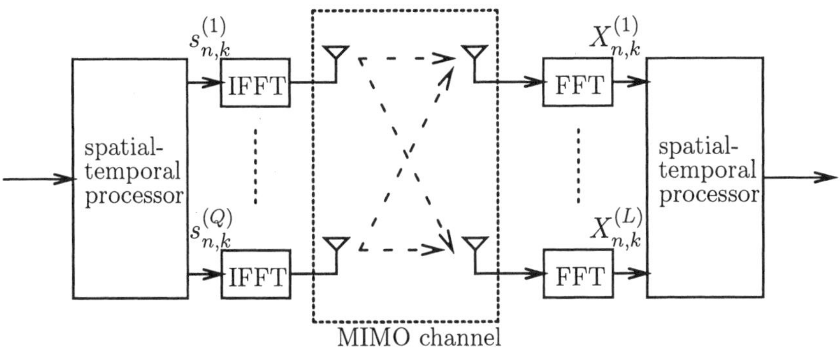

Figure 5.16. Multiple-input multiple-output OFDM system.

A MIMO-OFDM system can be shown in Figure 5.16. The signal from each receive antenna at the k-th subchannel of the n-th OFDM block can be expressed as,

$$X_{n,k}^{(l)} = \sum_{q=1}^{Q} H_{n,k}^{(ql)} s_{n,k}^{(q)} + N_{n,k}^{(l)},$$

where $H_{n,k}^{(ql)}$ is the channel frequency response at the k-th subchannel of the n-th OFDM block corresponding to the q-th transmit antenna and the l-th receive antenna, and $N_{n,k}$ is the additive white Gaussian noise. The crucial challenge here is that each received signal corresponds to several channel parameters.

5.4.1 Basic Channel Estimation

Channel estimation for OFDM can exploit correlation of channel parameters at different times and frequencies. A basic channel estimator has been introduced in [175]. Exploiting the correlation at different subchannels, channel frequency responses can be expressed as

$$H_{n,k}^{(q)} = \sum_{m=0}^{K_o-1} h_{n,m}^{(q)} W_N^{km}, \tag{5.4.1}$$

for $k = 0,\ 1, \cdots, N-1$ and $q = 1, \cdots, Q$. The K_o in the above expression depends on the ratio of the delay span of wireless channels and the symbol duration of OFDM, and $W_N = \exp(-\jmath\frac{2\pi}{N})$. Hence, to obtain $H_{n,k}^{(q)}$, we only need to estimate $h_{n,m}^{(q)}$. In the above discussion, we have omitted the index for receive antenna here since channel estimation for each receive antenna is performed independently.

During the training period, transmitted signals are known to the receiver; in the data transmission mode, transmitted signals can be recovered from the decided symbols. If the transmitted signals, $s_{n,k}^{(q)}$ from the q-th transmit antenna is known for $q = 1, \cdots, Q$. then $\tilde{h}_{n,m}^{(q)}$, a temporal estimation of $h_{n,m}^{(q)}$, can be found by minimizing the following cost function,

$$\sum_{k=0}^{N-1} \left| X_{n,k} - \sum_{q=1}^{Q} \sum_{m=0}^{K_o-1} \tilde{h}_{n,m}^{(q)} W_K^{km} s_{n,k}^{(q)} \right|^2 . \tag{5.4.2}$$

Direct calculation in [175] yields that

$$\begin{pmatrix} \mathbf{A}_n^{(11)} & \cdots & \mathbf{A}_n^{(Q1)} \\ \vdots & \cdots & \vdots \\ \mathbf{A}_n^{(1Q)} & \cdots & \mathbf{A}_n^{(QQ)} \end{pmatrix} \begin{pmatrix} \tilde{\mathbf{h}}_n^{(1)} \\ \vdots \\ \tilde{\mathbf{h}}_n^{(Q)} \end{pmatrix} = \begin{pmatrix} \mathbf{b}_n^{(1)} \\ \vdots \\ \mathbf{b}_n^{(Q)} \end{pmatrix}, \tag{5.4.3}$$

or

$$\begin{pmatrix} \tilde{\mathbf{h}}_n^{(1)} \\ \vdots \\ \tilde{\mathbf{h}}_n^{(Q)} \end{pmatrix} = \begin{pmatrix} \mathbf{A}_n^{(11)} & \cdots & \mathbf{A}_n^{(Q1)} \\ \vdots & \cdots & \vdots \\ \mathbf{A}_n^{(1Q)} & \cdots & \mathbf{A}_n^{(QQ)} \end{pmatrix}^{-1} \begin{pmatrix} \mathbf{b}_n^{(1)} \\ \vdots \\ \mathbf{b}_n^{(Q)} \end{pmatrix}, \tag{5.4.4}$$

where $\tilde{\mathbf{h}}_n^{(q)}$ is the temporal estimation of channel parameter vector, defined as

$$\tilde{\mathbf{h}}_n^{(q)} = \left(\tilde{h}_{n,0}^{(q)}, \cdots, \tilde{h}_{n,K_o-1}^{(q)} \right)^T,$$

and $a_{n,m}^{(ij)}$, $\mathbf{A}_n^{(ij)}$, $b_{n,m}^{(i)}$, and $\mathbf{b}_n^{(i)}$ are definded as

$$a_{n,m}^{(ij)} = \sum_{k=0}^{N-1} s_{n,k}^{(i)} s_{n,k}^{(j)\,*} W_N^{-km}, \tag{5.4.5}$$

$$\mathbf{A}_n^{(ij)} = \left(a_{n,m_1-m_2}^{(ij)} \right)_{m_1,m_2=0}^{K_o-1}, \tag{5.4.6}$$

$$b_{n,m}^{(i)} = \sum_{k=0}^{N-1} X_{n,k} s_{n,k}^{(i)\,*} W_N^{-km},$$

and

$$\mathbf{b}_n^{(i)} = \left(b_{n,0}^{(i)}, \cdots, b_{n,K_o-1}^{(i)} \right)^T,$$

respectively.

From the temporal estimation of channel parameters, robust estimation can be obtained using the approach developed in Section 5.2 and in [161], which exploits the correlation of channel parameters at different OFDM blocks. Then the robust estimation of channel parameter vectors at the n-th OFDM block can be obtained by

$$\widehat{\mathbf{h}}_n^{(i)} = \sum_{l\geq 0} c_l \tilde{\mathbf{h}}_{n-l}^{(i)},$$

where c_l's ($l \geq 0$) are the coefficients for the robust channel estimator described in Section 5.2.3 and [161], [175].

Figure 5.17 illustrates the block diagram of the basic channel estimator for a MIMO-OFDM system with two transmit antennas. It can be easily extended to a MIMO-OFDM system with any number of transmit antennas.

To calculate temporal estimation in the figure, a $2K_o \times 2K_o$ matrix inversion is need to get the temporal estimation of $h_{n,m}^{(1)}$ and $h_{n,m}^{(2)}$. In general, a $QK_o \times QK_o$ matrix inversion is required for a MIMO-OFDM system with Q transmit antennas, which is computationally intensive. In order to reduce the computational complexity, the significant-tap-catching estimator has been developed in [175].

Figure 5.18 shows the performance of a 2-input/2-output OFDM with space-time coding. The parameters of the simulated OFDM system are the same as the one described in Section 2.1.5. A 16-state space-time code with 4-PSK is used in the system. In brief, the overall system can transmit data

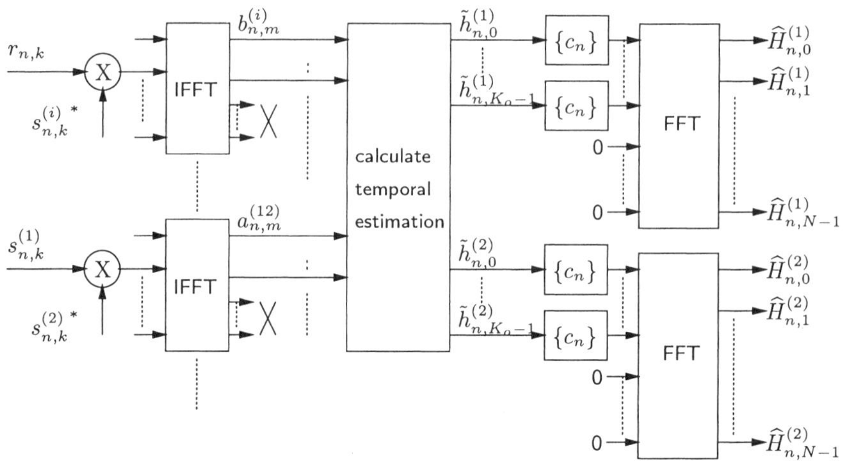

Figure 5.17. Basic channel parameter estimator for MIMO-OFDM with 2 transmit antennas.

at a rate of 1.18 Mbits/sec over an 800 kHz channel, i.e., the transmission efficiency is 1.475 bits/sec/Hz.

Figure 5.18 compares the performance between channels with the two-ray and the HT delay profiles with f_d=40 Hz. The system has the same performance when the ideal parameters of the previous OFDM block are used for decoding. However, when estimated parameters are used, the system has better performance for the two-ray delay profile than for the HT profile since the estimator has lower MSE for the two-ray delay profile as we can see from Figure 5.18(b). When a 7-tap or 9-tap significant-tap-catching techniques in [175] is used, the required SNR is 8 dB for a 10% WER, and 6 dB for a 1% BER, for the two-ray delay profile, and about 8.6 dB for a 10% WER and 6.6 dB for a 1% BER, respectively.

5.4.2 Optimum Training Sequences for Channel Estimation

Usually, training sequences are used at the beginning of each time slot for synchronization and channel estimation. In this section, we describe optimum training that can simplify initial channel estimation and optimize estimation performance.

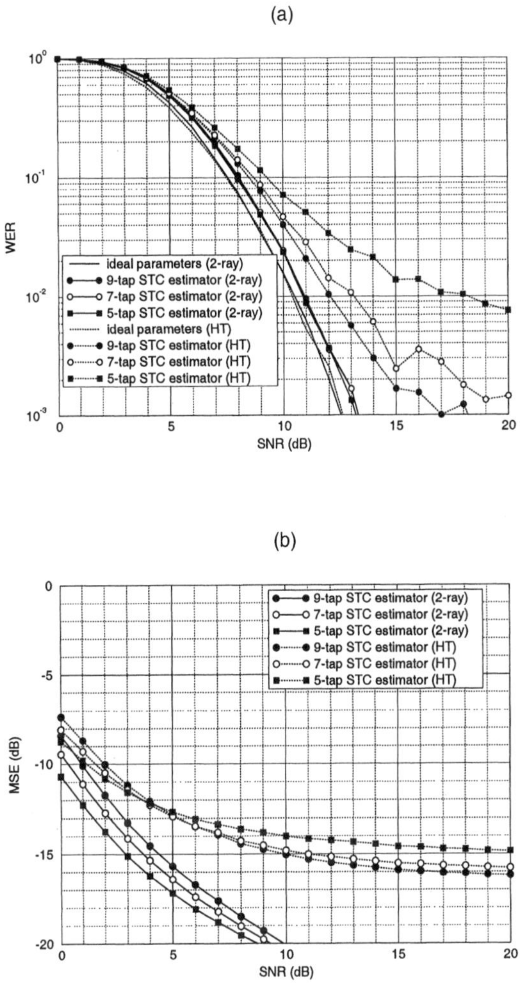

Figure 5.18. (a) WER and (b) MSE of a 2-input/2-output OFDM systems when a wireless channel with 40 Hz Doppler frequency and the two-ray and the HT delay profiles, respectively.

For simplicity, we assume that modulation results in constant-modulus signals, that is, $|s_{n,k}^{(q)}| = 1$. From (5.4.5),

$$a_{n,m}^{(ii)} = N\delta[m],$$

where $\delta[m]$ denotes the unit impulse function. Consequently, $\mathbf{A}_n^{(ii)} = N\mathbf{I}$, where $\mathbf{I}$ is a $K_o \times K_o$ identity matrix. If the training sequences, $\{s_{1,k}^{(q)}\}$'s, are chosen such that $\mathbf{A}_1^{(ij)} = \mathbf{0}$ for $i \neq j$, then, from (5.4.4), $\tilde{\mathbf{h}}_1^{(i)} = \frac{1}{N}\mathbf{b}_1^{(i)}$, and no matrix inversion is required for channel estimation.

To find $s_{1,k}^{(q)}$ for $q \geq 2$ with $|s_{1,k}^{(q)}| = 1$ and $\mathbf{A}_1^{(ij)} = \mathbf{0}$ for $i \neq j$, it is sufficient to find $a_{1,m}^{(ij)} = 0$ for $|m| \leq K_o - 1$ since $\mathbf{A}_1^{(ij)}$ only consists of $a_{1,m}^{(ij)}$'s for $|m| \leq K_o - 1$, where $m \equiv N + m$ if $m < 0$.

To construct training sequences such that $\mathbf{A}_1^{(ij)} = \mathbf{0}$, let the training sequence for the first antenna $\{s_{1,k}^{(1)}\}$ be any sequence that is *good* for time and frequency synchronization and other properties, such as low PAPR. For a MIMO-OFDM system with the number of transmit antennas, Q, less than or equal to $\frac{N}{K_o}$, let

$$s_{1,k}^{(q)} = s_{1,k}^{(1)} W_N^{-\bar{K}_o(q-1)k}, \tag{5.4.7}$$

for $q = 2, \cdots, Q$, where $\bar{K}_o = \lfloor \frac{N}{Q} \rfloor \geq K_o$ and $\lfloor z \rfloor$ denotes the largest integer no larger than z. Then for any $i \leq j$,

$$a_{n,m}^{(ij)} = N\delta[m - \bar{K}_o(j - i)]. \tag{5.4.8}$$

Note that $1 \leq j - i \leq p - 1$; therefore,

$$\bar{K}_o(j - i) \geq \bar{K}_o \geq K_o, \tag{5.4.9}$$

and

$$\bar{K}_o(j - i) \leq \bar{K}_o(p - 1) = \bar{K}_o p - \bar{K}_o \leq N - K_o. \tag{5.4.10}$$

Consequently, $a_{1,m}^{(ij)} = 0$ for $0 \leq m \leq K_o - 1$ or $N - K_o + 1 \leq m \leq N - 1$ (equivalent to $|m| \leq K_o - 1$), which results in $\mathbf{A}_1^{(ij)} = \mathbf{0}$ for all $i < j$. If $i > j$, $\mathbf{A}_1^{(ij)} = (\mathbf{A}_1^{(ji)})^H = \mathbf{0}$. Hence, $\mathbf{A}_1^{(ij)} = \mathbf{0}$ for all $i \neq j$.

It should be indicated that the above optimum training sequence design approach is not applicable to those MIMO-OFDM systems with over $\frac{N}{K_o}$ transmit antennas.

It is proved in [175] that the MSE of the basic temporal channel estimation reaches the low bound when $\mathbf{A}_1^{(ij)} = \mathbf{0}$ for $i \neq j$. Therefore, optimum training sequences can not only reduce the complexity of channel estimation but aslo improve the performance of temporal channel parameter estimation during training period.

5.4.3 Simplified Channel Estimation

In the previous section, we have introduced optimum sequences for channel estimation, which not only improve the initial channel estimation during the training period but also simplify channel estimator. During the data transmission period ($n > 1$), transmitted symbols are random; therefore, we cannot control $\mathbf{A}_n^{(ij)}$. Here, we introduce an approach that simplifies channel estimation during the data transmission mode.

From (5.4.3), for the n-th OFDM block, we have

$$\mathbf{A}_n^{(ii)}\tilde{\mathbf{h}}_n^{(i)} = \mathbf{a}_n^{(i)} - \sum_{j=1, j\neq i}^{Q} \mathbf{A}_n^{(ji)}\tilde{\mathbf{h}}_n^{(i)}, \tag{5.4.11}$$

for $i = 1, \cdots, Q$. In the above expression, the subscript n has been added to indicate that those vectors and matrices are related to the n-th OFDM block. From the discussion in the previous section, for a OFDM system with constant modulus modulation, $\mathbf{A}^{(ii)} = N\mathbf{I}$ for $i = 1, \cdots, Q$, and therefore,

$$\tilde{\mathbf{h}}_n^{(i)} = \frac{1}{N}(\mathbf{b}_n^{(i)} - \sum_{j=1, j\neq i}^{Q} \mathbf{A}_n^{(ji)}\tilde{\mathbf{h}}_n^{(i)}), \tag{5.4.12}$$

for $i = 1, \cdots, Q$. From the above equations, if $\tilde{\mathbf{h}}_n^{(i)}$'s for $j = 1, \cdots, i-1, i+1, \cdots, Q$ are known, then $\tilde{\mathbf{h}}_n^{(i)}$ can be estimated without any matrix inversion.

If robust estimation of channel parameter vectors at previous OFDM block, $\widehat{\mathbf{h}}_{n-1}^{(i)}$'s for $i = 1, \cdots, Q$ are used to substitute $\tilde{\mathbf{h}}_n^{(i)}$ on the right side of (5.4.12), then

$$\tilde{\mathbf{h}}_n^{(i)} = \frac{1}{N}(\mathbf{b}_n^{(i)} - \sum_{j=1, j\neq i}^{Q} \mathbf{A}^{(ji)}\widehat{\mathbf{h}}_{n-1}^{(j)}), \tag{5.4.13}$$

for $i = 1, \cdots, Q$, and the huge matrix inversion in (5.4.4) can be avoided. Consequently, computational complexity is reduced.

The simplified channel estimation described above significantly reduces the computational complexity of channel estimation; it may also cause some performance degradation. But, it is demonstrated by theoretical analysis and computer simulation in [176] that the performance degradation is negligible.

5.4.4 Enhanced Channel Estimation

In [175], [176], Sections 5.4.1, 5.4.2, and 5.4.3, we have introduced channel parameter estimators and optimum training sequences for OFDM with multiple transmit antennas. Furthermore, for a MIMO-OFDM system where many independent channels with the same delay profile are involved, the channel delay profile can be more accurately estimated. By exploiting the estimated channel delay profile, channel parameter estimation can be further improved.

From the above discussion, for the n-th OFDM block, channel parameters corresponding to the q-th transmit and the l-th receive antenna pairs, $h_{n,m}^{(ql)}$ in (5.4.1), can be estimated using the correlation of channel parameters at different times and frequencies. With $\widehat{h}_{n,m}^{(ql)}$, the estimated $h_{n,m}^{(ql)}$, the channel frequency response at the k-th tone of the n-th OFDM block can be reconstructed by

$$\widehat{H}_{n,k}^{(ql)} = \sum_{m=0}^{K_o-1} \widehat{h}_{n,m}^{(ql)} W_N^{km}. \tag{5.4.14}$$

The estimated channel parameter, $\widehat{h}_{n,m}^{(ql)}$, can be decomposed into the true channel parameter, $h_{n,m}^{(ql)}$, and an estimation error, $e_{n,m}^{(ql)}$, that is

$$\widehat{h}_{n,m}^{(ql)} = h_{n,m}^{(ql)} + e_{n,m}^{(ql)}. \tag{5.4.15}$$

From [176], $e_{n,m}^{(ql)}$ can be assumed to be Gaussian with zero-mean and variance σ^2, and independent for different q's, l's, n's, or m's. If the parameter estimation quality is measured by means of the normalized MSE (NMSE), which is defined as

$$\text{NMSE} = \frac{E\left|\widehat{H}_{n,k}^{(ql)} - H_{n,k}^{(ql)}\right|^2}{E\left|H_{n,k}^{(ql)}\right|^2},$$

then it can be calculated directly that the NMSE for the estimation in (5.4.14) is

$$\text{NMSE}_r = K_o\sigma^2, \tag{5.4.16}$$

where we have used the assumption that

$$\sum_{l=0}^{K_o-1} E\left|h_{n,m}^{(ql)}\right|^2 = \sum_{l=0}^{K_o-1} \sigma_m^2 = 1,$$

with $\sigma_m^2 = E\left|h_{n,m}^{(ql)}\right|^2$.

If channel's delay profile, that is, σ_m^2's for $m = 0, \cdots, K_o - 1$, is known, and can be used to reconstruct channel frequency response from $\widehat{h}_{n,m}^{(ql)}$, the

NMSE of $\widehat{H}_{n,k}^{(ql)}$ can be significantly reduced. In this case, if the α_m's are selected to minimize the NMSE of

$$\widehat{H}_{n,k}^{(ql)} = \sum_{m=0}^{K_o-1} \alpha_m \widehat{h}_{n,m}^{(ql)} W_N^{km}, \tag{5.4.17}$$

then it can be proved that the optimal α_m is

$$\alpha_m = \frac{\frac{\sigma_m^2}{\sigma_m^2+\sigma^2}}{\sum_{m'=0}^{K_o-1} \frac{\sigma_{m'}^4}{\sigma_{m'}^2+\sigma^2}} \tag{5.4.18}$$

and the NMSE is

$$\text{NMSE}_o = \frac{\sigma^2 \sum_{m=0}^{K_o-1} \frac{\sigma_m^2}{\sigma_m^2+\sigma^2}}{\sum_{m=0}^{K_o-1} \frac{\sigma_m^4}{\sigma_m^2+\sigma^2}}. \tag{5.4.19}$$

As indicated in [161], the channel delay profile depends on the environment, and therefore, is usually unknown. However, for MIMO-OFDM systems, channels corresponding to different transmit or receive antennas have the same delay profile. Therefore, $\sigma_m^2 = E\left|h_{n,m}^{(ql)}\right|^2$ can be estimated by

$$\widehat{\sigma}_m^2 = \frac{1}{QL}\sum_{q=1}^{Q}\sum_{l=1}^{L}\left|\widehat{h}_{n,m}^{(ql)}\right|^2.$$

With the estimated σ_m^2, enhanced channel frequency responses can be reconstructed by (5.4.17).

Figure 5.19 compares the MSE of the basic and enhanced channel estimation for a 4-input/4-output OFDM system. The MSE of the enhanced channel estimator is about 1.5 dB better for the TU channels and 1 dB better for the HT channels than the basic estimator described in Section 5.4.1 and [175].

5.5 Summary and Further Reading

We have presented various estimators for OFDM systems, which exploit the time and frequency correlations of the rapid dispersive fading wireless channels. These techniques can be used in OFDM systems with coherent detection and maximal ratio combining. The impact of channel estimation error on OFDM can be found in [178].

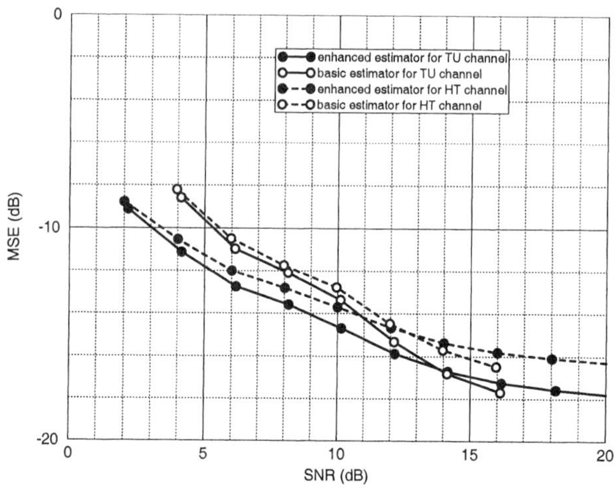

Figure 5.19. MSE comparison of the basic and the enhanced channel estimation techniques for a 4-input/4-output OFDM system.

Various decision-directed channel estimators have been discussed in this chapter. More materials on decision-directed channel estimation can be also found in [179], [180], [181], [182].

Pilot-symbol-aided channel parameter estimators are robust to high Doppler frequency. Therefore, it can be used for OFDM systems with high mobility. Besides the papers [162], [163], [170], [171], [183] that we have mentioned in this chapter, the pilot grid selection is also discussed in [184], [185], [186]. The polynomial and Gaussian filter interpolations are presented in [187], [188], respectively. In [189], an approach is introduced to estimate each path delay and gain of multipath channels. The performance of different pilot-symbol-aided estimators is analyzed and compared in [190], [191], [192]. The effect of frequency offset on the performance is analyzed in [191]. Channel estimation for OFDM with turbo codes is presented in [193]. The channel estimation for digital television systems is investigated in [194], [195], [196].

Blind signal processing techniques have been used in single-carrier systems for channel identification and equalization. Recently, blind techniques have been also proposed for OFDM systems. A sufficient condition for OFDM systems to be blindly identified has been obtained in [197]. For OFDM with a cyclic prefix, a subspace method for the blind and semi-blind identification of channel parameters has been proposed in [198]. For OFDM with precoding, deterministic channel estimation and symbol recovery algorithms have been derived in [199]. Channel parameters for OFDM can be also identified by exploiting the finite alphabet property of the signal constellations [200]. Recently, *the constant-modulus algorithm* (CMA) [201], [202] has been used to estimate channel parameters for OFDM systems with transmit diversity.

Channel equalization to reduce the inter-block interference in OFDM is discussed in [203], [204], [205]. Channel estimation for multi-carrier CDMA systems and for clustered OFDM are considered in [195], [206], [207] and in [208], respectively.

MIMO OFDM has been used in many wireless communication systems and proposed for numerous standards. In this chapter, we have only introduced some basic approaches for MIMO OFDM channel estimation. More on this topic can be found in [175], [176], [177], [209], [210], [211], [212], [213], [214], [215] and the reference there in. More on training design can be also found in [216], [217], [218], [219]. Channel estimation for MIMO systems with frequency and/or time offset is investigated in Chapter 7 of this book.

Appendix 5A: Derivation of MMSE Channel Estimator

By the *orthogonality principle* [160], the $\mathbf{C}_m$'s are determined by

$$\mathrm{E}\left\{(\widehat{\mathbf{H}}_n - \mathbf{H}_n)\tilde{\mathbf{H}}_{n-m}^H\right\} = \mathbf{0}. \tag{5A.1}$$

Direct calculation shows that (5A.1) is equivalent to

$$\sum_{m_1=0}^{\infty} \mathbf{C}_{m_1}\mathbf{R}_H[m-m_1] - \mathbf{R}_H[m] + \rho\mathbf{C}_m = \mathbf{0}, \tag{5A.2}$$

for $m = 0, 1, \cdots$, where $\mathbf{R}_H[m]$ is a $N \times N$ channel correlation matrix sequence defined as

$$\mathbf{R}_H[m] = (r_H[m, i-j])_{i,j=1}^{N},$$

with

$$r_H[m, l] = E\{H_{n+m,k+l}H_{n,k}^*\}.$$

Assuming the separation property holds for dispersive time-varying channels, that is,

$$r_H[m,l] = r_t[m]r_f[l],$$

where $r_t[m]$ is the channel's time-domain correlation that is defined as

$$r_t[m] = E\{H_{n+m,k}H^*_{n,k}\}.$$

Therefore,

$$\mathbf{R}_H[m] = (r_f[m]r_f[i-j])_{i,j=1}^N = r_t[m]\mathbf{R}_f,$$

and (5A.2) can be simplified to

$$\sum_{m_1=0}^{+\infty} r_t[m-m_1]\mathbf{C}_m\mathbf{R}_f - r_t[m]\mathbf{R}_f + \rho\mathbf{C}_m = \mathbf{0}. \tag{5A.3}$$

From the eigen-decomposition of $\mathbf{R}_f$ in (5.2.5), we have

$$\sum_{m_1=0}^{+\infty} r_t[m-m_1]\mathbf{C}_{m_1}\mathbf{U}\mathbf{D}\mathbf{U}^H - r_t[m]\mathbf{U}\mathbf{D}\mathbf{U}^H + \rho\mathbf{C}_m = \mathbf{0}, \tag{5A.4}$$

or equivalently,

$$\sum_{m_1=0}^{+\infty} \mathbf{U}^H\mathbf{C}_m\mathbf{U}(r_t[m-m_1]\mathbf{D} + \rho\delta[m-m_1]\mathbf{I}) - r_t[m]\mathbf{D} = \mathbf{0}, \tag{5A.5}$$

where $\mathbf{I}$ is a $N \times N$ identity matrix, and

$$\delta[m] = \begin{cases} 1 & \text{if } m = 0, \\ 0 & \text{otherwise.} \end{cases}$$

Since $\mathbf{D}$ is diagonal, both $r_t[m-m_1]\mathbf{D}+\rho\delta[m-m_1]\mathbf{I}$ and $r_t[m]\mathbf{D}$ are diagonal. Consequently, $\mathbf{U}^H\mathbf{C}_m\mathbf{U}$ must be diagonal. Let

$$\mathbf{U}^H\mathbf{C}_m\mathbf{U} = \mathbf{\Phi}_m = \text{diag}\{\phi_m^{(1)}, \phi_m^{(2)}, \ldots, \phi_m^{(N)}\}.$$

Then, (5A.5) can be simplified into

$$\sum_{m_1=0}^{+\infty} \phi_{m_1}^{(k)}(d_k r_t[m-m_1] + \rho\delta[m-m_1]) - d_k r_t[m] = 0, \tag{5A.6}$$

for $k = 1, 2, \cdots, N$.

For the k with $d_k \neq 0$, from Appendix A of [220],

$$\begin{aligned} \phi_k(\omega) &= \sum_{m_1=0}^{+\infty} \phi_n^{(k)} e^{-\jmath n\omega} \\ &= 1 - \frac{1}{M_k(\omega)\gamma_k[0]}, \end{aligned} \tag{5A.7}$$

where $M_k(\omega)$ is a stable one-sided Fourier transform

$$M_k(\omega) = \sum_{n=0}^{\infty} \gamma_n^{(k)} e^{-\jmath n\omega}, \tag{5A.8}$$

which is uniquely determined by

$$M_k(\omega)M_k(-\omega) = \frac{d_k}{\rho} p_t(\omega) + 1, \tag{5A.9}$$

with

$$p_t(\omega) = \sum_{n=-\infty}^{\infty} r_t[n] e^{-\jmath n\omega}.$$

The dc component $\gamma_k[0]$ in $M_k(\omega)$ can be found by

$$\gamma_k^2[0] = \exp\{\frac{1}{2\pi}\int_{-\pi}^{\pi} \ln[\frac{d_k}{\rho} p_t(\omega) + 1] d\omega\}. \tag{5A.10}$$

For the k with $d_k = 0$, we have $\phi_m^{(k)} = 0$ for $m = 0, 1, \cdots$, which can be also written as (5A.7).

Appendix 5B: MSE of MMSE Channel Estimator

The average MSE over different tones is defined as

$$\overline{\text{MSE}} = \frac{1}{N} E \left\| \widehat{\mathbf{H}}_n - \mathbf{H}_n \right\|^2.$$

Substituting (5A.2) into (5.2.11), the average minimum MSE for the estimation is

$$\begin{aligned} \overline{\text{MMSE}} &= \frac{1}{N}\rho \text{Tr}\{\mathbf{C}[0]\} \\ &= \frac{1}{N}\rho \text{Tr}\{\mathbf{U}\mathbf{\Phi}[0]\mathbf{U}^H\} \\ &= \frac{1}{N}\rho \sum_{k=1}^{N} \phi_k[0], \end{aligned} \tag{5B.1}$$

where Tr{ .} denotes the trace of a matrix that is defined as the summation of the diagonal elements of the matrix, and

$$\begin{aligned}\phi_k[0] &= \frac{1}{2\pi}\int_{-\pi}^{\pi} 1 - \frac{1}{M_k(\omega)\gamma_k[0]} d\omega \\ &= 1 - \frac{1}{\gamma_k^2[0]} \\ &= 1 - \exp\{-\frac{1}{2\pi}\int_{-\pi}^{\pi} \ln[\frac{d_k}{\rho}p_t(\omega)+1]d\omega\}. \end{aligned} \tag{5B.2}$$

Therefore,

$$\overline{\text{MMSE}} = \frac{\rho}{N}\sum_{k=1}^{N}\left(1 - \exp\{-\frac{1}{2\pi}\int_{-\pi}^{\pi} \ln[\frac{d_k}{\rho}p_t(\omega)+1]d\omega\}\right). \tag{5B.3}$$

For Jakes' model, $p_t(\omega) = p_J(\omega)$. Then by direct calculation,

$$\phi_k = 1 - (\frac{\alpha_k}{2})^{-\frac{\omega_d}{\pi}} \exp(-\frac{\omega_d(b(\alpha_k)+1)}{\pi}),$$

and

$$\overline{\text{MMSE}}_J(\omega_d) = \frac{\rho}{N}\sum_{k=1}^{N}\left(1 - (\frac{\alpha_k}{2})^{-\frac{\omega_d}{\pi}} \exp\{-\frac{\omega_d(b(\alpha_k)+1)}{\pi}\}\right), \tag{5B.4}$$

where

$$\alpha_k = \frac{2d_k}{\omega_d\rho},$$

and

$$b(\alpha_k) = \begin{cases} \frac{\pi}{2}\alpha_k - 1 - \sqrt{1-\alpha_k^2}\ln\frac{1+\sqrt{1-\alpha_k^2}}{\alpha_k} & \text{if } \alpha_k < 1, \\ \frac{\pi}{2}\alpha_k - 1 - \sqrt{\alpha_k^2 - 1}(\frac{\pi}{2} - \arcsin\frac{1}{\alpha_k}) & \text{if } \alpha_k \geq 1. \end{cases}$$

Appendix 5C: Mismatch Analysis

If an MMSE channel estimator is designed to match a channel with time- and frequency domain correlations $\bar{r}_t[m]$ and $\bar{r}_f[l]$, respectively, then its coefficients $\bar{c}_{m,l,k}$ are determined from (5.2.12) by

$$\bar{\mathbf{C}}(\omega) = \bar{\mathbf{U}}\bar{\Phi}_{(}\omega)\bar{\mathbf{U}}^H, \tag{5C.1}$$

where the definitions of $\bar{\mathbf{U}}$ and $\bar{\Phi}(\omega)$ are similar to those of $\mathbf{U}$ and $\Phi(\omega)$ except that $r_t[m]$ and $r_f[l]$ there are respectively substituted by $\bar{r}_t[m]$ and $\bar{r}_f[l]$. For a channel with time- and frequency domain correlations $r_t[m]$ and $r_f[l]$, rather than $\bar{r}_t[m]$ and $\bar{r}_f[l]$, from (5.2.11), the MSE for the designed channel estimator is

$$
\begin{aligned}
&\text{MSE}(\{\bar{\mathbf{C}}_m\}) \\
= \;& E\left\|\sum_{m=0}^{+\infty} \bar{\mathbf{C}}_m \tilde{\mathbf{H}}_{n-m} - \mathbf{H}_n\right\|^2 \\
= \;& \text{Tr}\left\{\sum_{m_1,m_2=0}^{+\infty} \left[(\bar{\mathbf{C}}_{m_1} - \delta[m_1]\mathbf{I})\mathbf{R}_f[m_2 - m_1](\bar{\mathbf{C}}_{m_2} - \delta[m_2]\mathbf{I})^H\right]\right. \\
&+\rho \sum_{m=0}^{+\infty} Tr\left\{\bar{\mathbf{C}}_m \bar{\mathbf{C}}_m^H\right\} \\
= \;& \frac{1}{2\pi}\int_{-\pi}^{\pi} p_t(\omega) Tr\{(\bar{\mathbf{C}}(\omega) - \mathbf{I})\mathbf{R}_f(\bar{\mathbf{C}}(\omega) - \mathbf{I})^H\}d\omega \\
&+\frac{\rho}{2\pi}\int_{-\pi}^{\pi} Tr\{\bar{\mathbf{C}}(\omega)\bar{\mathbf{C}}^H(\omega)\}d\omega. \qquad (5C.2)
\end{aligned}
$$

Substituting (5C.1) into (5C.2), we obtain a general formula for the MSE of the mismatched channel estimator,

$$
\begin{aligned}
\text{MSE}(\{\bar{\mathbf{C}}_m\}) \;=\; & \frac{1}{2\pi}\int_{-\pi}^{\pi} p_t(\omega) Tr\{\bar{\mathbf{U}}\mathbf{R}_f\bar{\mathbf{U}}^H(\bar{\Phi}(\omega) - \mathbf{I})^H(\bar{\Phi}(\omega) - \mathbf{I})\}d\omega \\
&+\frac{\rho}{2\pi}\int_{-\pi}^{\pi} Tr\{\bar{\Phi}(\omega)\bar{\Phi}(\omega)^H\}d\omega. \qquad (5C.3)
\end{aligned}
$$

For an estimator matching the channel's correlation in frequency domain and mismatching in time-domain, that is,

$$\bar{r}_f[k] = r_f[k], \text{ and } \bar{r}_t[m] = r_t[m].$$

Then

$$\bar{\mathbf{U}}\mathbf{R}_f\bar{\mathbf{U}}^H = \bar{\mathbf{D}} = \text{diag}\{\bar{d}_1, \cdots, \bar{d}_N\}$$

and

$$
\begin{aligned}
&\text{MSE}(\{\bar{\mathbf{C}}_m\}) \\
&= \sum_{k=1}^{N} \frac{\bar{d}_k}{2\pi} \int_{-\pi}^{\pi} p_t(\omega) \mid \bar{\phi}_k(\omega) - 1 \mid^2 d\omega + \sum_{l=1}^{N} \frac{\rho}{2\pi} \int_{-\pi}^{\pi} \mid \bar{\phi}_k(\omega) \mid^2 d\omega \\
&= \sum_{k=1}^{N} \frac{\bar{d}_k}{2\pi} \int_{-\pi}^{\pi} [p_t(\omega) - \bar{p}_t(\omega)] \mid \bar{\phi}_k(\omega) - 1 \mid^2 d\omega \\
&\quad + \frac{\rho}{2\pi} \sum_{k=1}^{N} \frac{\bar{d}_k}{2\pi} \int_{-\pi}^{\pi} \bar{p}_t(\omega) \mid \bar{\phi}_k(\omega) - 1 \mid^2 d\omega + \sum_{k=1}^{N} \frac{\rho}{2\pi} \int_{-\pi}^{\pi} \mid \bar{\phi}_k(\omega) \mid^2 d\omega \\
&= \sum_{k=1}^{K} \frac{\bar{d}_k}{2\pi} \int_{-\pi}^{\pi} [p_t(\omega) - \bar{p}_t(\omega)] \mid \bar{\phi}_k(\omega) - 1 \mid^2 d\omega + \overline{\text{MMSE}}. \qquad (5C.4)
\end{aligned}
$$

The first term in the above expression represents the MSE variation due to the mismatch. By means of (5A.10) and (5A.7), $\text{MSE}(\{\bar{\mathbf{C}}_m\})$ can be further simplified into

$$
\begin{aligned}
&\text{MSE}(\{\bar{\mathbf{C}}_m\}) \\
&= \sum_{k=1}^{N} \frac{\rho}{2\pi \bar{\gamma}_k^2[0]} \int_{-\pi}^{\pi} \frac{\bar{d}_k [p_t(\omega) - \bar{p}_t(\omega)]}{\bar{d}_k \bar{p}_t(\omega) + \rho} d\omega + \overline{\text{MMSE}} \\
&= \rho \sum_{k=1}^{N} \frac{1}{\bar{\gamma}_k^2[0]} \left(\frac{1}{2\pi} \int_{-\pi}^{\pi} \frac{\bar{d}_k p_t(\omega) + \rho}{\bar{d}_k \bar{p}_t(\omega) + \rho} d\omega - 1 \right) + \overline{\text{MMSE}}. \qquad (5C.5)
\end{aligned}
$$

If $\bar{p}_t(\omega) = p_B(\omega)$, that is, the time-domain correlation of the designed estimator is ideal ω_d-bandlimited, then, for any channels with $p_t(\omega)$ zero outside $[-\omega_d, \omega_d]$, and

$$
\frac{1}{2\pi} \int_{-\omega_d}^{\omega_d} p_t(\omega) d\omega = 1,
$$

we have

$$
\frac{1}{2\pi} \int_{-\pi}^{\pi} \frac{\bar{d}_k p_t(\omega) + \rho}{\bar{d}_k \bar{p}_t(\omega) + \rho} d\omega = 1.
$$

Therefore,

$$
\text{MSE}(\{\bar{\mathbf{C}}_m\}) = \overline{\text{MMSE}}_B(\omega_d).
$$

Hence, if an OFDM channel estimator is designed using $p_B(\omega)$ as the time-domain correlation, then the time-domain correlation mismatch of the estimator will not degrade its performance. This suggests that a robust channel estimator should use $p_B(\omega)$ as the time-domain correlation.

To analyze the frequency domain correlation mismatch, we assume that the time-domain correlation of the designed estimator is the same as that of the channel, that is, $\bar{p}_t(\omega) = p_t(\omega)$, and the frequency correlation matrix of the designed estimator has the same eigenvectors as that of the channel. That is, $\mathbf{R}_f$ can be eigen-decomposed into

$$\mathbf{R}_f = \bar{\mathbf{U}}^H \mathbf{D} \bar{\mathbf{U}}, \text{ or } \bar{\mathbf{U}} \mathbf{R}_f \bar{\mathbf{U}}^H = \mathbf{D},$$

where $\mathbf{D} = \text{diag}\{d_1, \cdots, d_N\}$ and $\sum_k d_k = K$. d_k and $\bar{d}_k$ for $k = 1, \cdots, N$ are generally different.

Applying the above two assumptions to (5C.3), we get

$$\begin{aligned}
&\text{MSE}(\{\bar{\mathbf{C}}_m\}) \\
= &\frac{1}{2\pi}\int_{-\pi}^{\pi} \bar{p}_t(\omega) Tr\{\mathbf{D}(\bar{\Phi}(\omega)-\mathbf{I})(\bar{\Phi}(\omega)-\mathbf{I})^H\} d\omega \\
&+\frac{\rho}{2\pi}\int_{-\pi}^{\pi} Tr\{\bar{\Phi}(\omega)\bar{\Phi}(\omega)^H\} d\omega \\
= &\sum_{k=1}^{N} \frac{1}{2\pi}\int_{-\pi}^{\pi} d_k \bar{p}_t(\omega) \mid \bar{\phi}_k(\omega) - 1 \mid^2 d\omega + \sum_{k=1}^{N}\frac{\rho}{2\pi}\int_{-\pi}^{\pi} \mid \bar{\phi}_k(\omega) \mid^2 d\omega \\
= &\sum_{k=1}^{N} (d_k - \bar{d}_k) \frac{1}{2\pi}\int_{-\pi}^{\pi} \bar{p}_t(\omega) \mid \bar{\phi}_k(\omega) - 1 \mid^2 d\omega + \overline{\text{MMSE}} \\
= &\sum_{k=1}^{N} (d_k - \bar{d}_k)\Gamma(\bar{d}_k) + \overline{\text{MMSE}},
\end{aligned} \tag{5C.6}$$

where

$$\Gamma(\bar{d}_k) = \frac{1}{2\pi\bar{\gamma}_k^2[0]}\int_{-\pi}^{\pi} \frac{\bar{p}_t(\omega)}{\frac{\bar{d}_k}{\rho}\bar{p}_t(\omega)+1} d\omega.$$

In the above derivation, we have used (5A.8) and (5A.7).

Appendix 5D: Derivation of MMSE Transform based Approach

For convenience of derivation, 2-D indexes, (n,k) and (ω_1, ω_2), are denoted as $\mathbf{n} = (n,k)^T$ and $\boldsymbol{\omega} = (\omega_1, \omega_2)$, respectively. Therefore, the channel parameter, $H_{n,k}$, coefficients for channel estimator, $c_{n,k}$, channel correlation,

$r_H[n,k]$ and its transform,$R_H(\omega_1,\omega_2)$, can be written as, $H_{\mathbf{n}}$, $c_{\mathbf{n}}$, $r_H[\mathbf{n}]$, and $R_H(\boldsymbol{\omega})$, respectively.

Let the pilot grid is determined by $\mathbf{V}$. Then according to Section 5.3.1, any index, $\mathbf{n}$ can be uniquely decomposed into $\mathbf{n} = \mathbf{r} + \mathbf{Vm}$, where $\mathbf{r} \in \mathcal{I}$.

Using the orthogonality principle [160, page 177], the coefficients for the MMSE estimator are obtained as the solution of

$$E\left\{\left(\sum_{\mathbf{l}\in\mathcal{Z}^2} c_{\mathbf{r}+\mathbf{V}(\mathbf{m}-\mathbf{l})}\tilde{H}_{\mathbf{Vl}} - H_{\mathbf{r}+\mathbf{Vm}}\right)\tilde{H}^*_{\mathbf{Vl}_1}\right\} = 0,$$

for all $\mathbf{l}_1 \in \mathcal{Z}^2$, or

$$r_H[\mathbf{r}+\mathbf{Vm}] = \sum_{\mathbf{l}\in\mathcal{Z}^2} c_{\mathbf{r}+\mathbf{V}(\mathbf{m}-\mathbf{l})}(r_H[\mathbf{Vm}_o] + \rho\delta[\mathbf{m}_o]), \tag{5D.1}$$

as defined before, where $r_H[\mathbf{n}] = r_H[n,k]$ is the correlation of channel parameters at different frequencies and times. Let

$$R_H(\mathbf{r},\boldsymbol{\omega}) \triangleq \sum_{\mathbf{m}\in\mathcal{Z}^2} r_H[\mathbf{r}+\mathbf{Vm}]e^{-j\boldsymbol{\omega}^T\mathbf{Vm}},$$

and

$$C(\mathbf{r},\boldsymbol{\omega}) \triangleq \sum_{\mathbf{m}\in\mathcal{Z}^2} c_{\mathbf{r}+\mathbf{Vm}}e^{-j\boldsymbol{\omega}^T\mathbf{Vm}}.$$

Then, (5D.1) can be expressed in the transform domain as

$$R_H(\mathbf{r},\boldsymbol{\omega}) = C(\mathbf{r},\boldsymbol{\omega})\left[R_H(\mathbf{0},\boldsymbol{\omega}) + \rho\right].$$

Therefore,

$$C(\mathbf{r},\boldsymbol{\omega}) = \frac{R_H(\mathbf{r},\boldsymbol{\omega})}{R_H(\mathbf{0},\boldsymbol{\omega}) + \rho}. \tag{5D.2}$$

which expresses the coefficients for the MMSE filtering in terms of $R_H(\mathbf{r},\boldsymbol{\omega})$.

Denote $C(\boldsymbol{\omega})$, the 2-D Fourier transform of $c_{\mathbf{n}}$, that is,

$$C(\boldsymbol{\omega}) = \sum_{\mathbf{n}\in\mathcal{Z}^2} c_{\mathbf{n}}e^{-j\boldsymbol{\omega}^T\mathbf{n}}.$$

then it can be also expressed in terms of $C(\mathbf{r},\boldsymbol{\omega})$ as

$$C(\boldsymbol{\omega}) = \sum_{\mathbf{r}\in\mathcal{I}} C(\mathbf{r},\boldsymbol{\omega})e^{-j\boldsymbol{\omega}^T\mathbf{r}}. \tag{5D.3}$$

Next we are trying to express $C(\boldsymbol{\omega})$ in terms of $R_H(\boldsymbol{\omega})$, the 2-D Fourier transform of $r_H[\mathbf{n}]$, defined as

$$R_H(\boldsymbol{\omega}) = \sum_{\mathbf{n}\in\mathcal{Z}^2} r_H[\mathbf{n}]e^{-\jmath\boldsymbol{\omega}^T\mathbf{n}} = \sum_{n,k=-\infty}^{\infty} r_H[n,k]e^{-\jmath(\omega_1 n+\omega_2 k)}.$$

From the definition of $R_H(\mathbf{r},\boldsymbol{\omega})$,

$$\begin{aligned}
&R_H(\mathbf{r},\boldsymbol{\omega})\\
&= \sum_{\mathbf{m}\in\mathcal{Z}^2} r_H[\mathbf{r}+\mathbf{V}\mathbf{m}]e^{-\jmath\boldsymbol{\omega}^T\mathbf{V}\mathbf{m}}\\
&= \sum_{\mathbf{m}\in\mathcal{Z}^2}\left\{\frac{1}{(2\pi)^2}\int_{-\pi}^{\pi}\int_{-\pi}^{\pi} R_H(\boldsymbol{\omega}_o)e^{\jmath\boldsymbol{\omega}_o^T(\mathbf{r}+\mathbf{V}\mathbf{m})}d\omega_1^o d\omega_2^o\right\}e^{-\jmath\boldsymbol{\omega}^T\mathbf{V}\mathbf{m}}\\
&= \frac{1}{(2\pi)^2}\int_{-\pi}^{\pi}\int_{-\pi}^{\pi} R_H(\boldsymbol{\omega}_o)\left\{\sum_{\mathbf{m}\in\mathcal{Z}^2} e^{-\jmath(\boldsymbol{\omega}-\boldsymbol{\omega}_o)^T\mathbf{V}\mathbf{m}}\right\}e^{\jmath\boldsymbol{\omega}_o^T\mathbf{r}}d\omega_1^o d\omega_2^o,
\end{aligned}$$

with $\boldsymbol{\omega}_o = (\omega_1^o, \omega_2^o)^T$. From [167, page 51],

$$\sum_{\mathbf{m}\in\mathcal{Z}^2} e^{-\jmath(\boldsymbol{\omega}-\boldsymbol{\omega}_o)^T\mathbf{V}\mathbf{m}} = (2\pi)^2\sum_{\mathbf{m}\in\mathcal{Z}^2}\delta(\mathbf{V}^T\boldsymbol{\omega}-\mathbf{V}^T\boldsymbol{\omega}_o-2\pi\mathbf{m}). \tag{5D.4}$$

Hence,

$$\begin{aligned}
&R_H(\mathbf{r},\boldsymbol{\omega})\\
&= \frac{1}{(2\pi)^2}\int_{-\pi}^{\pi}\int_{-\pi}^{\pi} R_H(\boldsymbol{\omega}_o)(2\pi)^2\sum_{\mathbf{p}\in\mathcal{Z}^2}\delta(\mathbf{V}^T\boldsymbol{\omega}-\mathbf{V}^T\boldsymbol{\omega}_o-2\pi\mathbf{p})e^{\jmath\boldsymbol{\omega}^T\mathbf{r}}d\omega_1^o d\omega_2^o\\
&= \frac{1}{|\det(\mathbf{V})|}\sum_{\boldsymbol{\omega}-2\pi\mathbf{U}\mathbf{m}\in[-\pi,\pi]^2} R_H(\boldsymbol{\omega}-2\pi\mathbf{U}\mathbf{m})e^{\jmath(\boldsymbol{\omega}-2\pi\mathbf{U}\mathbf{m})^T\mathbf{r}},
\end{aligned} \tag{5D.5}$$

with $\mathbf{U} = \mathbf{V}^{-1}$. From (5D.2), (5D.3) and (5D.5), we have that

$$\begin{aligned}
&C(\boldsymbol{\omega})\\
&= \sum_{\mathbf{r}\in\mathcal{I}}\frac{R_H(\mathbf{r},\boldsymbol{\omega})}{R_H(\mathbf{0},\boldsymbol{\omega})+\rho}e^{-\jmath\boldsymbol{\omega}^T\mathbf{r}}\\
&= \sum_{\mathbf{r}\in\mathcal{I}}\frac{1}{|\det(\mathbf{V})|}\sum_{\boldsymbol{\omega}-2\pi\mathbf{U}\mathbf{m}\in[-\pi,\pi]^2}\frac{R_H(\boldsymbol{\omega}-2\pi\mathbf{U}\mathbf{m})e^{\jmath(\boldsymbol{\omega}-2\pi(\mathbf{U}\mathbf{m})^T\mathbf{r})}}{R_H(\mathbf{0},\boldsymbol{\omega})+\rho}e^{-\jmath\boldsymbol{\omega}^T\mathbf{r}}\\
&= \frac{1}{|\det(\mathbf{V})|}\sum_{\boldsymbol{\omega}-2\pi\mathbf{U}\mathbf{m}\in[-\pi,\pi]^2}\frac{R_H(\boldsymbol{\omega}-2\pi\mathbf{U}\mathbf{m})}{R_H(\mathbf{0},\boldsymbol{\omega})+\rho}\sum_{\mathbf{r}\in\mathcal{I}} e^{-\jmath 2\pi(\mathbf{U}\mathbf{m})^T\mathbf{r}}.
\end{aligned}$$

It can be proved from the property of $\mathcal{I}$ [165] that

$$\sum_{\mathbf{r}\in\mathcal{I}} e^{-j2\pi(\mathbf{Um})^T\mathbf{r}} = \begin{cases} |\det(\mathbf{V})|, & \text{if } \mathbf{m}=(0,0)^T, \\ 0, & \text{otherwise.} \end{cases}$$

Consequently,

$$C(\boldsymbol{\omega}) = \frac{R_H(\boldsymbol{\omega})}{R_H(\mathbf{0},\boldsymbol{\omega})+\rho}.$$

If the pilot symbols are dense enough to make $R_H(\mathbf{r},\boldsymbol{\omega})$ alias-free, then $R_H(\mathbf{0},\boldsymbol{\omega}) = \frac{R_H(\boldsymbol{\omega})}{|\det(\mathbf{V})|}$ within the non-zero support of $R_H(\boldsymbol{\omega})$ and

$$C(\boldsymbol{\omega}) = \frac{R_H(\boldsymbol{\omega})}{\frac{R_H(\boldsymbol{\omega})}{|\det(\mathbf{V})|}+\rho}. \tag{5D.6}$$

From (5D.6), the coefficients for the MMSE estimator can be determined once the channel statistics are known.

The MSE for the estimator can be found by

$$\begin{aligned}
&\overline{\text{MMSE}} \\
&= E\left|H_{\mathbf{r}+\mathbf{Vm}} - \sum_{\mathbf{l}\in\mathcal{Z}^2} c_{\mathbf{r}+\mathbf{V}(\mathbf{m}-\mathbf{l})}\tilde{H}[\mathbf{Vm}_o]\right|^2 \\
&= E\left\{\left(H_{\mathbf{r}+\mathbf{Vm}} - \sum_{\mathbf{l}\in\mathcal{Z}^2} c_{\mathbf{r}+\mathbf{V}(\mathbf{m}-\mathbf{l})}\tilde{H}_{\mathbf{Vl}}\right)H^*_{\mathbf{r}+\mathbf{Vm}}\right\} \\
&= r_H[\mathbf{0}] - \sum_{\mathbf{l}\in\mathcal{Z}^2} c_{\mathbf{r}+\mathbf{Vl}}r^*_H[\mathbf{r}+\mathbf{Vl}] \\
&= 1 - \frac{1}{(2\pi)^2}\int_{-\pi}^{\pi}\int_{-\pi}^{\pi}\frac{|R_H(\mathbf{r},\boldsymbol{\omega})|^2}{R_H(\mathbf{0},\boldsymbol{\omega})+\rho}d\omega_1 d\omega_2.
\end{aligned}$$

If the pilot symbols are dense enough to satisfy the alias-free condition, then $|R(\mathbf{r},\boldsymbol{\omega})|^2 = |R(\mathbf{0},\boldsymbol{\omega})|^2$ for all $\mathbf{r}\in\mathcal{Z}^2$. Furthermore, there are $|\det(\mathbf{V})|$ replicas of $1/|\det(\mathbf{V})|R(\boldsymbol{\omega})$ within $[-\pi,\ \pi]\times[-\pi,\ \pi]$. Let the nonzero support of $R(\boldsymbol{\omega})$ be D, then

$$
\begin{aligned}
& \overline{\mathrm{MMSE}} \\
= \; & 1 - \frac{|\det(\mathbf{V})|}{(2\pi)^2} \iint_D \frac{|R_H(\mathbf{0},\boldsymbol{\omega})|^2}{R_H(\mathbf{0},\boldsymbol{\omega}) + \rho} d\omega_1 d\omega_2 \\
= \; & \frac{1}{(2\pi)^2} \iint_D R_H(\boldsymbol{\omega}) d\omega_1 d\omega_2 - \frac{|\det(\mathbf{V})|}{(2\pi)^2} \int\int_D \frac{\left|\frac{R_H(\boldsymbol{\omega})}{\det(\mathbf{V})}\right|^2}{\frac{R_H(\boldsymbol{\omega})}{\det(\mathbf{V})} + \rho} d\omega_1 d\omega_2 \\
= \; & \frac{\rho}{(2\pi)^2} \iint_D \frac{R_H(\boldsymbol{\omega})}{\frac{R_H(\boldsymbol{\omega})}{\det(\mathbf{V})} + \rho} d\omega_1 d\omega_2. \qquad (5D.7)
\end{aligned}
$$

Chapter 6

PEAK POWER REDUCTION TECHNIQUES

Chintha Tellambura and Mathias Friese

6.1 Introduction

This chapter discusses methods to reduce large envelope variations of OFDM signals. The envelope variations are one of the most frequently cited drawbacks of OFDM, because in practice any transmission system reveals some nonlinear characteristics and, moreover, is peak-power limited. Nonlinearities cause spectral widening of the transmit signal resulting in unwanted out-of-band (OOB) noise, a major concern especially in radio applications where possibly large differences in signal strength from mobile transmitters impose stringent requirements on adjacent channel interference (ACI) suppression. The transmit signal itself is degraded by nonlinearities, which results in increased bit error rates in the receiver.

In the literature, the envelope variations are often described in terms of the crest-factor (CF), peak-to-average power ratio (PAPR), peak-to-mean envelope power ratio (PMEPR) or simply peak-to-average power ratio (PAPR). Not all of these terms are used consistently, rendering the comparison of absolute numbers sometimes difficult. We adopt the terms PAPR [221], [222] and CF [223] and their precise mathematical definitions will be given later.

High peak values in OFDM result from the superposition of a large number of usually statistically independent subchannels that can constructively sum up to high peaks. Because of this well-known mechanism the appearance of high peak values had been considered unavoidable for some time. However, the recent interest in the application of OFDM to wireless networks has resulted in the development of methods to combat these problems.

There are several different classes of approaches. For example, some researchers accept a high PAPR for the transmit signal and propose new amplifier concepts [224] for increasing the transmitter power efficiency. The optimization of existing amplifiers is under consideration [225]. Others attempt to find intelligent methods for modulation, in order to generate transmit signals with reduced PAPR. Consequently, existing amplifier technology can be used.

This chapter discusses techniques for PAPR reduction as well as a fundamental theory on the PAPR problem. We consider the basic concepts in order to give an understanding that helps to rank all the different approaches with their different properties. More details on specific issues can be found in the references. Finally, comparing theory with the state of the art shows that there is still research required in this new field in order to approach the theoretical limits with low complexity technology.

6.2 PAPR-Properties of OFDM Signals

Since our focus is on radio applications, throughout this chapter we only consider the complex envelope of the passband signal. Nevertheless, many important PAPR reduction techniques were initially intended for baseband applications, such as digital subscriber lines [222], [226]. Most of these techniques can be translated easily for bandpass applications.

For all PAPR-related issues of OFDM signals, we consider a single modulation interval (OFDM symbol) only. Since the guard interval or cyclic prefix of duration T_g repeats a preceding part of the signal, it does not affect PAPR and we set $T_g = 0$. Therefore, the transmitted signal can be written as

$$s(t) = \sum_{n=0}^{N-1} s_n e^{j2\pi n \Delta f t}, 0 \leq t < T_s \tag{6.2.1}$$

where $j = \sqrt{-1}$ and Δf is the frequency separation between any two adjacent subchannels. The OFDM-symbol duration T_s is equal to $[\Delta f]^{-1}$ to ensure the orthogonality among the subchannels.

The OFDM-coefficients s_n are typically taken from a fixed modulation constellation (alphabet) of size $M = 2^m$. We use the term data symbol to refer to each s_n and other synonymous terms are information symbol, modulation symbol and so on. For notational convenience, we introduce the vector $s = (s_0, s_1, \cdots, s_{N-1})$[1]. We assume that $E\{s_n\} = 0$ (which is usually

[1]Note however that not every s_n carries data. Some s_n are permanently set to zero (known as virtual carriers) and some are permanently set to fixed values which are perfectly known to the receiver (pilot symbols).

the case). In the case of uncoded transmission, m bits of information are transported per data symbol. Each data symbol s_n is chosen from the set $\mathcal{Q} = \{\lambda_1, \lambda_2, \ldots, \lambda_M\}$ of M distinct elements. The set $\mathcal{Q}$ is called the signal constellation or signalling alphabet and M (the number of elements of $\mathcal{Q}$) is called the order of the constellation. Binary phase shift keying (BPSK), quadrature phase shift keying (QPSK), M-ary phase shift keying (MPSK) and M-ary quadrature amplitude modulation (MQAM) constellations can be described as follows:

$$\begin{aligned}
\mathcal{Q}_{\text{BPSK}} &= \{\exp(j\pi x), x = 0,1\} = \{(-1)^x,\ x = 0,1\}, \\
\mathcal{Q}_{\text{QPSK}} &= \left\{\exp\left[j\left(\frac{\pi x}{2}\right)\right], x = 0,1,2,3\right\} = \{j^x,\ x = 0,1,2,3\}, \\
\mathcal{Q}_{\text{MPSK}} &= \left\{\exp\left[j\left(\frac{2\pi}{M}x\right)\right], x = 0,1,\ldots,M-1\right\}, \\
\mathcal{Q}_{\text{MQAM}} &= \{[\pm(2x_1+1) \pm j(2x_2+1)], x_1 = x_2 = 0,1,\ldots,\log_2(M)-2\}.
\end{aligned} \tag{6.2.2}$$

Note that for a PSK constellation $\mathcal{Q}$, for any $u \in \mathcal{Q}$, $|u|^2 = 1$. This constant-modulus condition does not hold for QAM constellations. For an MQAM constellation, a normalization factor may be used to make $E[|u|^2] = 1$, where $E[.]$ is the expectation. This normalization factor is $1/\sqrt{10}$ and $1/\sqrt{42}$ for 16QAM and 64QAM respectively. In general, all elements of $\mathcal{Q}$ occur with equal probability $1/M$.

Higher-order signal constellations can be decomposed into lower-order signal constellations, leading to the construction of low PAPR sequences. For example, [227] shows that the 16QAM constellation can be decomposed as

$$\mathcal{Q}_{16\text{QAM}} = \sqrt{2}\exp\left(\frac{\pi j}{4}\right) z_1 + \frac{\sqrt{2}}{2}\exp\left(\frac{\pi j}{4}\right) z_2 \tag{6.2.3}$$

where $z_1, z_2 \in \mathcal{Q}_{\text{QPSK}}$. Using (6.2.2), this can be further simplified as

$$\mathcal{Q}_{16\text{QAM}} = \left\{\sqrt{2}\exp\left(\frac{\pi j}{4}\right) j^{x_i} + \frac{\sqrt{2}}{2}\exp\left(\frac{\pi j}{4}\right) j^{y_k}\right\} \tag{6.2.4}$$

for some $x_i, y_k \in \{0,1,2,3\}$. Similarly, the QPSK constellation can be written as a summation of two BPSK constellations [228]

$$\mathcal{Q}_{\text{QPSK}} = \frac{\sqrt{2}}{2} z_1 + j\frac{\sqrt{2}}{2} z_2 \tag{6.2.5}$$

where $z_1, z_2 \in \mathcal{Q}_{\mathrm{BPSK}}$. Thus QPSK constellation points can be written as

$$\frac{\sqrt{2}}{2}(-1)^{x_i} + j\frac{\sqrt{2}}{2}(-1)^{y_k} \tag{6.2.6}$$

for some $x_i, y_k \in \{0, 1\}$. These representations will be used later in this chapter.

We define magnitude $m(t)$ and phase $\phi(t)$ of $s(t)$ according to $s(t) = m(t)\exp(j\phi(t))$ where $m(t) \geq 0$ and real. The time average of the power of an individual OFDM symbol is given by $\bar{P}$, with

$$\bar{P} = \frac{1}{T_s}\int_0^{T_s} |s(t)|^2\, dt = \sum_{n=0}^{N-1} |s_n|^2, \tag{6.2.7}$$

where the right hand side of (6.2.7) can be obtained by direct calculation (and is a consequence of the well-known Parseval formula). The time average clearly depends on the magnitudes of the data symbols s_n. Therefore, if the s_n are modelled as random variables with varying magnitude, $\bar{P}$ is also a random variable. The time average of the total OFDM signal, consisting of a large number of OFDM symbols, is equal to the ensemble average of $\bar{P}$. Therefore, the transmit power P_{av} of the total OFDM signal is given by the statistical expectation

$$P_{av} = E\{\bar{P}\}. \tag{6.2.8}$$

The root-mean-square value (RMS) of the complex envelope is defined by $RMS = \sqrt{P_{av}}$.

In the following, as a measure of the height of the peak values of an individual OFDM symbol, the crest factor C will be considered, where

$$C = \frac{\max\limits_{t\in[0,T_s)} \{m(t)\}}{\sqrt{P_{av}}}, \text{with } m(t) = |s(t)|\,. \tag{6.2.9}$$

Note that with this definition C can be smaller than 1, because the peak power of an individual symbol may be smaller than the ensemble average power. In the frequently considered special case of fixed average power P_{av} (this is in particular true for PSK-modulation of the subchannels), however, it is lower bounded by 1.

The worst case CF depends also on the modulation constellation. Let the CF of the constellation be defined by $C_{alphabet} = \max|s_n|/\sqrt{E\{|s_n|^2\}}$. In the worst case, all coefficients have maximal magnitude and their corresponding time-domain components sum up with the same phase at a certain time instant (classical example: all s_n are real and positive, $t = 0$). In this case, $C_{max} = \sqrt{N}.C_{alphabet}$. For example, in the case of PSK modulation, $C_{alphabet} = 1$, with $C_{max} = \sqrt{N}$.

In many publications, instead of the time-continuous waveform $s(t)$, its N samples $s(kT/N), k = 0, \cdots, N-1$ are considered for CF assessment. As shown in Chapter 1, the N samples can be generated by IFFT of the sequence s. Therefore, another possible definition of the CF is

$$C_{sample} = \frac{\max\limits_{n} \{m_k\}}{\sqrt{P_{av}}} \text{ with } m_k = |s(kT/N)|. \tag{6.2.10}$$

The advantage of this model is mainly that some analytical calculations are possible that give theoretical insight. For practical purposes, however, it can be inaccurate to consider the time samples only, because the real signal is always time-continuous. Possible overshoot in between the samples is not considered in (6.2.10). Therefore, $C \geq C_{sample}$ and CF estimations based on (6.2.10) can be inaccurate. Finally, note that C as well as C_{sample} relate to the complex envelope signal. The radio frequency signal $Re\{s(t)\exp(j2\pi f_c t)\}$ with large carrier frequency f_c will have a CF that is about $\sqrt{2}$ times larger.

For random input data, the N samples $s(kT/N), k = 0, \cdots, N-1$ can be considered as zero-mean unit-variance complex Gaussian random variables. Thus $m_k = |s(kT/N)|$ is Rayleigh distributed. The CF is determined by the largest magnitude of all the samples. Therefore, the probability that the CF is less than a desired value $C_d = m_d/\sqrt{P_{av}}$ can be expressed as

$$\Pr(C_{sample} \leq C_d) = \prod_{k=0}^{N-1} \Pr(m_k \leq m_d) = (1 - e^{-C_d^2})^N. \tag{6.2.11}$$

Note that this is only an approximation.

The PAPR or the peak-to-mean envelope power ratio (PMEPR) is

$$\xi = \frac{\max\limits_{t\in[0,T_s)} |s(t)|^2}{P_{av}} = \frac{\max\limits_{t\in[0,T_s)} |s(t)|^2}{E\{|s(t)|^2\}} \tag{6.2.12}$$

where $\max |s(t)|^2$ denotes the maximum instantaneous power of the OFDM signal. Throughout this chapter the term **PAPR** will refer to the baseband

PAPR. The definition in (6.2.12) is also termed **PAPR** [229], [230] and peak-to-average power (**PAPR**) ratio as well [231]. PAPR is the most frequently used term. The square root of the PAPR is the crest factor discussed above. Strictly speaking, PMEPR is a more precise term for ξ, because $|s(t)|$ is the envelope but not the transmitted signal itself. Nevertheless, we refer to ξ as PAPR for brevity.

For some problems, a polynomial viewpoint can also be helpful; define an associated polynomial $P_s(z)$ of degree $N-1$ as

$$P_s(z) = \sum_{n=0}^{N-1} s_n z^n \tag{6.2.13}$$

in which the coefficients are the data symbols s_n. Comparing (6.2.1) and (6.2.12), the PAPR can also be defined as

$$\xi = \max_{|z|=1} \frac{|P_s(z)|^2}{P_{av}}. \tag{6.2.14}$$

That is, PAPR relates to the peak amplitude of the associated polynomial on the unit circle.

A statistical definition of PAPR is also useful, noting that PAPR is randomly distributed for random input data symbols. An OFDM symbol is said to have a peak at ξ_p at probability P_c if

$$\Pr[\xi > \xi_p] = P_c. \tag{6.2.15}$$

That is, the probability that PAPR exceeds ξ_p is P_c. The PAPR is less than ξ_p for $100(1-P_c)\%$ of the OFDM symbols. We refer to ξ_p as the $100(1-P_c)$-th percentile PAPR.

6.2.1 Maximum PAPR of an N Subcarrier OFDM Signal

The PAPR of an MPSK OFDM signal is always less than or equal to N, where N is the number of subchannels. To see this, consider the OFDM signal given in (6.2.1) where the input data symbols are chosen from an

MPSK constellation such that $|s_n| = 1$. The PAPR is bounded as

$$
\begin{aligned}
\xi &= \max_{t\in[0,T_S)} |s(t)|^2/N \\
&= \max_{t\in[0,T_s)} \left| \frac{1}{\sqrt{N}} \sum_{n=0}^{N-1} s_n e^{j2\pi\Delta ft} \right|^2 \\
&\leq \frac{1}{N} \left[\sum_{n=0}^{N-1} \left| s_n e^{j2\pi\Delta ft} \right| \right]^2 \\
&\leq N.
\end{aligned}
\tag{6.2.16}
$$

This is the maximum of the PAPR of a given N subchannel MPSK OFDM system. There are only M^2 sequences having maximum PAPR of N[232].

The PAPR is a function of the input data frame and recall that there are M^N distinct data frames for a N subchannel OFDM signal with M-ary modulation. For any input data frame, we have

$$1 < \xi \leq N. \tag{6.2.17}$$

For example, for $N = 256$ the PAPR can be as high as 24 dB $(10\log_{10}(256))$. Fortunately, very high PAPR values are very rare. For example, with BPSK, only four sequences 0000..., 1111..., 0101... and 1010... achieve $\xi = N$. For randomly distributed data, the probability of an occurrence of this is $4/2^N = 2^{2-N}$. This probability is negligible when N is large, as is the case in practice.

6.2.2 Estimating True PAPR from Discrete Time Signals

The rate $\frac{LN}{T_s}$ sampled version of the OFDM signal given in (6.2.1) is

$$x[k/L] = x\left(k\frac{T_s}{LN}\right) = \sum_{n=0}^{N-1} s_n e^{j2\pi kn/N}, \quad k = 0, 1, \ldots, LN-1. \tag{6.2.18}$$

When $L = 1$, these samples $x[k/L]$ are called Nyquist-rate samples. The continuous time signal is generated from these samples by low pass filtering at the transmitter. For this generation, N samples are sufficient. However, the Nyquist-rate samples do not necessarily coincide with the peaks of the continuous time signal. Therefore, it is essential to oversample the OFDM signal to estimate the true PAPR.

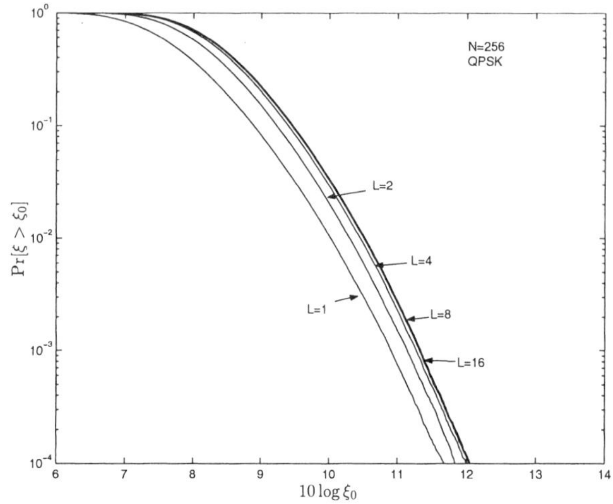

Figure 6.1. CCDF of PAPR with different oversampling ratios.

If the OFDM symbol is divided into LN samples (6.2.18), where $L > 1$ is an integer referred to as the oversampling factor, these samples can be computed by a zero-padded IDFT. The PAPR using such samples is given by

$$\xi_L = \max_{k\in[0,NL)} \frac{|x[k/L]|^2}{E\{|x[k/L]|^2\}}. \tag{6.2.19}$$

The PAPR estimate for the special case $L = 1$, i.e., the PAPR estimated using Nyquist-rate samples, is defined as

$$\xi_{Nyq} = \max_{k\in[0,NL)} \frac{|x[k/1]|^2}{E\{|x[k/1]|^2\}}. \tag{6.2.20}$$

We clearly have

$$\xi \geq \xi_L > \xi_{Nyq} \tag{6.2.21}$$

where ξ_{Nyq} is the PAPR estimated using Nyquist-rate samples. Theoretically, ξ_L approaches ξ as L becomes sufficiently large. Figure 6.1 shows the PAPR CCDF for several values of the oversampling factor. Note that the curves do not change much when L increases above $L = 4$, suggesting that an oversampling ratio of 4 is sufficient for accurate results.

PAPR distributions can accurately be derived using the distributions and bounds for the discrete time samples with $L > 1$. According to Theorem 2 in [233], if $h(z)$ is a complex polynomial of degree $N-1$, then the maximum of $|h\left(e^{j\theta}\right)|$ is bounded by the maximum of its samples on the unit circle as

$$\max_{\theta}\left|\sum_{n=0}^{N-1} s_n e^{jn\theta}\right| \leq \frac{k}{k-\pi} \max_{0\leq k\leq LN}\left|\sum_{n=0}^{N-1} s_n e^{j\frac{2\pi nk}{LN}}\right| \tag{6.2.22}$$

where L is the oversampling factor and $L > \pi$. An upper bound for the CCDF of PAPR is derived using the above relationship

$$\Pr(\xi > \xi_0) \leq L_{opt} N e^{-\xi_0(1-\pi/L_{opt})^2} \tag{6.2.23}$$

for $\forall\xi > 0$, L_{opt} is the optimum oversampling factor given by

$$L_{opt} = \pi\xi_0\left(1+\sqrt{1-\frac{1}{\xi_0}}\right). \tag{6.2.24}$$

Equation (6.2.24) is obtained considering the fact that $L_{opt} > \pi$ [233].

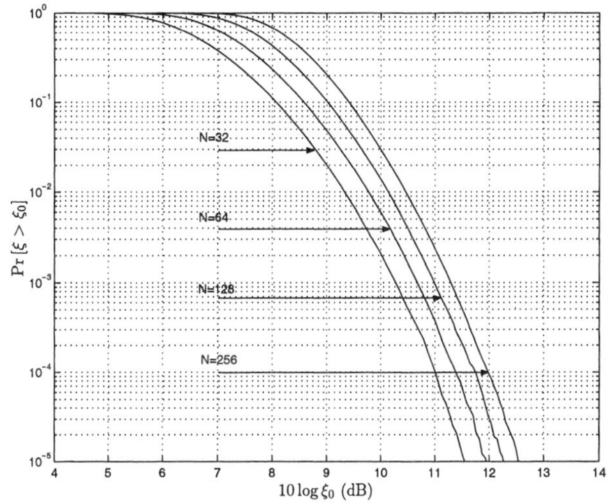

Figure 6.2. The CCDF for N QPSK subchannels.

Although high peaks are possible in principle, they occur rarely: even for 256 subchannels the PAPR rarely exceeds 12.5 dB (Figure 6.2), whereas the theoretical maximum PAPR is about 24 dB. Therefore, for practical purposes it is appropriate to consider as the maximum PAPR a value that is not exceeded with some predefined probability, e.g., 10^{-5}, denoted by

$100(1 - 10^{-5})$th percentile PAPR. This 99.999th percentile PAPR will be much below the theoretically possible maximum value and, from Figure 6.2, for up to 256 subchannels this percentile PAPR is always below 14 dB. The rarity of the large peaks gives rise to the hope that with some intelligent signal processing the highest peaks might be avoided completely.

Finally, Figure 6.2 shows that not only very high, but also very low crest factors are rare. For example, in 256-carrier-OFDM the CF will be almost always greater than 7 dB, since below that the ccdf approaches the value 1. This is no surprise, but we will return to this important observation later when discussing PAPR reduction schemes.

The design of multicarrier signals with low PAPR is not a completely new field of research. Wideband signals, i.e., signals with prescribed flat frequency spectrum and low PAPR, have been of interest for measurement purposes for several decades. Also (6.2.1) can be considered as a trigonometric polynomial of order $N-1$ in z with $z = e^{2\pi\Delta ft}$ and has therefore also been investigated from a pure mathematical point of view. For further information see, e.g., [234], [235], [236], [237]. However, in all these cases the objective was to find a single waveform only. In contrast to that, for data transmission it is necessary to find a large set of such waveforms. For instance, for transmitting 100 bits of information we need 2^{100} different waveforms. Furthermore, these waveforms must be generated in real time.

6.3 PAPR-Reduction with Signal Distortion

PAPR-reduction approaches can be divided into two classes. The first one distorts the transmitted signal by clipping the signal peaks, increasing the bit error rate (BER). The other class is characterized by the fact that, although the PAPR of the transmit signal is reduced, the signal remains undistorted. While this costs bandwidth efficiency, the BER remains unchanged.

6.3.1 Peak-Clipping Effect on System Performance

Since peak-clipping is the most pragmatic PAPR reduction method and is used (implicitly) in most of today's system implementations, a brief overview of its impact on system performance will be given here. We consider two types of nonlinearities: a complex baseband hard clipper and a model for transistor high-power amplifiers (HPA). Both these models leave the phase $\phi(t)$ undistorted. We consider these two models only because we assume that any phase distortion (AM/PM conversion), as introduced by travelling wave tube amplifiers, can be eliminated almost completely by a suitable pre-distorter [238], [239]. However, for the amplitude distortion (AM/AM conversion) this is not perfectly possible because the limiting effect always remains. Models for transistor HPAs can be found, e.g., in [240] and [241].

Nonlinear amplitude distortions can be described by a nonlinear characteristic $g(m)$. Complex baseband hard clipping is given by

$$g(m) = \begin{cases} m & m < m_{\max}, \\ m_{\max} & m \geq m_{\max}, \end{cases} \tag{6.3.1}$$

where m_{max} denotes the clipping magnitude. The behavior of a real solid state HPA is better described by

$$g(m) = \frac{m}{\left(1 + (m/m_{\max})^{2p}\right)^{1/2p}} \tag{6.3.2}$$

where the parameter p controls the smoothness of the transition from the linear region into saturation. A typical value is $p = 3$ [241]. For $p \to \infty$, (6.3.2) approximates (6.3.1). Note that hard clipping of the complex baseband signal according to (6.3.1) is not equivalent to hard clipping of the RF-signal, as studied for example in [242]. If the RF-signal is subject to hard clipping, then the equivalent baseband nonlinear characteristic $g(.)$ is a smoother limiter, similar to the HPA-model given in (6.3.2). The physical interpretation of hard clipping of the complex baseband signal is a HPA according to (6.3.2) plus a pre-distortion device, which maintains a linear amplitude characteristic up to the peak power (for example, see [238]).

The nonlinear distortion of $s(t)$ leads to intermodulation (IM) noise among the subchannels. IM-noise at frequencies within the OFDM transmission band degrades the SNR at the demodulator output and hence increases the bit error rate. IM-noise at frequencies outside the OFDM transmission band is regarded as unwanted out-of-band (OOB) energy and may degrade the BER of transmissions in adjacent channels and, in typical radio applications, the level of OOB radiation must be below the limits specified by regulatory bodies.

The modulation of the HPA can be described by either the input back-off (IBO) or the output back-off (OBO). Both figures are defined relative to the peak power $P_{\max}$ of the HPA,

$$\text{IBO} = 10 \cdot \log_{10}\left(\frac{P_{\max}}{P_{\text{av,in}}}\right), \quad \text{OBO} = \log_{10}\left(\frac{P_{\max}}{P_{\text{av,out}}}\right) \tag{6.3.3}$$

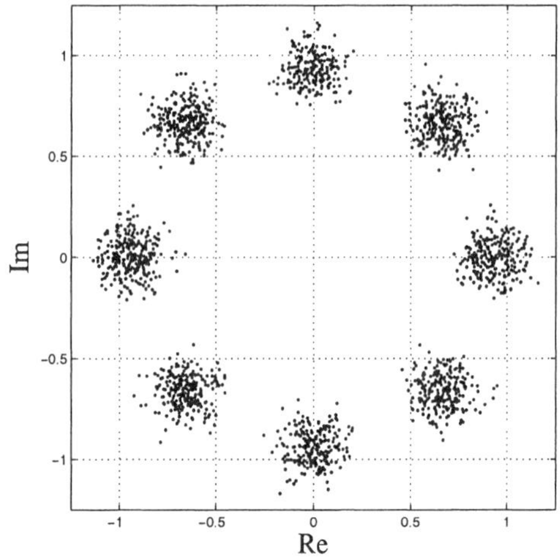

(a)

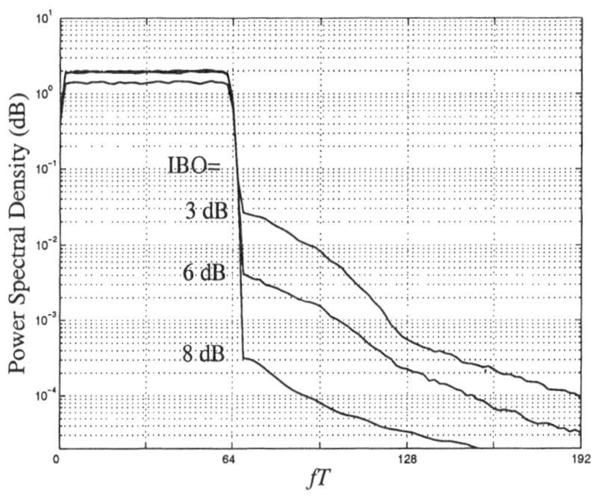

(b)

Figure 6.3. (a) Receiver signal space (8-PSK) at an IBO of 3 dB. (b) Estimated average PSD when peak clipping is present for an IBO of 3,6 and 8 dB (128 subchannels; 100 OFDM symbols evaluated; because of symmetry right half of spectrum shown only).

where $P_{\text{av,in}}$ is equal to P_{av} in (6.2.8) and $P_{\text{av,out}}$ is $E\{g^2(m)\}$. Hence, for a given nonlinearity and a given transmit signal, IBO and OBO are directly related to each other. Figure 6.3 visualizes the effects of IM-noise for the case of hard baseband clipping according to (6.3.2). Figure 6.3a shows the demodulator output for an 8-PSK OFDM. The IM results in a noise-like disturbance. Note also that because the peak values are clipped, the signal power is decreased. On average, the original signal points, with unit magnitude, are moved slightly towards the origin [243], [244].

In Figure 6.3b the estimated power spectral density (PSD) of the distorted OFDM signal is shown. In the case without a nonlinearity, the PSD decays slowly according to the $\sin^2(\pi f T_s)/(\pi f T_s)^2$ spectrum of each carrier. In practice, however, the decay rate is increased by applying additional filtering to the OOB components with negligible linear distortions to the useful signal components. Because of cost considerations, such filtering is usually located before the HPA. Distortions due to the HPA will then increase the OOB power to the level shown in Figure 6.3b. Therefore, Figure 6.3b gives a quite realistic view of the effect of spectral spreading. An important figure is the level of the noise PSD at the edge of the transmission band (here $fT = 64$) relative to the inband signal level. For example, for an IBO of 6 dB this value is only about 32 dB down and, furthermore, its rate of decay is very slow, leading to adjacent channel interference (ACI).

Next we compare the loss in energy efficiency due to peak-clipping for OFDM with that of constant-envelope (CE) modulation. A constant envelope has two advantages. First, the average transmit energy could be made equal to the peak power of the HPA. Second, because there is no IM-noise due to peak-clipping, the transmission also would not suffer from any BER degradation.

In contrast, with OFDM, for a given transmitter peak power, the average power is reduced and IM-noise is generated. The back-off of the transmitter must be chosen carefully. A small back-off achieves a high average transmit power. But, large back-off achieves low IM noise. Clearly, there is an optimum back-off value, where the total degradation

$$TD = OBO + \Delta SNR \tag{6.3.4}$$

is minimized. In this equation ΔSNR denotes the SNR-degradation due to IM at a given BER. Since the precise optimization is quite complicated, the reader is referred to [240], [243] for details. Figure 6.4 summarizes the major results from [243]. With an optimal back-off, depending on the modulation constellation, only 4-7 dB are lost when compared to a CE modulation

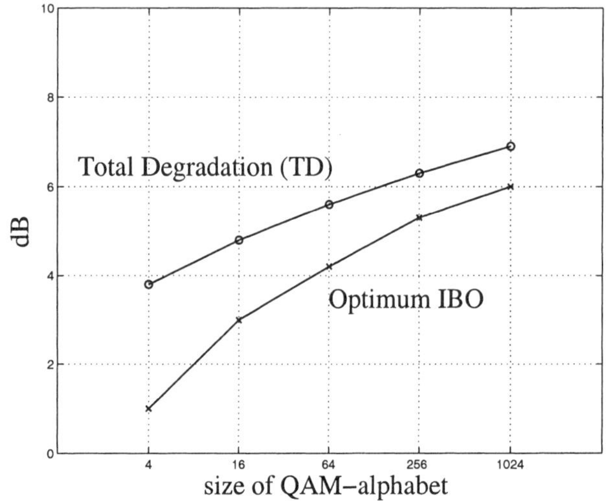

Figure 6.4. Resulting total degradation (TD) in power efficiency of OFDM compared to a CE modulation for various sizes of the constellation for optimally chosen IBO at a BER of 10^{-4}.

scheme. Corresponding values for the transistor HPA model according to (6.3.2) instead of peak-clipping can be found in [240]. The figures reported here for the case of ideal baseband clipping are about 2 dB better than those for the transistor HPA.

An interpretation of these results is warranted. First, considering the loss up to 7 dB, the CE modulation assumed here is hypothetical and for reference only. In practice, the spectral efficiency of CE modulation would be much smaller than that of OFDM. If the spectral efficiency were increased to the values of, e.g., 16-QAM OFDM, then the modulation with CE would suffer from an even higher energy inefficiency.

6.3.2 PAPR-Reduction by Clipping and Filtering

As mentioned before, the most obvious approach for reducing OFDM peaks is simply to clip $s(t)$ so that $m(t)$ is limited to a certain desired maximum value m_d. Clipping can be considered also as a simple PAPR reduction method, although it introduces several undesired effects. Fortunately, it is possible to perform clipping already before the HPA, most desirably in digital domain [242], [245]. If an ideal pre-distorter is assumed for the HPA, then

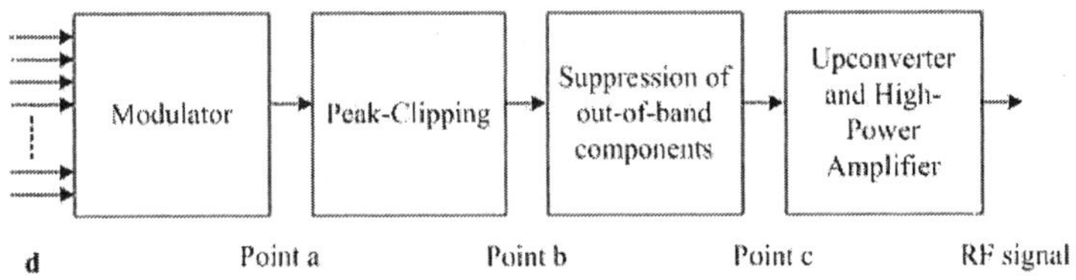

Figure 6.5. System configuration for clipping and filtering.

the HPA will not introduce any additional distortions as long as $m_d \leq m_{max}$, i.e., the signal peaks do not exceed the peak power limit of the HPA. The advantage of this method is that distortions may be controlled by intelligent signal processing, because they are introduced much earlier in the transmitter's signal chain instead of at the output stage. Figure 6.5 shows a typical configuration. The conventional OFDM modulator is followed by a peak-clipping device, and a linear filter for removing the OOB components. Finally, the up-converter translates the complex baseband signal into a real-valued RF signal. The task of the linear filter is to reduce the unwanted OOB noise. Here, complete removal of the OOB components by an ideal low-pass filter is assumed. Of course, in practice, an ideal low-pass filter can only be approximated. Removal of the OOB noise by filtering leads to a regrowth of some of the signal peaks. The amount of CF regrowth may be determined by computer simulation.

While we have described the clipping and filtering approach for the time-continuous baseband signal $s(t)$ here, it might be realized in several variations. For example, clipping and filtering can be applied to the transmit signal at IF instead of baseband. In this case only a real-valued signal needs to be processed. Because digital filtering at IF has much higher computational complexity, other filter realizations, like surface acoustic wave (SAW) filters could be suitable solutions.

Instead of the time-continuous signal (which, in practice, means a sampled signal with a sufficiently high over-sampling rate), clipping and filtering could be applied to the equidistant samples $s(kT/N)$ only (cf. (6.2.10)) [242]. In this case the low-pass filtering can be considered part of the interpolation process in the digital to analog conversion.

The results of the clipping and filtering approach can also be obtained directly [246], [247]. Instead of clipping with subsequent filtering, the time-domain signal $s(t)$ is multiplied with some multiplication signal that drops down at time-instants where large peaks occur. Consequently, the peaks

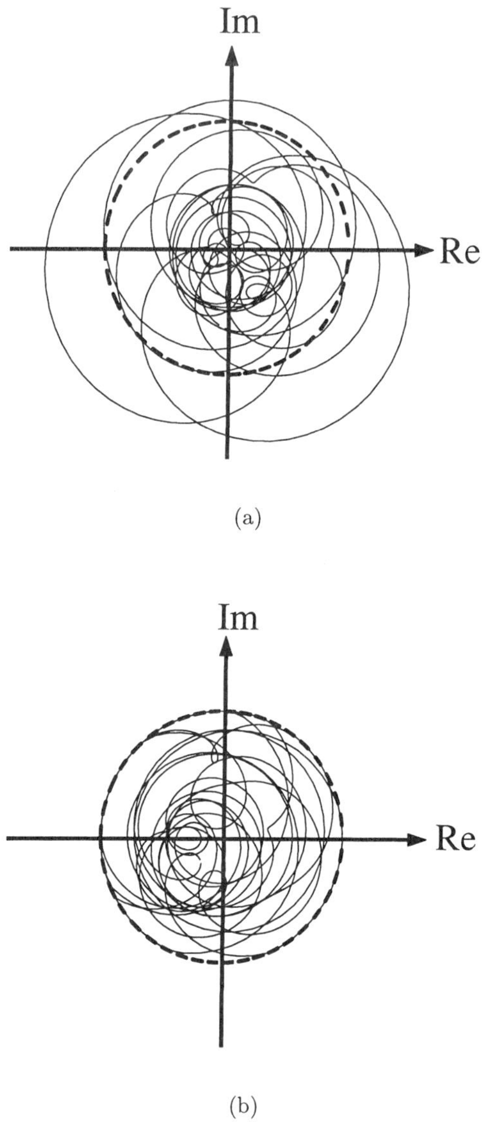

Figure 6.6. Polar plot of complex envelope $s(t)$ of a 128 subchannel OFDM signal. (a) Conventional signal; (b) PAPR reduced distortionless below 3.2 dB.

are reduced. The dips in the multiplication signal are smooth and chosen according to some window function that results in small spectral spreading only. In case of a few peaks only, this method proposes lower computational complexity since no additional filtering operation is required.

6.4 Limits for Distortionless PAPR-Reduction

The last section showed that the high peak values of OFDM signals can be reduced by clipping at the cost of a certain level of distortions, in particular increased BER and additional out-of-band noise.

We can design an OFDM system so that only waveforms with relatively small peak power are generated. With a perfectly predistorted HPA whose peak power is at least equal to the peak power of the transmit signal, no OOB components are generated and also the BER performance is not degraded. Hence, this PAPR reduction method is distortionless. For illustration, Figure 6.6 shows the polar plot of OFDM signals with $N = 128$ subchannels. No PAPR reduction is performed in (a) and the signal in (b) is processed (distortionless) so that a low CF of 3.2 dB is obtained. Hence it can be transmitted with a HPA at an IBO of only 3.2 dB without clipping.

However, with distortionless methods, the PAPR can be reduced only at the cost of a (slightly) reduced bandwidth efficiency and higher computational complexity for modulation and/or demodulation.

The overall PAPR can be reduced simply by not using high PAPR waveforms for transmission. This motivates a generic PAPR reduction principle, as depicted in Figure 6.7. Symbol sequences $s = (s_0, s_1, ..., s_{N-1})$ leading to high PAPR are excluded from transmission. Undesired sequences can be excluded, in principle, by using a look-up table. The percentage of ex-

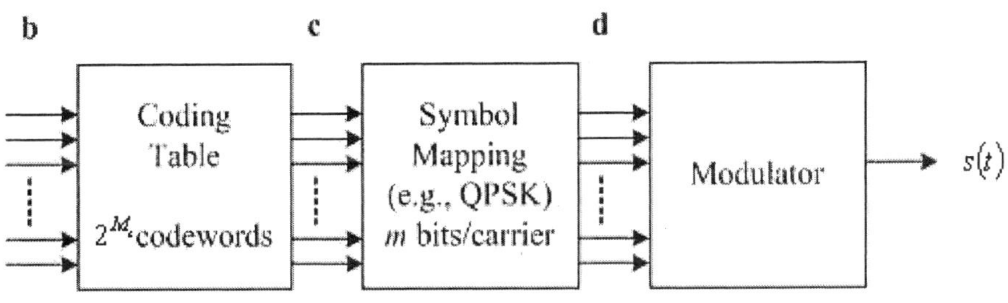

Figure 6.7. Principle of redundant block coding for PAPR reduction.

cluded sequences s is determined by the limit on the PAPR and amounts to

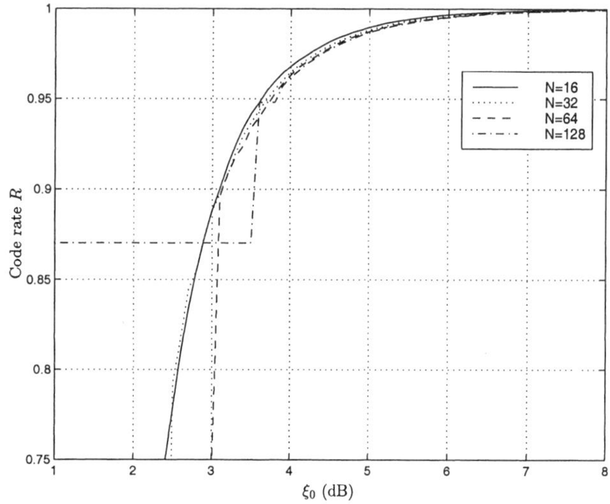

Figure 6.8. Code rate for limiting the PAPR.

a bandwidth efficiency loss. The PAPR of individual OFDM-symbols only rarely exceeds 12 dB (Figure 6.2), indicating that the reduction in bandwidth efficiency will be rather small. For a quantitative example, consider N-subchannel MPSK OFDM, which can accommodate $N \log_2(M)$ information bits at most. Suppose that we need a code rate $R = 1 - K/(N \log_2(M))$ to limit the PAPR to ξ_0; that is, out of all M^N sequences, $M^{RN \log_2(M)}$ will have PAPR less than ξ_0. By definition, we have

$$\Pr(\xi < \xi_0) = \frac{M^{N \log_2(M) - K}}{M^N} \tag{6.4.1}$$

which can be rearranged as

$$R = 1 + \frac{1}{N \log_2(M)} \log_2 \Pr(\xi < \xi_0). \tag{6.4.2}$$

This coding rate can be estimated by simulation. Figure 6.8 shows the required code rate to limit the PAPR. For $N = 128$, to reduce PAPR to 7 dB from 21 dB for the uncoded case, the required code rate can be as high as 0.98, suggesting that the PAPR can be much reduced by a small amount of redundancy. Unfortunately, such codes have so far been elusive.

6.5 Techniques for Distortionless PAPR-Reduction

This section describes distortionless schemes for generating OFDM signals with low PAPR. In practice, the key problem in PAPR reduction is computational complexity. In principle, PAPR can be reduced with little performance penalty (i.e., bandwidth efficiency loss). All that is required is a table of sequences that exceed a PAPR limit. However, for a large number of subchannels, the table grows too large and is not feasible. Hence, the coding table in Figure 6.7 (eventually including also the mapping) should be replaced by an implementable algorithm.

Distortionless PAPR reduction techniques include the selective mapping approach, optimization techniques like the partial transmit sequence approach and algebraic coding techniques. Several of these techniques are based on a very simple premise: For a given data vector s, generate multiple OFDM signals and choose the minimum PAPR signal for transmission. Formally, this approach can be stated as follows.

> For each information bit-word $\mathbf{b}^i$ define a set of associated signals $x^l(\mathbf{b}^i, t)$, $l = 1, \cdots, M_a$. Then select that signal $x^{l_{opt}}(\mathbf{b}^i, t)$ for transmission that has the lowest peak power.

This very general principle comprises two separate problems. First, for each $\mathbf{b}^i$ we must generate associated signals $x^l(\mathbf{b}^i, t)$, $l = 1, \cdots, M_a$ so that a selection can reduce the peak power of the resulting multicarrier signal at all. The second problem is performing the selection with reasonable complexity. Selecting the best signal may sound trivial, but as we will see, the definition of associated signals $x^l(\mathbf{b}^i, t)$ typically has to be done in frequency domain in terms of associated coefficient sequences $\mathbf{d}^l(\mathbf{b}^i)$. Since from $\mathbf{d}^l(\mathbf{b}^i)$ the peak-power in time-domain is not visible, the selection typically requires multiple transformations in time-domain, i.e. modulation processes.

In fact, a major challenge is that there is no simple rule known that helps in accurately determining the peak power that corresponds to a particular coefficient sequence $\mathbf{d}^l$. Only some very basic rules for avoiding the worst case peaks are known.

6.5.1 Selective Mapping

The basic principle of PAPR reduction becomes most obvious with selective mapping (SLM). The method has been developed independently by several authors [248], [249], [250]. In selective mapping, M_a statistically independent signals $x^l(\mathbf{b}^i, t)$, with $l = 1, \cdots, M_a$ are generated from the same information $\mathbf{b}^i$ and that signal with the lowest peak power is selected for transmission. Figure 6.9 shows an SLM OFDM system. The information word

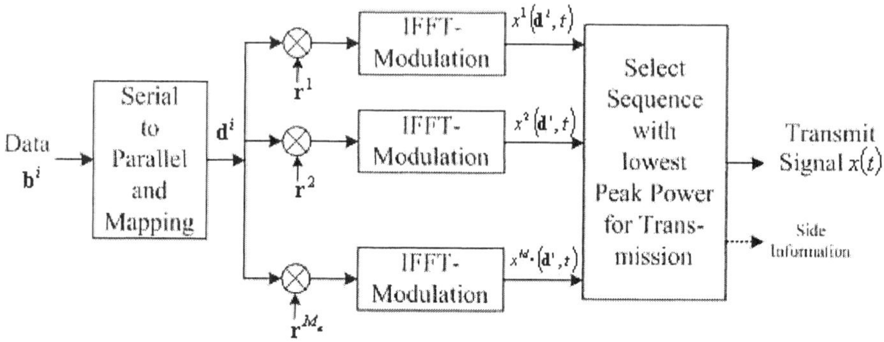

Figure 6.9. Selective mapping approach.

$\mathbf{b}^i$ is mapped onto a vector of OFDM coefficients $\mathbf{d}^i$ conventionally. From this vector, M_a statistically independent coefficient sequences are derived by elementwise multiplication with M_a fixed, but statistically independent vectors $\mathbf{r}^1, \cdots, \mathbf{r}^{M_a}$. Each coefficient sequence is transformed in time-domain by conventional modulation. Then, for each resulting time domain waveform $x^l(\mathbf{d}^i, t)$, the peak power is measured and that signal $x^{l_{opt}}(\mathbf{d}^i, t)$ with lowest corresponding peak power is selected for transmission. Since the peak powers are assumed to be statistically independent, selecting the best signal will reduce PAPR.

For detection, the receiver needs to know which vector $\mathbf{r}^{opt}$ has been used in the transmitter in order to divide the demodulator output elementwise by $\mathbf{r}^{opt}$. This knowledge could be transmitted as side information without significantly reducing the data rate, since only $\log_2(M_a)$ bits are needed. Typically, the number of sequences, M_a, must be kept small, because for each associated sequence, a complete OFDM symbol (6.2.1) must be generated.

Of course, it is of greatest interest to know how many sequences are necessary to achieve a significant CF reduction and how much reduction is possible. Since for obtaining a CF $C^{C_{M_a}}$ larger than a certain desired value C_d it is a necessary requirement that the CF of all signals $x^l(\mathbf{d}^i, t)$ is larger than C_d, C^{M_a} is determined by the following relationship [2]

$$\begin{aligned}
\Pr(C^{M_a} > C_d) &= \Pr((C(r^1) > C_d), \cdots, (C(r^{M_a}) > C_d)) \\
&= \prod_{l=1}^{M_a} \Pr(C(r^l) > C_d) \\
&= (\Pr(C > C_d))^{M_a} \qquad (6.5.1)
\end{aligned}$$

The second line of (6.5.1) is derived assuming that the probabilities are independent. Figure 6.10 plots simulation results for $N = 128$ subchannels and $M_a = 1, 2, 4$ and 16. With an increasing number M_a of associated sequences, the PAPR reduces significantly.

However, although with a small number of associated sequences, e.g., $M_a = 2$ and $M_a = 4$, the percentile PAPR (e.g., at a probability of 10^{-4}) can be reduced, further significant reduction requires a dramatically increasing M_a, thereby limiting the practical applicability of SLM. In conventional OFDM, there is a lower limit for the CF, that is exceeded by almost every OFDM transmit symbol (the CCDF approaches 1). With selective mapping, it is very difficult to reduce the CF below that value, because according to (6.5.1) selective mapping benefits from reducing the CCDF by exponentiation. This mechanism works best, if the CCDF for individual symbols is already low. Figure 6.10 also shows estimations of the CCDF based on the analytic model (6.2.11) of the CCDF of time-samples that has been inserted in (6.5.1). As expected, the model is optimistic, because it neglects possible missed peak between the sampling instants. For higher probabilities of the CCDF, the modelling error becomes even more significant.

For implementing the SLM approach, we need to define the M_a 'independent' vectors $\mathbf{r}^l$, $l = 1, \cdots, M_a$. To ensure independence, it is sufficient to take vectors that are generated by a pseudo-noise generator. Furthermore, it is desirable to limit the elements of $\mathbf{r}^l$ to unit magnitude, i.e. $r_i^l = \exp(j\phi_{i,l})$, possibly with a limited number of discrete phases only, e.g. 4 or 8 phases [249]. Restricting the phase values to a discrete set also has the advantage that in typical, rotational invariant constellations, the constellation points are maintained, an advantage that serves for synchronization purposes. Also the complex multiplications reduce to a much simpler modulo arithmetic of the discrete phase values.

[2] Assuming PSK-OFDM here, because strictly (6.5.1) is valid only if the selection process has no impact on the average power. Also, the elements of the sequences r shall have unit modulus.

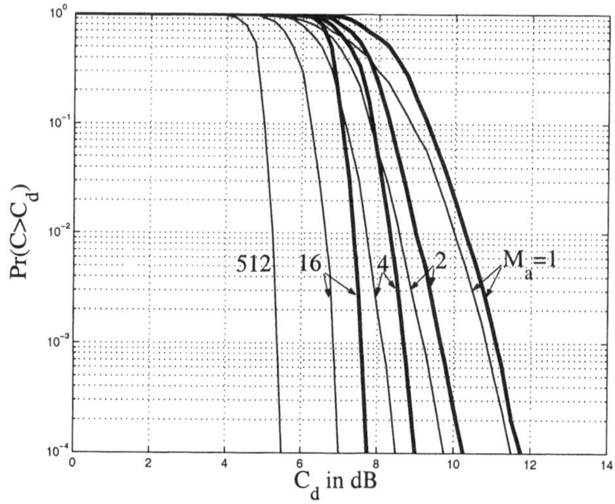

Figure 6.10. CCDF of the CF for $N = 128$ and 8PSK OFDM with SLM. Thin lines are computed using (6.2.11) and (6.5.1).

The method proposed in [250] works directly on the binary words $\mathbf{b}^i$ to be transmitted. Here $\mathbf{b}^i$ is scrambled with one of M_a possible different (binary) m-sequences. The result is mapped subsequently onto conventional modulation constellations. The advantage is that, also in the case of more general QAM constellations, the multiplications with the vectors $\mathbf{r}^l$ is replaced by the simpler modulo arithmetic. However, the main source of computational complexity in the selective mapping approach remains, i.e. performing a M_a-fold modulation.

Therefore, in [251] the set of associated signals $x^l(\mathbf{d}^i, t)$ is defined immediately in time-domain. In particular, it is proposed that for each vector of coefficients s, the corresponding time-domain signal $s(t)$ according to (6.2.1) is decomposed into its time-symmetric and antisymmetric component. Then, in a second step, some low-complexity transformations are applied directly to the time-domain components, thereby defining several associated signals $x^l(\mathbf{d}^i, t)$. The best one is selected for transmission. Essentially, there is only one IFFT required to obtain the time-domain components; therefore computational complexity is reduced significantly.

6.5.2 Optimization Techniques

The basic idea of all the optimization techniques is to define the transmit signal $s(t)$ as a function of a set of parameters. Some of the parameters are determined by the information to be transmitted, whereas the remaining parameters are free for optimization. The optimization is performed so that the peak values in the time-domain waveform are as small as possible. As a simple definition of parameters consider the OFDM-coefficients $s = (s_0, s_1, ..., s_{N-1})$. Some of them could be used for transporting information and the remaining ones could be used for optimization. Since some parameters (subchannels) are not available for transferring information anymore, the bandwidth efficiency is reduced.

Optimization techniques can also be regarded as a generalization of the SLM principle because, in effect, to each information word $\mathbf{b}^i$ a large set of possible waveforms $x^l(\mathbf{b}^i, t)$ is associated. The optimization procedure selects the best transmit signal. Because the number of associated sequences increases exponentially with the number of optimization parameters, these methods can reduce PAPR much more than SLM. But the optimization procedure can be computationally expensive. Furthermore, the loss in bandwidth efficiency typically is higher than with SLM.

Cancellation of Peaks

As already mentioned, it is possible to reserve some of the elements of the coefficient vector $s = (s_0, s_1, ..., s_{N-1})$ as optimization parameters for reducing the PAPR of $s(t)$. The sum of all sub carriers that are reserved for optimization can be considered as a compensation signal to reduce the peaks created by the contribution from the information [222], [252]. Specifically,

$$s(t) = \underbrace{\sum_{n \in I_{\mathrm{inf\,o}}} s_n e^{j2\pi n\Delta f t}}_{\text{information signal}} + \underbrace{\sum_{n \in I_{opt}} s_n e^{j2\pi n\Delta f t}}_{\text{compensation signal}} . \tag{6.5.2}$$

In (6.5.2), I_{info} and I_{opt} denote disjoint sets of indices for the information carrying subchannels and optimization subchannels, respectively. Since the optimization parameters are the coefficients of individual subchannels or tones they are called optimization tones.

The amount of achievable PAPR reduction increases with increasing size, $|I_{opt}|$ of the set I_{opt}. However, increasing $|I_{opt}|$ will also reduce the bandwidth efficiency. Since the compensation tones are unavailable for information transfer a relative redundancy, r $= I_{opt}/N$ can be defined for this approach.

The specific location of compensation tones has significant impact on the achievable amount of PAPR reduction. In [253], it is shown that for example using all tones with even index n for compensation, leads to a scheme that can not reduce the peak value of $s(t)$ at all. But better results are obtained if the locations of compensation tones are selected pseudo-randomly [222].

Note that, in (6.5.2), the information and optimization components are added linearly. This is an important property, because the PAPR is then convex on the optimization parameters. In particular this problem can be cast as a quadratically constrained linear program (QCLP)[222].[3] In practice, however the computational complexity for determining the optimal compensation tones may be too high. Therefore, in [222] a suboptimal iterative procedure for determining the optimal compensation tones has been proposed. PAPR can also be reduced by reserving some of the subchannels for peak reduction purpose and optimizing their values. Although this approach reduces the efficiency of the OFDM it does not need side information. The general idea follows from [254], [255], [256], [257], [258], [259].

Adaptive subchannel selection is presented in [260]. Subcarriers with low SNR are excluded at the transmitter, which improves the BER of the signal. These excluded subchannels are used to reduce the PAPR of the signal, thereby achieving low BER and PAPR. An iterative technique to find a suitable reduction function corresponding to excluded subchannels is also presented.

The use of parity subchannels to reduce the PAPR is reported in [259]. A PAPR suppressing signal is generated at parity subchannels when ever the PAPR is above a threshold level. These parity subchannels are based on the coding scheme prosed in [261]. The process of generating suppressing signal is iterated till peak power reduce to the given threshold. A PAPR reduction of 5 dB is observed for 8 subchannel OFDM signals. This method may not be suitable for OFDM systems with large number of subchannels.

Partial Transmit Sequences (PTS)

As already mentioned, the proper definition of the optimization parameters is crucial for the achievable amount of PAPR reduction. PTS is mainly motivated by the idea that an individual scaling and phase rotation of appropriately defined 'partial transmit signals' should be a powerful means

[3]In the special case of real-valued signal such as in DSL applications, this problem becomes linear program(LP)

to avoid the constructive addition of the components to high peak values [221], [223], [262], [263]. Figure 6.11 shows a possible implementation of this approach in the transmitter. The OFDM signal is partitioned into M_{PTS}

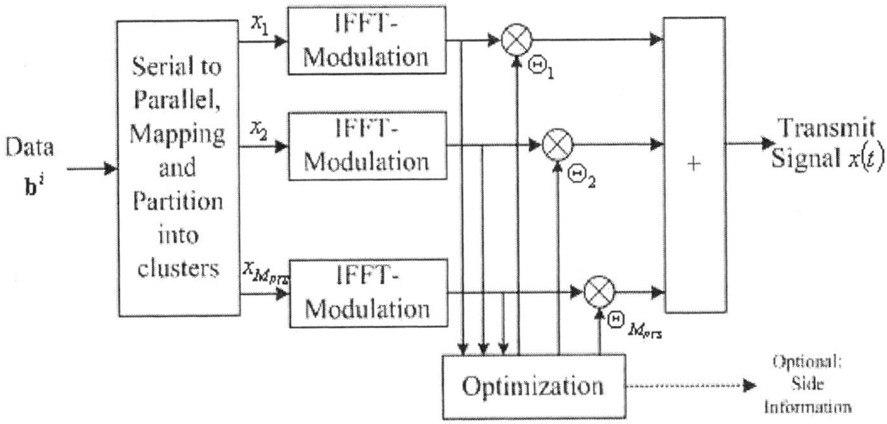

Figure 6.11. Partial transmit sequence approach.

disjoint sets of subchannels

$$s(t) = \sum_{g=1}^{M_{PTS}} \Theta_g x_g(t) \tag{6.5.3}$$

with

$$x_g(t) = \sum_{n \in I_g} s_n . e^{j2\pi n \Delta f t}. \tag{6.5.4}$$

The components $x_g(t)$ are called the 'partial transmit sequences' (PTS) or 'group signals'. The latter implies that the sets I_g consist of adjacent subchannels but this is not necessarily the case. Each group signal is transformed in time-domain by conventional modulation. In time-domain, all group signals are weighted with an individual factor Θ_g, and added to result in the transmit signal $s(t)$. The complex factors Θ that usually have unit modulus allow for a phase rotation of each group signal. They are chosen so that the peak power of $s(t)$ is minimal.

As with SLM, the receiver must have knowledge about the generation process of the OFDM signal, i.e. the chosen factors $(\Theta_1, \Theta_2, \cdots, \Theta_{M_{PTS}})$. These factors could be transmitted as side information. Alternatively within each group signal differential encoding can be applied. In any case (neglecting the possibly required side information), the relative redundancy of this scheme is $r = M_{PTS}/N$, since we spend M_{PTS} complex factors for optimization.

In principle, the group signals may consist of different numbers of subcarriers, but the best PAPR reduction capability is expected, when they are the same for all groups, i.e. $|I_g| = N/M_{PTS}$. where N is assumed to be an integer multiple of M_{PTS}. Also the particular selection of subchannels has an impact on the achievable amount of CF reduction [221]. For brevity, we will consider only the simple case of adjacent subchannels with $I_g = \{(g-1)\cdot N_g, \cdots, g\cdot N_g - 1\}$. For the same reasons as mentioned in the selective mapping approach, also for PTS it is useful to restrict possible phase values $\theta_g = arg(\Theta_g)$ to a finite set. Figure 6.12 shows the achieved PAPR

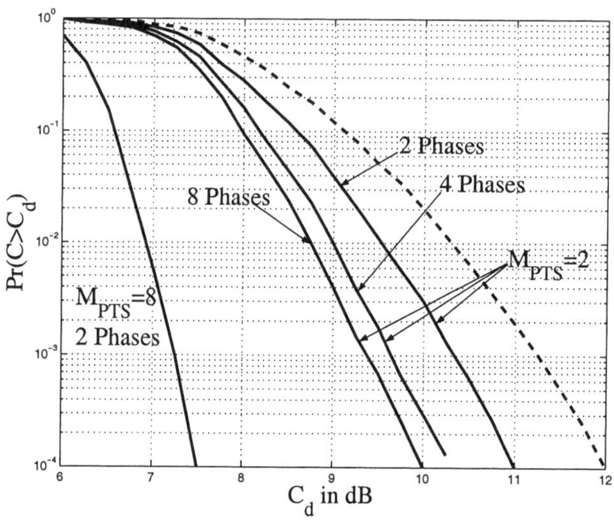

Figure 6.12. PAPR reduction with PTS for 128 subchannels. Dashed line denotes conventional OFDM.

reduction for various numbers of groups M_{PTS} and two to eight phases, i.e. $\theta_g \in \{0, \pi\}, \theta_g \in \{0, \pi/2, \pi, 3\pi/2\}$ and $\theta_g \in \{0, \pi/8, 2\pi/8, \cdots, 7\pi/8\}$. The results have been obtained by simulations with 8-PSK subchannel modulation and a full search over all parameter tuples. With an increasing number of groups M_{PTS} as well as with increasing number of possible phase values, the achievable PAPR reduction increases. With 8 phase values however, most of the possible PAPR reduction for a certain value of M_{PTS} is achieved. A finer granularity has little effect only. Also, in addition to the fact that the effective CF is reduced, also the true maximum peaks can be significantly reduced [223].

A disadvantage of the method is that the number of possible time-domain signals $s(t)$ rapidly increases with the number of group signals M_{PTS} and phase values. Therefore, applying a full search procedure for finding the best time-domain waveform will often be impossible. There are several ways to combat this problem. In [231] it is proposed not to perform the full search optimization, but to try a certain number of sets of optimization parameters $(\Theta_1, \Theta_2, \cdots, \Theta_{M_{PTS}})$. We may note however that this approach is similar to SLM. In particular, because of the linearity of the Fourier transform, the factors Θ_g can be considered also in frequency domain. Hence, the rotation vectors $\mathbf{r}$ of SLM can be expressed in terms of the PTS optimization variables according to

$$r = \left(\underbrace{\Theta_1, \Theta_1, \cdots,}_{N_g \text{ variables}} \cdots, \underbrace{\Theta_{M_{PTS}}, \Theta_{M_{PTS}}, \cdots}_{N_g \text{ variables}} \right). \tag{6.5.5}$$

Consequently, the limits of the SLM approach are also valid for this variation of the PTS approach. The next section explores another approach to reduce the computational complexity.

Iterative PTS

The PAPR reduction method presented in this subsection can be regarded as an optimized and specialized implementation of the PTS approach. Its main property is that it contains a powerful iterative procedure for the optimization of the free parameters in PTS. This procedure allows for the reduction of the signal peaks close to the theoretically possible values and reduces the complexity of PTS, which is exponential in N down to order $N \log N$.

We may use the notation of the PTS approach (6.5.3) and (6.5.4) here, but we will consider only the special case of equal numbers of subchannels $N_g = N/M_{PTS}$ per group signal $x_g(t)$. Within each group signal $x_g(t)$, information is carried in form of complex factors δ_n, that are differentially encoded in frequency direction [4] according to

$$s_n = \delta_n s_{n-1}, \quad n = (g-1)N_g + 1, \cdots, gN_g - 1. \tag{6.5.6}$$

[4]This is not a necessary requirement of the iterative approach.

It is assumed that $|\delta_n| = 1$, which implies DPSK modulation among the subchannels, but in general also DAPSK is possible.[5]

In what follows, also $|\delta_n| = 1$ shall be assumed. The absolute phase of each group signal is represented by $\theta_g = \arg(s_g N_g)$ and shall be optimized to reduce the CF. Also with this scheme, the price to be paid for the CF reduction is a loss in data rate, because, in comparison with conventional DPSK, each N_g-th phase difference is not available for information transfer. The relative redundancy r is calculated again by dividing the number of 'lost' parameters (used for optimization) by the number of information-carrying parameters, when no CF reduction scheme where applied,

$$r = \frac{(N-1) - M_{PTS}(N_g - 1)}{N-1} \approx \frac{1}{N_g} \tag{6.5.7}$$

For large N, it depends mainly on the number of subchannels per group. For example, in the case of $N_g = 4$ rate loss is about 25 percent.

The optimization algorithm is based on an iterative procedure that alternates between the time and frequency representations of the OFDM signal. It has already been used for the reduction of the peak values of multi-tone signals [264]. For the application in OFDM, the algorithm is generalized to the reduction of peak values of information-carrying signals. Its principle relies on the linearity property of the Fourier transform. Each group signal $x_g(t)$, is regarded as a basis function $\xi_g(t)$ of a function subspace,

$$x_g(t) = \underbrace{\sum_{n=(g-1)N_g}^{gN_g-1} \Delta_n . e^{j2\pi n\Delta f t}}_{\xi_g(t)} \tag{6.5.8}$$

with the information dependent factors $\Delta_g N_g = 1$ and

$$\Delta_n = \prod_{n'=(g-1)N_g+1}^{n} \delta_{n'}, \quad n = gN_g + 1, \cdots, gN_g - 1. \tag{6.5.9}$$

In (6.5.8), the first carrier of the group signal has been chosen to be the optimization parameter, but other definitions are possible. According to (6.5.8), the basis functions $\xi_g(t), g = 1, \cdots, M_{PTS}$, have an information-dependent shape, but nevertheless they are always mutually orthogonal (because they consist of a weighted sum of mutually orthogonal functions). The

[5]The transmission of additional side information is not required if non-coherent detection is applied at the receiver.

linear combination of these basis functions can form a subspace only, where the information to be transmitted defines that particular subspace.

The optimization algorithm starts with generating $s(t)$ according to (6.5.3)-(6.5.6), with the optimization parameters $\theta = \arg(\Theta_g)$ set to pseudo-randomly selected starting values.

In each iteration, the components of the time-domain signal $s(t)$, whose magnitude is greater than a certain threshold m_t, are defined as an error signal $e(t)$,

$$e(t) = \begin{cases} 0 & |s(t)| \leq m_t, \\ (|s(t)| - m_t).e^{j\arg(s(t))} & |s(t)| > m_t. \end{cases} \tag{6.5.10}$$

In order to reduce the undesired peaks, this error signal is subtracted from $s(t)$. This subtraction is done in the frequency domain. Therefore, the error signal $e(t)$ is expressed as a series, according to the given set of basis functions $\xi_g(t)$, $g = 1, \cdots, M_{PTS}$. The mean square approximation of $e(t)$ by a weighted sum of basis functions,

$$e(t) \approx \sum_{g=1} M_{PTS} e_g \cdot \xi_g(t) \tag{6.5.11}$$

leads (because of the mutual orthogonality of the $\xi_g(t)$) to the following expression for the coefficients e_g

$$e_g = \frac{1}{N_g T_s} \int_{t=0}^{T_s} e(t).\xi_g^*(t)dt, \; g = 1, \cdots, M_{PTS}. \tag{6.5.12}$$

Using the definition of $\xi(t)$ in (6.5.8), it can be shown that (6.5.12) is a weighted sum of Fourier coefficients $\epsilon(t)$ of $e(t)$, with $\epsilon_n = \frac{1}{T_s} \int_{t=0}^{T_s} e(t) \cdot e^{-j2\pi n \Delta f t} dt$. Consequently, the calculation of the e_g can be performed by an efficient IFFT algorithm, extended by the following simple linear superposition for each group,

$$e_g = \frac{1}{N_g} \sum_{n=gN_g}^{(g+1)N_g - 1} \Delta_n^* \epsilon_n. \tag{6.5.13}$$

In the next step, the error coefficients e_g are subtracted from the coefficients s_{gN_g}. In general, the resulting value $s_{gN_g} - e_g$ will not meet the condition $s_{gN_g} - e_g = 1$. Therefore a new coefficient is calculated according to

$$s_{gN_g}^{i+1} = 1 \cdot e^{j\arg(s_{gN_g}^i - e_g^i)}, \tag{6.5.14}$$

where i denotes the number of iterations. This nonlinear mapping ensures that the Euclidean distance between the new value $s_{gN_g}^{i+1}$ and the desired value $s_{gN_g}^{i} - e_g^i$ is always smaller than or equal to the old distance between $s_{gN_g}^{i}$ and the desired value. This property guarantees the convergence of the algorithm. More precisely, it can be proved that the RMS value of $|e(t)|$ is monotonically decreasing [253]. Then, with a suitable selection of the threshold m_t, the CF is almost always decreasing.

Adaptive PTS

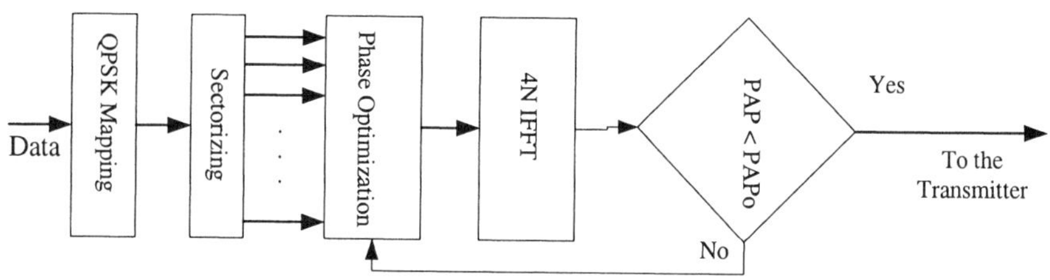

Figure 6.13. Adaptive PTS approach.

We next present another reduced-complexity variant of PTS, which is called adaptive PTS. The key idea is that most input vectors intrinsically have low PAPR (Figure 6.2) and PTS optimization can be terminated early when such inputs occur. As above, we only consider binary (i.e., $\Theta_g \in \{1, -1\}$) weighting factors and we divide the input data block into M_{PTS} subblocks. As a first step, set all the weighting factors to 1 and compute the PAPR of the combined signal. If it is less than a set threshold PAPR_{T_s}, then stop optimization immediately. If not, invert the first weighting factor ($\Theta_1 = -1$) and recompute the resulting PAPR. If it is less than the threshold, retain Θ_1 as part of the final phase sequence and stop optimization. The algorithm continues in this fashion until PAPR is less than PAPR_{T_s} or all or part (K) of the $2^{M_{PTS}-1}$ combinations are searched. The corresponding block diagram is shown in Figure 6.13.

The adaptive algorithm can now be written as follows:

step 1 Set $[\Theta_1, \dots, \Theta_{M_{PTS}}] = [1, 1, \dots, 1]$,

step 2 Set IterCount = 1,

step 3 While PAPR($\mathbf{X}$') > PAPR$_{T_s}$ or IterCount < K,

step 4 Change the weight vector by one bit,

step 5 IterCount ++

Here, K can be set to $2^{M_{\text{PTS}}-1}$ (then this reverts to full PTS) or a lesser value. The maximum number of iterations of this technique is K, and the minimum is 1. The actual number of iterations varies from one input frame to another. We characterize the complexity of this scheme by the average number of iterations per input frame.

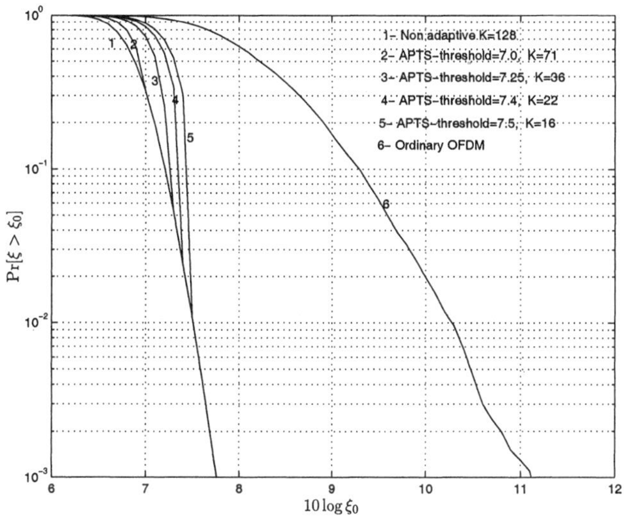

Figure 6.14. CCDF of an OFDM signal with APTS ($M_{\text{PTS}} = 8$).

In the following results, 10^5 random OFDM symbols were generated to obtain CCDFs. We assume 256 subchannels throughout and use QPSK data symbols with the energy normalized to unity. Figure 6.14 shows the CCDF for $M_{\text{PTS}} = 8$. The 99.9th percentile PAPR of the original OFDM signal is about 11.2 dB. The ordinary PTS technique improves it by 3.75 dB. Curve 2 shows results for APTS with a threshold value PAPR$_{T_s}$ of 7.5 dB. In the region CCDF< 10^{-2}, both the techniques provide identical performance. Ordinary PTS requires 128 iterations per OFDM symbol while

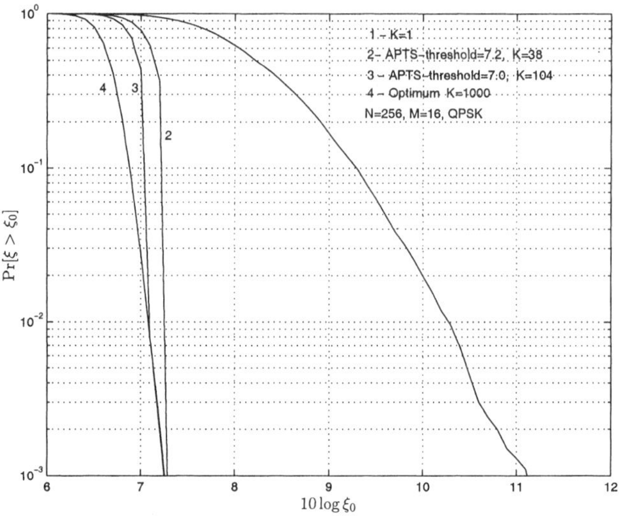

Figure 6.15. Comparison of performance of optimum PTS with APTS for $M_{\text{PTS}} = 16$.

APTS requires only 16 (on average) iterations per OFDM symbol. This amounts to an 87% reduction in complexity.

Results are also shown for APTS with PAPR_{T_s} of 7.4 dB, 7.25 dB and 7.0 dB. These require 22, 36 and 71 iterations (on average) and reduce complexity up to 83%, 72% and 44% respectively. Therefore, the complexity of APTS can be reduced greatly by limiting the number of iterations by selecting a suitable threshold value. Lower threshold values yield better performance but result in higher complexity.

Figure 6.15 shows results for 16 subblocks (M_{PTS}=16). The optimal PTS curve requires 32768 iterations per OFDM symbol, which is too time consuming for simulations. We therefore use randomly-generated 2000 binary weighting patterns, compute the PAPR of each and subsequently take the minimum PAPR as the optimal PAPR. Curves 2 and 3 show the performance of APTS with K=1000 and threshold limits of 7.2 dB and 7.0 dB respectively. Curves 2 and 3 correspond to an average of 38 and 104 iterations, yielding complexity reduction of 98% and 94.8% respectively, when compared with the optimum result. The performance loss is only about 0.1 dB at the 99th percentile PAPR. By contrast, the flipping algorithm [265] requires only 16 iterations in this case, but results in a performance loss of

1 dB. The adaptive approach promises to considerably reducing the PTS complexity.

6.5.3 Modified Signal Constellation

Another optimization approach is to modify a standard constellation such as MPSK and MQAM by adding more points. These additional points can be used as free optimization variables.

The PAPR of an N-subchannel MPSK-OFDM system can be as high as $10\log_{10}(N)$. If some of these subchannels (say, N_n) are made to zero, the theoretical maximum PAPR reduces to $10\log_{10}(N - N_n)$. This fact has been exploited by several authors to reduce the PAPR. Parallel combinatory OFDM (PC-OFDM) is proposed in [266], [267]. In the transmitter first, which subchannels to be zero and nonzero are chosen. Then $N - N_n$ nonzero subchannels are modulated by MPSK. Now another set of bits called parallel combinatory bits are mapped to the zero positions of the OFDM symbol. There are $\binom{N}{N_n}$ different ways to choose N_n zeros out of N subchannels. Therefore, the total number of bits carried by a PC-OFDM signal can be expressed as

$$m_{tot} = (N - N_n)\log_2(M) + \left\lfloor \log_2 \binom{N}{N_n} \right\rfloor \tag{6.5.15}$$

where $\lfloor x \rfloor$ is the largest integer smaller than or equal to x.

PC-OFDM uses an MPSK signal constellation extended with an additional zero amplitude point. Using an MPSK signal constellation, an $(M+1)$-ary amplitude and phase shift keying (APSK) constellation is constructed. The bandwidth efficiency of PC-OFDM may be higher compared to conventional OFDM depending on N_n. The complexity of the PC-OFDM system increases when N and N_n increase. The difficult part is to select the zero positions of the OFDM signal for given parallel combinatory bits. Introduction of zero amplitude signal point further reduces the Euclidian distances between signal points. However, simulation results show that PC-OFDM performs better than OFDM in an AWGN channel but not in fading channels.

Another approach is known as Tone Injection (TI) [268]. This approach expands the constellation to a bigger one, where one of several values that carry the same information can be selected. This extra degree of freedom is used to generate multicarrier signals with lower PAPR. The process of constellation expansion can be explained as follows. If X_n carries b_k bits, then it could take one of 2^{b_k} discrete values. If a QAM constellation with minimum distance between points s_k is assumed, the real and imaginary parts of X_n can take values $\pm s_k/2, \pm 3s_k/2, \ldots, \pm(M_k - 1)s_k/2$, where $M_k = 2^{b_k/2}$ is the number of levels per dimension. Real and imaginary parts of constellation points are modified such that PAPR is reduced. Note that no side information is needed. If we assume the constellation point is A then the modified point is expressed as $\widehat{A} = A + pD + jqD$, where p and q are any integer value and D is a positive real number known at the receiver. An iterative approach is presented to find proper values for p and q. A 6 dB reduction in PAPR is observed with moderate complexity.

6.5.4 PAPR-Reduction Effect on the System Performance

An OFDM system with 256 carriers, 16QAM and a guard interval of 12.5% is simulated. The OFDM signal is passed through nonlinear devices. Figure 6.16 depicts the CCDF of PAPR of an OFDM signal. It also shows the PAPR reduction using SLM. When $K = 2$, two OFDM symbols are generated from the same information sequence and one with the minimum PAPR is chosen for the transmission. Similarly, when $K = 16$, 16 different OFDM symbols using the same information sequence is generated. These different sequences can be generated in different ways as shown in References [269], [270], [271]. When $K = 16$, PAPR reduction on the order of 4 dB at the 99.9th percentile PAPR is obtained. This PAPR reduction is obtained with the increase in the system complexity. An OFDM system with 256 subchannels needs highly complex systems at the transmitter to reduce the PAPR below 7 dB. PAPR reduction codes are capable of achieving low PAPR when N is small. But these schemes reduce the system throughput significantly as N gets large.

Inband distortion

The effect of inband distortion can be observed by the BER performance. Figure Figure 6.17 depicts the BER performance of an OFDM signal passing through a soft limiter nonlinear device. This figure shows BER performance of both ordinary OFDM and PAPR reduced OFDM signals for different back-off values of the nonlinear device. P-OFDM refers to the PAPR reduced OFDM. PAPR of the signal is reduced by SLM. The BER improves significantly even in an AWGN channel depending on the back-off value of the non-linear device. No significant improvement in BER is observed when the back-off is very low, 0 dB for example. In this case both OFDM and

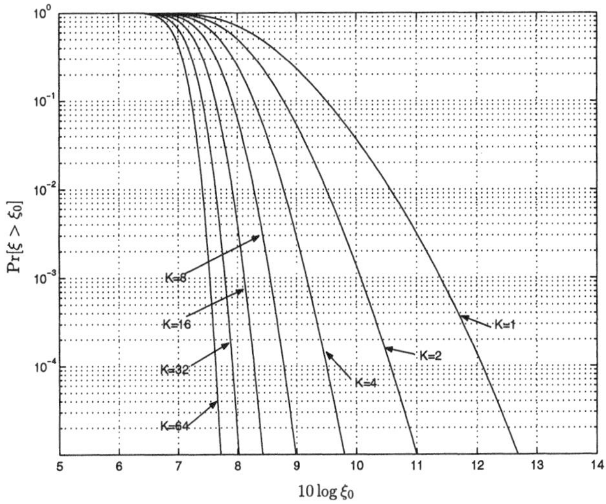

Figure 6.16. CCDF of PAPR of an OFDM signal.

PAPR reduced OFDM experience severe signal distortions. When the back off greater than 2 dB a clear improvement in BER is observed. In these cases distortion in OFDM is high while the distortion in PAPR reduced OFDM is low. When the back-off is increased to 5 dB an improvement of about 10 dB is observed at 10^{-6} BER. If the back-off is increased further both schemes do not experience signal distortion thus would not show any BER degradation. Results shown in Figure 6.17 are obtained without using any forward error correction (FEC) codes. Typically, wireless systems need FEC to combat channel impairments. These FEC codes can also be used to improve the BER caused by peak limiting. Figure 6.18 shows the BER of an QPSK OFDM signal coded with Golay complementary sequences (6.5.27). Number of subchannels in the coded system is 32 and PAPR is 3 dB at most. BER is significantly improved even for a clipping level of 0 dB.

Out of band radiation

OFDM in general has sharp signal transitions at the signal boundaries, giving out high OBR even without nonlinearity effect. This can be alleviated by using extended guard intervals at both ends of the OFDM signal and a smoothing window function (raised cosine window) [272]. Figure 6.19 depicts the spectrum of an OFDM signal when passed through a SSPA nonlinear device. Curves 1 to 7 represent the different back-off values of the amplifier.

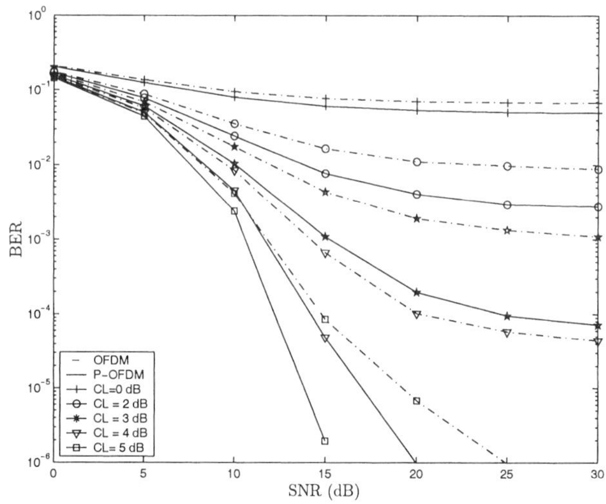

Figure 6.17. Bit error rate in an AWGN channel, 16QAM with SLM (K=16).

For example, for a back-off of 9 dB, to provide -80 dB adjacent channel separation requires a channel spacing of about three times the symbol rate for normal OFDM, with the PAPR reduced OFDM the channel spacing is nearly the symbol rate. Thus, PAPR reduced OFDM provides a spectrally-efficient solution for the adjacent channel interference problem, which can arise in portable cellular applications. This performance also reveals that depending on the permissible OBR level, the amplifier back-off can be reduced by 2 to 3 dB when the PAPR of an OFDM is reduced considerably. When the back-off of the amplifier is increased, OBR level is reduced.

6.5.5 Algebraic Coding

The last sections describe efficient PAPR reduction techniques. However, from a more theoretical point of view it is still unsatisfying that, despite the ingenious approaches taken, we still cannot explicitly state data sequences $s = (s_0, s_1, \ldots, s_{N-1})$ that will result in low PAPR. We succeeded in finding sufficiently large sets of such sequences for transferring information, but the detailed definition of each particular sequence is left to some optimization algorithm, which means that the structure of the resulting code is unknown to us.

A much more direct way to obtain coefficients s that leads to signals $s(t)$

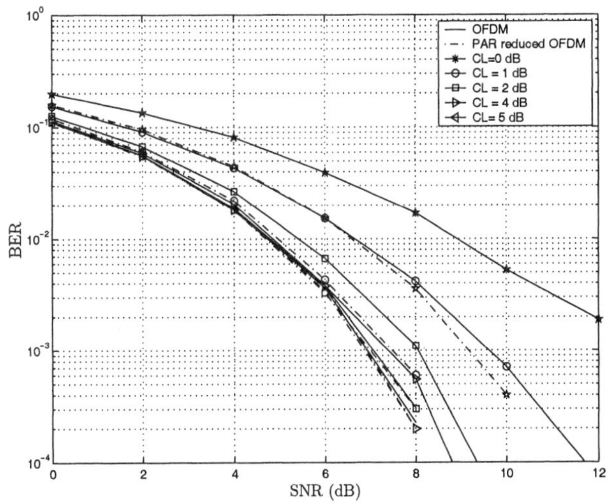

Figure 6.18. Bit error rate in an AWGN channel (PAPR reduced to 3 dB); $CL = 20\log_{10}(m_{\max}/\sqrt{N})$.

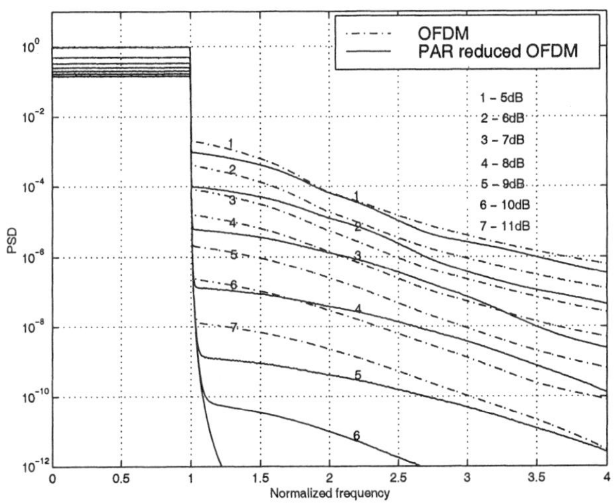

Figure 6.19. Spectrum of an OFDM signal after passing through a SSPA.

with low PAPR is by capturing the properties of low-PAPR signals in an appropriately structured coding scheme. Such a method would not require any time-consuming optimization or selection procedures. Also, the presence and knowledge of a certain structure is a key requirement for efficiently exploiting the redundancy of the code for error correction as will be discussed later on. n principle block coding schemes to reduce the PAPR have many advantages. First, the transmission of side information to the receiver is unnecessary. Second, codes can include error control capability, which can reduce the BER. Good coding schemes, which reduce PAPR and provide good error correction capabilities having higher code rates have been elusive to date.

Few definitions are required for this discussion. A M-ary code $\mathcal{C}$ is a given set of N-tuples $= (s_0, s_1, \ldots, s_{N-1})$, where each $s_n \in \mathcal{Q}$. The Hamming distance between two sequences $(s, e\ )$ or codewords is defined as

$$s_H(d, e) = \sum_{n=0}^{N-1} \delta(s_n - e_n) \tag{6.5.16}$$

where $\delta(\cdot)$ is the Kronecker delta function. A critical parameter of a code $\mathcal{C}$ is the minimum Hamming distance, or just minimum distance, which measures how good it is at error-correcting. This is defined to be the smallest of the distances between any two distinct codewords. That is

$$s_{\min} = \min\{s_H(d, e) \mid d, e \in \mathcal{C}, d \neq e\}. \tag{6.5.17}$$

The PAPR of a code $\mathcal{C}$ is defined as

$$\text{PAPR}(\mathcal{C}) = \max\{\text{PAPR}(d) \mid \forall d \in \mathcal{C}\}. \tag{6.5.18}$$

Finally, the code rate of a code $\mathcal{C}$ is defined as

$$R = \frac{\log_2 \sharp(\mathcal{C})}{N \log_2 M} \tag{6.5.19}$$

where $\sharp(\mathcal{C})$ denotes the number of elements of set $(\mathcal{C})$. The following notation will be used where necessary. An $[n, k, s_{\min}, \eta]$-code is a code of length n, containing k information symbols, with minimum distance $s_{\min}$ and $\text{PAPR}_{\max}$ η. The aperiodic autocorrelation function (AACF) of a data vector $s = (s_0, s_1, \ldots, s_{N-1})$ is defined as

$$C_s(k) = \sum_i s_i s^*_{i+k} \tag{6.5.20}$$

where * denotes complex conjugation and here and below the summations are understood be over only those integer values for which both i and $i + k$ lie within $\{0, 1, \cdots, N-1\}$.

Encoding Principles

In order to give an overview of the algebraic coding techniques, here we focus on the basic relations and describe the more advanced methods in examples only. The instantaneous envelope power $P_s(t)$ of $s(t)$ (6.2.1) can be expressed by

$$P_s(t) = |s(t)|^2 = \sum_i \sum_k s_i s_k^* e^{j2\pi(i-k)\Delta f t}. \tag{6.5.21}$$

By substituting $k = i + u$ in (6.5.21), we obtain

$$P_s(t) = N + \sum_{u \neq 0} \sum_i s_i s_{i+u}^* e^{-j2\pi u \Delta f t}. \tag{6.5.22}$$

Using (6.5.20)in (6.5.22), the instantaneous power can be related to the AACF as

$$P_s(t) = N + \sum_{u \neq 0} C_s(u) e^{-j2\pi u \Delta f t}. \tag{6.5.23}$$

According to (6.5.23), the AACF and the instantaneous power $P_s(t)$ are a Fourier pair. Consequently, for achieving a near constant envelope with a low PAPR, the AACF should be impulse-like.

Two vectors $s = (s_0, s_1, \ldots, s_{N-1})$ and $e = (e_0, e_1, \ldots, e_{N-1})$ are said to be a Golay complementary pair if

$$C_s(u) + C_e(u) = 0 \quad \text{for} \quad u \neq 0. \tag{6.5.24}$$

Any member of a Golay pair is called a Golay sequence. Let all coefficients s_n and e_n have constant modulus 1 (PSK-modulation of the subchannels). Then a PSK Golay pair satisfies

$$C_s(u) + C_e(u) = \begin{cases} 2N & u = 0, \\ 0 & u \neq 0. \end{cases} \tag{6.5.25}$$

Golay-sequence codes for quadrature amplitude modulation are considered in [273], [228], [227]. If two sequences s and e are complementary, from (6.5.23) and (6.5.25) it follows that the sum of the corresponding instantaneous powers is

$$P_s(t) + P_e(t) = 2N. \tag{6.5.26}$$

Furthermore, since each power is a non-negative function, it can be concluded that $P_s(t) \leq 2N$ as well as $P_e(t) \leq 2N$. Hence the peak power of any signal

$s(t)$, derived from a complementary sequence s is limited by $2N$. Since the average power of $s(t)$ is $\overline{P} = N$, the CF follows to be not greater than 3 dB.

The fact that complementary sequences translate into time-domain waveforms with PAPR or CF not greater than 3 dB was noticed first in [236], which can be regarded as a generalization of earlier work on Shapiro-Rudin sequences in [234]. Because these publications focused on measurement applications instead of data transmission, no attempts were made to generate a larger set of complementary sequences.

Golay complementary sequences and Reed-Muller codes

Many of the early code proposals were comprehensively generalized by Davis and Jedwab when they proposed an infinite family of binary Golay complementary sequences with parameters $[2^m, \log_2(m!)+m, 2^{m-2}, 2.0]$, and defined them as certain Reed-Muller (RM) $\text{RM}(2,m)$ cosets of $\text{RM}(1,m)$ [274], [275]. This gave a set of codes that had predictable and guaranteed error control and envelope limitation properties. These codes might not have optimal error correcting capability when compared with standard alternatives, but a reduction in coding gain could be quite easily outweighed by a greater reduction in envelope fluctuation.

We call this family DJ, where DJ comprises codewords, $c(\mathbf{x})$, which is defined in corollary 4 in [276]. For any permutations π of the symbols $\{1, 2, \ldots, m\}$ and for any c, $c_k \in \mathcal{Z}_{2^h}$

$$s_i = 2^{h-1} \sum_{k=1}^{m-1} x_{\pi(k)} x_{\pi(k+1)} + \sum_{k=1}^{m} c_k x_k + c \qquad (6.5.27)$$

gives the i-th symbol of a Golay sequence over $\mathcal{Z}_{2^h}$ of length 2^m, where $(x_1, x_2, \ldots, x_m)$ is the binary representation of symbol position $i, i \in 2^m$ $(i = \sum_{j=1}^{m} x_j 2^{m-j})$. Here $\mathcal{Z}_{2^h} = \{0, 1, \ldots, 2^h - 1\}]$ and each subchannel contains exactly $h, (h > 1)$ code bits. This explicitly determines $2^{h(m+1)}.m!/2$ Golay sequences over $\mathcal{Z}_{2^h}$ of length 2^m. This is the DJ construction generalized to higher-order constellations and to PAPRs which are a multiple of 2. The construction is optimal for low N. For instance, for $m = 3$ and binary sequences we can construct an optimal $[8, \log_2(48), 2, 2.0]$ DJ code. There are only 16 more sequences with PAPR ≤ 2.0 which are not included in the DJ set, and the inclusion of any of these sequences would reduce s from 2 to 1. Unfortunately the rate, k/N, of the DJ construction vanishes rapidly for $N > 32$. Therefore the DJ construction is only practically useful in the context of OFDM for systems requiring no more than 32 subchannels. It remains an open problem to discover low PAPR error-correcting code constructions for $N > 32$ with an acceptable rate. Unfortunately, many OFDM systems require anything from 8 to 8192 subchannels.

Rudin-Shapiro Recursion

Rudin-Shapiro (RS) polynomials are defined recursively by the formulas [277]

$$\begin{aligned} P_{n+1}(z) &= P_n(z) + z^{2n} Q_n(z) \\ Q_{n+1}(z) &= P_n(z) - z^{2n} Q_n(z), \quad n \geq 0, \end{aligned} \tag{6.5.28}$$

where $P_0(z) = Q_0(z) = 1$. RS sequences belong to a larger family of Golay sequences of length 2^k [236]. Construction of RS sequences is equivalent to the recursive construction of Golay complementary pairs [234]. However, in Golay complementary pairs the number of tones is not restricted to power of two as in the case of RS sequences. Binary and multi-phase Golay complementary sequences can be used to construct multitone signals with flat envelopes rendering a PAPR of at most 3 dB [236], [278]. Based on this results a code was proposed by van Nee [279] to build a special class of block codes that guarantee a PAPR of 3 dB at most. Such codes can be constructed for any number of subchannels N, provided N is a power of two. When employing MPSK to modulate the subchannels, M-ary source vectors X_m of length $(\log_2 N + 1)$ are encoded into M-ary codewords C_m of dimension N. These codes can be regarded as linear block codes with the generator matrix G_N for the first-order Reed-Muller code $\mathrm{RM}(1, \log_2 N)$, translated by a constant offset D_N, called the kernel, is a row vector of length N with elements representing the phases of Golay complementary sequences, while G_N has dimension $N(\log_2 N + 1)$. These are complementary codes of length N. These codes are a special case of DJ codes described above.

The complementary code has a rate of $(\log_2 N + 1)/N$ which decreases with the length of the code. This is due to the higher amount of redundancy needed to limit the PAPR to 3 dB as the number of subchannels increases. The code rate and the bandwidth efficiency decay when the number of subchannels is increased.

Here, for the purpose of illustration, we will only show an example [280]. Assume 8-PSK modulation and $N = 8$ subchannels. In this case the coefficient vector s of length 8 is obtained from an vector of eight integers $\nu = (\nu_1, \cdots, \nu_7)_s^T$ with $\nu \in 0, 1, 2, 3, 4, 5, 6, 7$, representing the phase values according to $s_n = e^{j2\pi\nu/8}$, with $n = 0, \cdots, 7$. The vector ν is defined by the matrix equation

$$\nu = Gh + g \mod 8 \tag{6.5.29}$$

where h denotes a vector of integers $h = (h_0, h_1, h_2, h_3)_s^T$, representing the information to be transmitted, with elements $h_i \in 0, l, 2, 3, 4, 5, 6, 7$ and

$$G = \begin{pmatrix} 1000 \\ 0100 \\ 0010 \\ 7110 \\ 0001 \\ 7101 \\ 7011 \\ 6111 \end{pmatrix}, \quad g = \begin{pmatrix} 0 \\ 0 \\ 0 \\ 4 \\ 0 \\ 0 \\ 4 \\ 0 \end{pmatrix}. \tag{6.5.30}$$

Computer simulation shows that all transmit signals generated according to (6.2.1) and (6.5.29) have a CF $\leq$ 3 dB. Note also that G acts as a generator matrix, so that here we have a direct link to the traditional theory of linear block codes. This feature will be of importance when considering error correction capabilities.

In this particular example with a CF of 3 dB, the relative redundancy is exactly $r = 50\%$. This case is close to the theoretical limit. However, despite the good PAPR reduction here, in a typical practical application it is desirable to increase the number of subchannels and to increase the spectral efficiency.

In principle, the method described above can be generalized to other PSK constellations and an increased numbers of sub carriers. However, because the construction rule is based on complementary sequences, the CF is always limited by $\sqrt{2}$. For such a low CF it will not be possible to reduce the redundancy below approximately 30%. What is much more important, however, is the fact that, because the construction is based on concatenation, with increasing number of subchannels, the code redundancy, $r = 1 - (\log_2(N) + 1)/N$ very quickly approaches 100%. For example, in the case of 64 and 512 subchannels it is $r = 89.1\%$ and 98%, respectively.

Decoding and Error Correction Capabilities

For the algebraic PAPR-reduction coding methods, suitable decoding algorithms have been derived. For example, the matrix G (6.5.30) is a code generator matrix. Consequently, the resulting set of vectors Gh forms a linear code. The addition of the constant vector g does not change distance properties of this code. For linear codes the minimum distance is given by the minimum weight over all nonzero codewords. For G in (6.5.30) the generated code has a minimum Hamming distance of 4. Hence, in addition to the PAPR reduction, it also provides increased immunity against channel noise. Furthermore, from standard coding theory, errors can be detected by applying a parity check matrix, defined by $H.G = 0$. The parity check matrix corresponding to G in (6.5.30) is

$$H = \begin{pmatrix} 17710000 \\ 17007100 \\ 10707010 \\ 27707001 \end{pmatrix}. \tag{6.5.31}$$

The following procedure is possible for performing simple hard decision decoding at the receiver. Assume that the receiver's decision unit outputs the possibly erroneous vector ν' corresponding to ν at the transmitter. In the first step the vector g is subtracted to obtain $h' = G \cdot \nu'$. Then the syndrome is obtained by calculating $H \cdot h'$. Using the syndrome, error detection and correction is possible. However, in overall system simulations, the increased Hamming distance translates to lower bit error probability only if maximum likelihood (ML) decoding is applied.

Efficient and optimal (i.e., soft-decision ML) decoding of Golay sequences requires computing the Euclidean distance or (equivalently) the correlation between the received vector and all valid codewords. Those correlations can be efficiently computed using the fast Hadamard transform (FHT); a FHT decoder for first-order RM codes is developed in [281]. While the FHT decoder efficiently computes the correlations, its complexity remains high. Another decoding approach is to reorder the received symbols according to a reliability measure; see [282] for a review of such techniques. Ordered statistic decoding (OSD) is an example of this approach that has been extended to Golay sequences in OFDM [283]. OSD is a low complexity decoding technique for codes with moderate or high rates. However, for codes with low rates (Golay sequence codes are low rate), this technique requires extensive computation. Finally, classical hard-decision RM decoding algorithms include majority logic and threshold decoding schemes. These decoders are simple and fast, but suboptimal. A class of sub-optimal soft decision decoding algorithms for RM codes using majority logic decoding has also been developed [284].

6.6 PAPR Reduction for Multicarrier CDMA - (MC-CDMA)

Recently much attention has been given for combining OFDM and code division multiple access (CDMA) techniques, resulting multi-carrier CDMA (MC-CDMA). MC-CDMA sends a single data symbol in multiple narrowband subchannels. Each subchannel is encoded with a phase offset of 0 or π based on an orthogonal spreading code. For a spreading code of length N there are N subchannels. This is also a multiple access scheme since different users use the same set of subchannels but with a different spreading codes that are orthogonal to each other. OFDM structure reduces the intra-cell interference and increases the capacity within a cell in a cellular environment using many parallel subchannels. In contrast, direct sequence CDMA (DS-CDMA) is specially robust against inter-cell interference by spreading each user's signal across the system bandwidth using a unique high chip rate spreading code. MC-CDMA combines both both techniques, thereby reducing both inter and intra cell interferences by spreading every data bit across a set of orthogonal subchannels. By doing this MC-CDMA provides an efficient method of frequency diversity, especially suitable for fading channels. The lowpass equivalent transmitted signal $x_{mc}(t)$ of an MC-CDMA system can be expressed as [285]

$$x_{mc}(t) = \sqrt{\frac{eE_b}{NT_b}} \sum_{m=0}^{M-1} a_m[k] \sum_{n=0}^{N-1} c_m[n] e^{\frac{j2\pi n}{T_b}} P_{T_b}(t - kT_b), \; kT_b \leq t < kT_b + T_b \tag{6.6.1}$$

where N is the number of subchannels, $a_m[k]$ is the m-th user's binary antipodal input data symbol in k-th bit interval and $c_m[n], n = 0, 1, \ldots, N-1$ is the n-th chip of the spreading code assigned to the i-th user. E_b and T_b are the bit energy and the bit duration of the considered system respectively. The shaping pulse $P_{T_b}(t) = 1$ for $0 \leq t < T_b$ and zero otherwise. The subchannels are spaced by $1/T_b$. MC-CDMA also suffers from high PAPR problem. Several techniques for PAPR reduction are reported to date [229], [262], [286], [287], [288], [289].

6.7 PAPR Reduction for Multicode CDMA

Multicode CDMA [290] is one of the techniques that has been proposed to support multirate operation in a packetized CDMA network. It provides easy integration of data streams having variable rates into a common unified architecture, with all the transmissions modulating the same carrier, occupying the same bandwidth and having the same processing gain. In a multicode CDMA system if a user needs K times the basic source rate, K spreading codes are assigned to it. The user converts its packet stream into K basic-rate parallel packet streams which are spread using the K different codes, respectively. Intuitively, with multicode CDMA, the different data rate requirements of the users can be easily catered for by assigning enough number of codes to each user. In multicode CDMA, a Walsh-Hadamard transform of input data is taken before spreading. The output of the operation is not binary, rather multilevel. Thus multicode signal can have a large variation in the envelope power giving out high PAPR. PAPR reduction in multicode CDMA can be seen in recent reach efforts [291], [292], [293], [294], [295]. The PAPR of milti-code CDMA can be defined as

$$\xi = \frac{1}{N} \max_{0 \leq n \leq N} |S_n|^2 \tag{6.7.1}$$

where N is the block length and S_n's are multilevel output signals of the Hadamard transform.

6.8 Concluding Remarks

The main part of the chapter was devoted to methods for reducing the PAPR without distorting the transmitted signal. Special attention has been given to selected mapping, optimization and algebraic coding techniques. Starting with an investigation of the properties of OFDM signals, it has been shown that high peak values are extremely rare: A PAPR greater than 14 dB is almost impossible. Consequently, in typical wireless applications, clipping techniques can reduce the PAPR down to about 10 dB without causing unacceptable high out-of-band radiation or signal degradation. Further PAPR reduction can be achieved by sophisticated modulation and coding techniques. In theory, e.g., for QPSK transmission, a PAPR less than 6 dB can be achieved with only a 4% code redundancy.

Advanced versions of selective mapping can reduce PAPR at the cost of small computational complexity. Another advantage of this technique is that high bandwidth efficiency is maintained. Optimization techniques can achieve some PAPR reduction with fairly small complexity. However, optimization techniques can be designed as well for high PAPR reduction at high complexity. Therefore they cover the largest range of applications. In particular, the cancellation approach seems to provide low PAPR with acceptable complexity.

Algebraic coding seems to be the most elegant method, because it allows the straightforward exploitation of code redundancy for error correction. For practical applications, however, the bandwidth efficiency is low. For large numbers of subchannels, bandwidth efficiency decreases dramatically. The major breakthrough that would be required for algebraic coding is a method for enlarging the set of low-PAPR sequences.

While the approaches can provide significant reductions in PAPR, computational complexity is one of the key problems in PAPR reduction. Most algorithms allow for a reasonable trade-off between the key performance metrics: PAPR reduction, bandwidth-efficiency and computational complexity. However, at this time, none of the three major approaches selected mapping, optimization and algebraic coding, is capable of achieving high bandwidth efficiency and a PAPR close to the theoretical limits together without increasing computational complexity significantly. However, compromises are possible. For example, for point-to-point applications, low complexity methods can be suggested with a small amount of PAPR reduction. But for broadcasting applications, high complexity algorithms are allowed at the transmitter side. Hence, PAPR can be reduced significantly.

In this chapter, the latest results have been summarized, but PAPR reduction is still an active research area. Further developments are expected especially in the standardization bodies concerned with digital subscriber line technologies and mobile radio systems. Interestingly, the investigations and results on PAPR-reduction techniques for multicarrier systems stimulated the development of similar PAPR reduction techniques for single-carrier systems [296].

Chapter 7

SYNCHRONIZATION FOR MIMO OFDM

Gordon Stüber and Apurva Mody

7.1 Introduction

A MIMO-OFDM receiver operating in the acquisition mode must perform time synchronization, radio frequency (RF) oscillator offset estimation and correction, sampling frequency offset estimation and correction, and initial channel estimation. Once the acquisition phase is complete, receiver enters a tracking mode. For systems that operate under high Doppler conditions, the channel must be tracked continuously. Even for applications where the users are stationary, variations in the effective channel will still occur due to drift in the oscillators used in the RF chain and the oscillators used at baseband to clock the digital-to-analog (D/A) and analog-to-digital (A/D) converters in the transmitter and receiver, respectively.

This chapter focuses on open loop sampling frequency offset and RF oscillator frequency offset tracking schemes for MIMO-OFDM systems that operate in a frame mode, consisting of a preamble followed by data. Examples of such systems include IEEE 802.11a and IEEE 802.16a. We also consider design issues associated with the MIMO-OFDM preamble, as well as the pilot signal structures, that enable the suggested algorithms to work efficiently.

Notation: bold upper case letters denote matrices, and upper case letters with an underline denote vectors; $(\cdot)^*$ and $(\cdot)^H$ denote conjugate and conjugate transpose operations, respectively; $E[\cdot]$ and $\mathrm{Var}(\cdot)$ are the expectation and variance operators; Tr denotes the trace of the matrix; $\mathrm{Diag}(\cdot)$ denotes a diagonal matrix; $\underline{X}_i$ denotes the ith entry of a vector; $\mathbf{X}_{1:N}^{(1:Q,1:L)}$ represents a 3-D matrix of dimension $Q \times L \times N$ and $\mathbf{X}_k^{(q,1:L)}$ represents the qth row vector of the sub-matrix $\mathbf{X}_k^{(1:Q,1:L)}$.

7.2 MIMO System Model

The individual OFDM modulators and demodulators are implemented using an N-point inverse fast Fourier transform (IFFT) and fast Fourier transform (FFT), respectively. A block of N data symbols specifies the frequency domain coefficients input to the IFFT. The block of N time domain coefficients at the IFFT output constitute an OFDM block. Let $q; 1 \leq q \leq Q$ index the transmit antenna and let $d = 0, 1, \ldots$ be the running OFDM symbol index. Then the time-domain samples are $s_n^{d,q} = \frac{1}{\sqrt{N}} \sum_{k'=0}^{N-1} S_{k'}^{d,q} \exp\{j2\pi nk'/N\}$, $0 \leq n \leq N-1$, where $S_{k'}^{d,q}$ is the data symbol transmitted from the qth antenna, for the dth OFDM symbol at the k'^{th} sub-carrier. The total average energy transmitted from all Q antennas, E_s, is normalized to unity such that $E_s = \frac{1}{2}\mathrm{E}\left\{\sum_{q=1}^{Q} \left|s_n^{d,q}\right|^2\right\} = 1$, $\forall d,\ n$.

To mitigate the effects of channel delay spread, a cyclic prefix (CP) consisting of the last G coefficients are inserted in front of the OFDM block to form an OFDM symbol. To fully remove ISI the cyclic prefix duration, $T_g = GT_{\text{sa}}$, must equal or exceed the maximum delay of the channel, T_h, where $1/T_{\text{sa}}$ is the sample rate. The time required to transmit one OFDM symbol is the OFDM symbol time, $T = T_s + T_g$, where $T_s = NT_{\text{sa}}$. For each antenna, the corresponding real and imaginary components of the OFDM symbol samples are applied to a pair of D/A converters operating at a sampling frequency of $1/T_{\text{sa}}$ Hz to generate the set of complex envelopes. The set of complex envelopes on the interval $(dT, (d+1)T)$ is

$$s^{d,q}(t) = \sum_{k'=0}^{N-1} S_{k'}^{d,q} \cdot \exp\left\{j\frac{2\pi k'}{T_s}(t - dT)\right\}, q = 1, \ldots, Q \ . \tag{7.2.1}$$

The complex envelopes are up-converted to RF with a carrier frequency of f_T Hz. It is assumed that the same RF oscillator is used for all Q RF chains at the transmitter, so the frequency f_T is the same for all transmit antennas. Similarly, all D/A converters in the transmitter chain are driven by a common sample clock oscillator. Similar assumptions are made for the RF and sample clock oscillators in the receiver chain. The waveforms are transmitted over a $Q \times L$ MIMO channel characterized by the time-variant impulse responses

$$h(\tau, t)^{(q,\ell)} = \sum_{m=0}^{M-1} h_m^{(q,\ell)}(t)\delta(\tau - \tau_m^{(q,\ell)}) \ , \tag{7.2.2}$$

where $\{h_m^{(q,\ell)}(t)\}$ are the time varying complex path gains between the qth transmit and the ℓth receive antenna, $\{\tau_m^{(q,\ell)}\}$ are the corresponding path time delays, and M is the total number of paths for each channel. Assuming that the channel quasi-static, meaning that it stays essentially constant over the OFDM symbol duration T, the channel transfer functions are given by

$$\tilde{H}_{d,k'}^{(q,\ell)} = \sum_{m=0}^{M-1} h_m^{(q,\ell)}(dT) \exp\left\{-j2\pi\tau_m^{q,l}k'/T_s\right\}, k' = 0, \ldots, N-1 \ . \quad (7.2.3)$$

The received bandpass signals are down-converted to complex baseband using a local oscillator with carrier frequency f_R Hz. Since the transmitter and the receiver RF oscillators have different frequencies, the received complex envelope is affected by an RF oscillator frequency offset of $k_o + \epsilon = (f_R - f_T)T_s$ sub-carrier spacings, where k_o and ϵ refer to the integer and the fractional components of the RF oscillator frequency offset in the units of sub-carrier spacing, respectively. The received complex envelope is also affected by RF oscillator phase noise $\vartheta(t)$. Since k_o has the net effect of shifting the down-converted OFDM signal by an integer number of sub-carrier spacings, it is ignored for the time being in the system equation. We later show a way to estimate and correct this integer offset. At the ℓth receiver antenna, the received complex baseband signal is

$$\begin{aligned} r^{\ell}(t) &= \frac{1}{\sqrt{N}} \sum_{q=1}^{Q} \sum_{k'=0}^{N-1} S_{k'}^{(d,q)} \tilde{H}_{d,k'}^{q,\ell} \\ &\quad \times\exp\left\{\frac{j2\pi k't}{T_s}\right\} \exp\left\{j\left(\frac{2\pi\epsilon t}{T_s} + \vartheta(t)\right)\right\} \\ &\quad +w^{\ell}(t) \ , \ell = 1, \ldots, L \ , \end{aligned} \quad (7.2.4)$$

where $w^{\ell}(t)$ represents the complex additive white Gaussian noise (AWGN) with a variance of N_0. The real and imaginary components of the received complex baseband waveform are applied a pair of balanced A/D converters with a sampling frequency of $1/T'_{\text{sa}}$ Hz. To counteract the aliasing introduced by the frequency offsets, and for improved parameter estimation, the waveforms can be over-sampled at the receiver by factor of u followed by a uN-point FFT which acts as an OFDM demodulator. However, we will not consider over-sampling in this chapter. Since the sampling frequencies at the transmitter and receiver differ, the received discrete-time samples are affected by the normalized SF offset coefficient $\beta = (T'_{\text{sa}} - T_{\text{sa}})/T_{\text{sa}}$. The RF oscillator frequency offset ϵ drifts relatively slowly with time, whereas the sampling clock offset and phase noise may change more rapidly [297]. Some of the techniques discussed later in this chapter will assume that β and ϑ only take on new values every Q OFDM symbols.

The nth sample for the dth OFDM symbol may be obtained from $r^{\ell}(t)$ by substituting $t = d(N+G)T'_{\text{sa}} + nT'_{\text{sa}} + \epsilon = d(N+G)(\beta+1)T'_{\text{sa}} + n(\beta+1)T'_{\text{sa}} + \epsilon$, where ϵ is the unknown phase component of the A/D converter. Afterwards, the guard interval is removed and an N-point FFT is applied to the samples $\{r_n^{\ell}\}_{n=0}^{N-1}$ for the dth OFDM symbol. Assuming that the timing synchronization window is accurate as discussed in Chapter 4, the demodulated samples at the kth sub-carrier are

$$
\begin{aligned}
R_k^{d,\ell} &= \frac{1}{N}\sum_{q=1}^{Q}\sum_{k'=0}^{N-1}\sum_{n=0}^{N-1} S_{k'}^{d,q}\tilde{H}_{k'}^{q,\ell} \\
&\quad \times\exp\left\{\frac{j2\pi k'}{N}\left[d(N+G)(\beta+1)+n(\beta+1)+\epsilon\right]\right\} \\
&\quad \times\exp\left\{\frac{j2\pi\epsilon}{N}\left[d(N+G)(\beta+1)+n(\beta+1)+\epsilon\right]\right\} \\
&\quad \times\exp\left\{\frac{-j2\pi nk}{N}\right\}\exp\left\{j\vartheta^{d}\right\} \\
&\quad +W_k^{d,\ell} \qquad (7.2.5) \\
&= \sum_{q=1}^{Q}\left(\sum_{k'=0}^{N-1} S_{k'}^{(d,q)}H_{k'}^{q,\ell}\right. \\
&\quad \left.\times\ \exp\left\{\frac{j2\pi}{N}\left[d(N+G)(k'\beta+\epsilon+\epsilon\beta)\right]\right\}\right) \\
&\quad \times\frac{1}{N}\frac{\sin[\pi(k'+k\beta+\epsilon+\epsilon\beta-k)]}{\sin[\frac{\pi}{N}(k'+k\beta+\epsilon+\epsilon\beta-k)]} \\
&\quad \times\exp\left\{j\pi(k'+k\beta+\epsilon+\epsilon\beta-k)(1-\frac{1}{N})\right\} \\
&\quad \times\exp\left\{j\vartheta^{d}\right\} \\
&\quad +W_k^{d,\ell} \qquad (7.2.6)
\end{aligned}
$$

where $W_k^{d,\ell}$ is the AWGN term in the frequency domain. The above equation can be separated into the useful terms when $k' = k$ and the inter-carrier interference (ICI) terms when $k' \neq k$. This gives

$$
\begin{aligned}
R_k^{d,\ell} &= \sum_{q=1}^{Q} S_k^{(d,q)} H_k^{q,\ell} \exp\left\{\frac{j2\pi}{N}\left[d(N+G)(k\beta+\epsilon+\epsilon\beta)\right]\right\} \\
&\times \frac{1}{N}\frac{\sin[\pi(k\beta+\epsilon+\epsilon\beta)]}{\sin[\frac{\pi}{N}(k\beta+\epsilon+\epsilon\beta)]}\exp\left\{j\pi(k\beta+\epsilon+\epsilon\beta)\left(1-\frac{1}{N}\right)\right\} \\
&\times \exp\left\{j\vartheta^d\right\} + W_{k\neq k'}^{d,\ell \text{ ICI}} + W_k^{d,\ell},
\end{aligned}
\tag{7.2.7}
$$

If the RF oscillators are stable and the phase noise fluctuations are relatively small then the approximation $\exp\{j\vartheta^d\} \approx 1 + j\vartheta^d$ may be made, where the imaginary part of the phase noise $j\vartheta^d$ introduces additional noise $W_k^{d,\ell \text{ PN}}$ into the system. The term $\epsilon\beta$ is assumed to be small and is ignored.

If the frame structure of the MIMO-OFDM system is organized into groups of Q OFDM symbols, starting with the dth OFDM symbol and extending to the $(d+Q-1)^{\text{th}}$ OFDM symbol, then (7.2.7) has the matrix representation

$$
\mathbf{R}_k^d = A_k^d C_k \mathbf{\Lambda}_k \mathbf{S}_k^d \cdot \mathbf{H}_k^d + \mathbf{W}_k, \tag{7.2.8}
$$

where

$$
A_k^d = \exp\left\{j\left[2\pi(k\beta+\epsilon)d(1+G/N)+\vartheta^d\right]\right\} \tag{7.2.9}
$$

is a scalar function of d, β, ϵ and ϑ;

$$
C_k = \frac{1}{N}\frac{\sin[\pi(k\beta+\epsilon+\epsilon\beta)]}{\sin[\frac{\pi}{N}(k\beta+\epsilon+\epsilon\beta)]}\exp\left\{j\pi(k\beta+\epsilon+\epsilon\beta)\left(1-\frac{1}{N}\right)\right\} \tag{7.2.10}
$$

is a scalar function of β and ϵ only;

$$
\mathbf{\Lambda}_k = \begin{bmatrix}
\exp\left\{\frac{j2\pi(k\beta+\epsilon)0(N+G)}{N}\right\} & 0 & \cdots \\
0 & \ddots & 0 \\
0 & \cdots & \exp\left\{\frac{j2\pi(k\beta+\epsilon)(Q-1)(N+G)}{N}\right\}
\end{bmatrix}
\tag{7.2.11}
$$

is a $Q \times Q$ diagonal matrix; $\mathbf{S}_k^d$ is the $Q \times Q$ signal transmission matrix, where the $(d,q)^{\text{th}}$ element represents the symbol transmitted from the qth transmit antenna, for the dth OFDM symbol, at the kth sub-carrier. The channel matrix $\mathbf{H}_k^d$ has dimensions $Q \times L$ and its $(q,\ell)^{\text{th}}$ element represents the frequency domain channel coefficient between the qth transmit and ℓ^{th} receive antenna, at the dth OFDM symbol and k^{th} sub-carrier. Note that the channel is assumed to be quasi-static over a frame duration, typical of fixed wireless access. Finally, $\mathbf{W}_k$ is the noise matrix representing the effects of the AWGN, ICI, and ISI.

7.3 Preamble and Pilot Structures

7.3.1 Preamble

The signal transmission matrices $\mathbf{S}_k$, for the preamble and pilot sub-carriers may assume different forms ranging from diagonal to unitary. Fig. 7.1 shows a typical MIMO-OFDM frame structure for Q transmit and L receive antennas. The frame consists of preamble and data portions. The data portion is comprised of length-N data sequences also arranged in a $Q \times Q$ matrix structure, each preceded by a length-G guard interval. The parameter P controls the preamble insertion rate. The preamble is comprised of length $N_I = N/I$ sequences arranged in a $Q \times Q$ matrix structure. Each length-N_I sequence is preceded by a suitable guard interval of length G. Although the training sequences shown in Fig. 7.1 all have length-N_I, the preamble could consist of a mixture of training sequences having different lengths. Also, it is not necessary that training sequences be transmitted from all antennas simultaneously. The IEEE 802.16a standard, for example, specifies the preamble and pilot tone matrix structure for 2×2 OFDM systems using Alamouti' s [298] space-time code. In the standard, a training sequence of the form CP + 2×128 (similar to the case when $I = 2$ in this chapter) is transmitted from one antenna at a time, to maintain medium access control (MAC) conformity and co-existence with single-input single-output (SISO) OFDM systems. This results in a diagonal signal transmission matrix for the preamble.

Since a diagonal structure requires that the preamble sequences be transmitted with a higher power to achieve the same performance, a unitary structure may be preferred that can be constructed based on orthogonal designs for binary space-time block codes. The orthonormal property of the transmission matrix also has the advantage of making the matrix easily invertible. For $Q = 2$, we can use the Alamouti [298]' s signal transmission matrix

$$\mathbf{S} = \frac{1}{\sqrt{2}} \begin{bmatrix} \underline{S}_1 & \underline{S}_1 \\ -\underline{S}_1 & \underline{S}_1 \end{bmatrix}. \qquad (7.3.1)$$

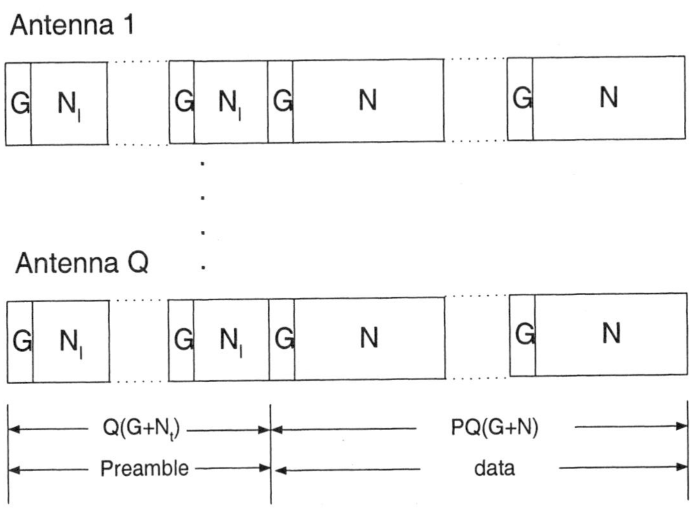

Figure 7.1. Frame structure for a $Q \times L$ MIMO OFDM system.

Likewise, for $Q = 4$,

$$\mathbf{S} = \frac{1}{\sqrt{4}} \begin{bmatrix} \underline{S}_1 & \underline{S}_1 & \underline{S}_1 & \underline{S}_1 \\ -\underline{S}_1 & \underline{S}_1 & -\underline{S}_1 & \underline{S}_1 \\ -\underline{S}_1 & \underline{S}_1 & \underline{S}_1 & -\underline{S}_1 \\ -\underline{S}_1 & -\underline{S}_1 & \underline{S}_1 & \underline{S}_1 \end{bmatrix}. \tag{7.3.2}$$

where $\underline{S}_1$ is a length-N vector.

When the number of transmit antennas is not of the form $Q = 2^\ell$ for some ℓ, e.g., $Q = 3$, either a diagonal matrix may be used or Gram-Schmidt orthonormalization procedure can be used to obtain the signal transmission matrix as shown in Appendix A. However, this procedure yields transmitted symbols in the frequency domain that are not from a finite alphabet.

7.3.2 Pilots

In order to perform tracking functions, pilot tones may be inserted at known positions in the 2-D time-frequency grid in the data portion of the frame. See, for example, Fig. 5.9 in Chapter 4. Similar to the preamble, the pilot tones have a $Q \times Q$ matrix structure. The IEEE 802.16a standard with

$N = 256$, for example, recommends the insertion of 8 pilot tones at fixed sub-channels [12 36 60 84 172 196 220 244], as shown in Fig. 7.2. Fig. 7.2 also shows the method for generating pilot sequences in the frequency domain for the standard. At the beginning of each frame, in the Downlink (DL) and Uplink (UL), the shift register is initialized with sequences as shown in the figure. In the DL, the shift register output is given by the sequence [11111111111000000000110...], where the 3rd 1, i.e., $P_3 = 1$ is used to generate the pilot tones for the first DL OFDM symbol following the preamble. The pilot matrix elements of each OFDM symbol in the DL are initialized as, $\mathbf{S}_{60}^{1,1} = \mathbf{S}_{84}^{1,1} = \mathbf{S}_{172}^{1,1} = \mathbf{S}_{220}^{1,1} = 1 - 2P_d$ and $\mathbf{S}_{12}^{1,1} = \mathbf{S}_{36}^{1,1} = \mathbf{S}_{196}^{1,1} = \mathbf{S}_{244}^{1,1} = 1 - 2\bar{P}_d$, where $\bar{P}_d$ is the complement of P_d. For the UL, the pilot tones are assigned as $\mathbf{S}_{12}^{1,1} = \mathbf{S}_{36}^{1,1} = \mathbf{S}_{60}^{1,1} = \mathbf{S}_{84}^{1,1} = \mathbf{S}_{172}^{1,1} = \mathbf{S}_{220}^{1,1} = 1 - 2P_d$ and $\mathbf{S}_{196}^{1,1} = \mathbf{S}_{244}^{1,1} = 1 - 2\bar{P}_d$. For MIMO-OFDM, the pilot symbols are inserted as known signal transmission matrices $\mathbf{S}_k^p$ with the value of P_d determining the entire pilot matrix when Q is of the form 2^ℓ.

Some work has optimized the preamble sequences and structures for OFDM systems [216], [300]. Preamble sequences must have desirable properties in both the time and frequency domains. These include constant envelope properties in the frequency domain, good autocorrelation properties in the frequency and time domain, and low peak-to-average power ratio in the time domain. Often the preamble sequence is designed in the frequency domain with elements chosen from the modulation alphabet or a subset thereof, converted to the time domain, and stored for direct modulation. For best channel estimation performance, the sequences must have a flat spectrum, where the spectral flatness can be measured using the spectral max-min ratio (SMMR) given by

$$\Xi(\underline{S}) = \frac{\{\max|\underline{S}_k| : 0 \leq k \leq N\}}{\{\min|\underline{S}_k| : 0 \leq k \leq N\}} . \tag{7.3.3}$$

For good synchronization performance, the sequences transmitted from different antennas should be orthogonal in time, must have good correlation properties, and low peak to average power ratio (PAPR) [301]. Here, we use the sequences suggested in the IEEE 802.16a Standard to construct the preamble. The standard suggests a sequence for the value of $N_I = 128$. For other values of N_I we have generated our own sequences using computer search. The sequences are generated with BPSK alphabet in the frequency domain and the sequence having the lowest PAPR is selected to form the preamble.

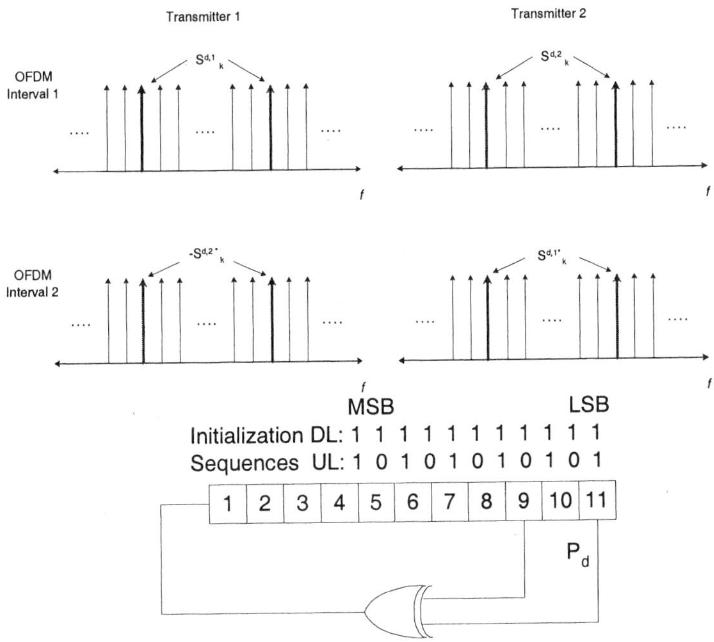

Figure 7.2. An example of pilot mapping for a $2 \times L$ OFDM system and pilot generation using a shift register [299].

7.4 Time Synchronization and Sample/RF Frequency Offset Estimation

Signal acquisition as defined in this chapter consists of time synchronization, RF oscillator frequency offset estimation and correction, sampling frequency offset estimation and correction, and initial channel estimation. Fig. 7.3 depicts the overall synchronization functionality of the MIMO-OFDM system. In our described method synchronization takes place in the following steps: *Step A* coarse time synchronization, *Step B* Sample/RF oscillator frequency offset estimation, *Step C* Integer Frequency Offset correction, *Step D* fine time synchronization, *Step E* Sample/RF Frequency Offset Estimation, and *Step F* initial channel estimation.

The MIMO-OFDM receiver first performs coarse time synchronization to obtain an approximate range of samples over which the OFDM frame is

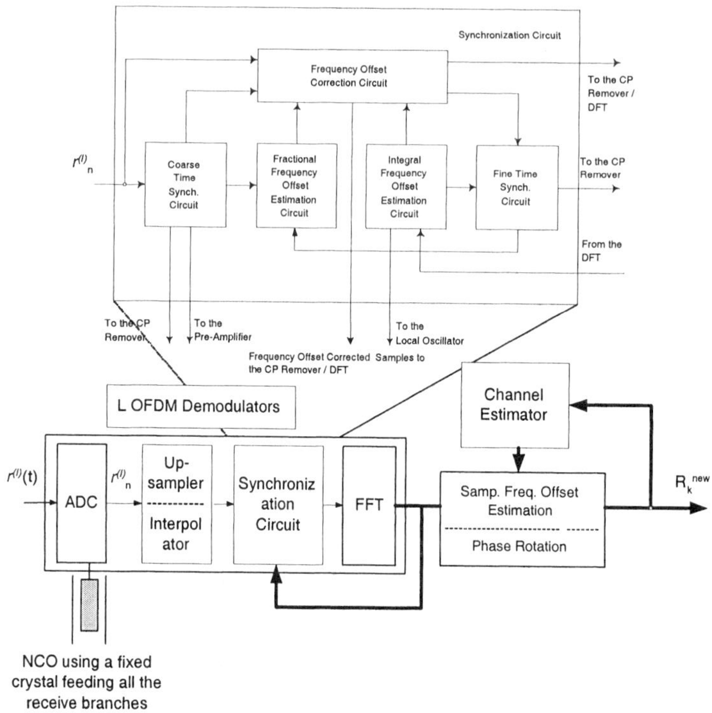

Figure 7.3. Receiver Implementation of a MIMO-OFDM System.

likely to begin. This is followed by frequency offset estimation and correction in the time and the frequency domains. Finally, fine time synchronization determines the optimal sampling instant to within a small range of samples as discussed in Chapter 4.

Step A. Coarse Time Synchronization- Coarse time acquisition can be performed by exploiting the repeated samples in the guard interval. The discrete autocorrelation of the sampled received complex envelope is computed over a window of G samples that are at a distance of N_I from each other. Müller-Weinfurtner [302], has analyzed the various metrics proposed in the literature for time synchronization based on the cyclic prefix in OFDM systems. It is found that a simple threshold comparison between the autocorrelation of the sampled received complex envelope and the energy of the complex envelope over the autocorrelation window yields an ML estimate and is an adequate metric to obtain the coarse time synchronization instant. A threshold ρ_{coarse} is introduced to reduce the probability of false lock or false alarm (P_{FA}). The autocorrelation can be calculated using the signals received over one or more receiver branches. Here we calculate the autocorrelation using all the receive branches. Our coarse time synchronization metric is

$$n^{\ell}_{\text{coarse}} = \left\{ \arg\underbrace{\max}_{n} \left\{ \phi^{\ell}_{n} \right\} : |\phi^{\ell}_{n}| \geq \rho_{\text{coarse}} \cdot (P^{\ell}_{n} + P^{\ell}_{n+N_I}) \right\}. \quad (7.4.1)$$

where $\phi^{\ell}_{n} = \sum_{k=0}^{G-1} (r^{\ell\,*}_{n+k} \cdot r^{\ell}_{n+k+N_I})$ is the autocorrelation function and $P^{\ell}_{n} = \sum_{k=0}^{G-1} |r^{\ell}_{n+k}|^2$ is the signal energy calculated over G samples. We empirically choose the threshold ρ_{coarse} as 10% of the incoming signal energy over the correlation window.

Step B. Simple RF Frequency Offset Estimation - Frequency offset estimation can be carried out based on the inherent periodicity in the OFDM signal and the preamble due to the cyclic prefix. The cyclic prefix and the last G samples of the useful part of the OFDM symbol are exactly the same in the absence of a frequency offset and are affected by the channel in a similar fashion if the channel changes slowly enough. If any frequency offset exists between the transmitter and receiver RF oscillators, it is reflected in the cyclic prefix and the posterior part of the OFDM symbol as a proportional phase shift $\theta = 2\pi\epsilon\frac{N_I}{N}$, where ϵ is the fractional portion of the RF frequency offset. Moose [303] showed that a relative frequency offset estimate of up to $\pm I/2$ subchannel spacings can be generated based on the phase of the autocorrelation function ϕ^{ℓ}_{n} at time instant n^{ℓ}_{coarse} as:

$$\widehat{\epsilon}^{\ell} = \frac{I}{2\pi} \angle \{\phi^{\ell}_{n^{\ell}_{\text{coarse}}}\}, \quad (7.4.2)$$

where $I = N/N_I$. Moose [303] further showed this to be an ML estimator for the frequency offset. The frequency offset can then be removed from the received sample stream at the ℓth receive antenna by multiplying it by $\exp\left\{-j2\pi\widehat{\epsilon}^{\ell} n/N_I\right\}$ during the training sequence and $\exp\left\{-j2\pi\widehat{\epsilon}^{\ell} n/N\right\}$ during the data sequence. Note that a reduction in the length of the preamble sub-sequences by a factor of I increases the fractional frequency offset estimation range by a factor of I. Hence, in terms of frequency offset estimation, short training sequences in the preamble may be useful.

Step C. Residual Frequency Offset Correction - A frequency offset of an integral number of subchannel spacings may still exist and it must be removed. If a diagonal signal transmission matrix is used, or if a non-diagonal signal transmission matrix is constructed such that the same sequence is transmitted from all the transmit antennas for $d = 1$, a cyclic cross-correlation of the demodulated OFDM symbol with the original sequence can be performed to estimate the residual frequency offset k_o [304]. The received frequency corrected time domain sequence from the first stage of frequency offset compensation $\{r_n^{1,\ell\ c}\}_{n=0}^{N_I-1}$ is repeated I times and its N-point FFT yields the sequence $\{R_k^{1,\ell\ c}\}_{k=0}^{N-1}$. Then the cyclic cross-correlation (in the frequency domain) with the known transmitted sequence is

$$\chi_k = \sum_{\ell=1}^{L} \left| \sum_{n=0}^{N-1} S_{(k+n)_N}^{1,1\ *} R_n^{1,\ell\ c} \right| \quad k = 0, 1, \ldots, N-1, \tag{7.4.3}$$

and frequency offset of integral number of subchannel spacings is estimated using $\widehat{k_o} = \text{argmax}\{\chi_k\}$. In summary, the fractional part of the relative frequency offset within $\pm I/2$ sub-carrier spacings can be estimated in the time domain, while the integer part beyond $\pm I/2$ sub-carrier spacings can be estimated in the frequency domain.

Step D. Fine Time Synchronization - Fine time acquisition consists of finding the start of the useful portion of the OFDM frame to within a few samples. Once the frequency offset is removed, fine time synchronization can be performed by cross-correlating the frequency compensated samples of the received complex envelope with the transmitted time domain preamble sequences. Fine time synchronization ensures detection of the training sequence, especially when $N_I = N$, and reduces P_{FA}. For a MIMO-OFDM system with Q transmitters, Q cross-correlators can be employed for each receive antenna and the fine time synchronization metric computed as

$$n_{\text{fine}}^{\ell} = \left\{ \arg \underbrace{\max}_{n} \left\{ \psi_n^{\ell} \right\} : \psi_n^{\ell} \geq \rho_{\text{fine}} \cdot P_n^{\ell} \right\}, \tag{7.4.4}$$

where the cross-correlation sequence is $\psi_n^{\ell} = \sum_{q=1}^{Q} \left| \sum_{k=0}^{N_I-1} (s_k^{1,q\ *} \cdot r_{n+k}^{\ell\ c}) \right|$, $P_n^{\ell} = \sum_{k=0}^{N_I-1} |r_{n+k}^{\ell\ c}|^2$ and the $r_{n+k}^{\ell\ c}$ are the received frequency offset corrected samples. If the same sequences are transmitted from all antennas then only one cross-correlator is needed per receiver antenna. Once again, the threshold ρ_{fine} is set at 10% of the energy contained in the N_I received samples.

Since fine time synchronization is a computationally expensive process, it is conducted for a small window centered around the coarse time synchronization instant n^{ℓ}_{coarse}. The net time synchronization instant is chosen as $n_{\text{opt}} = \sum_{\ell=1}^{L} \alpha_\ell n^{\ell}_{\text{fine}}$ where non-equal weights may be applied to put more emphasis on the stronger signal. A negative offset of a few samples is applied to n_{opt} to ensure that the fine time synchronization instant lies within the guard interval. Finally, the sampling frequency at the receiver may be increased by an integer factor to obtain a more accurate time and frequency synchronization. Our simulation results show an improvement in the BER floor if the sampling frequency is doubled at the receiver.

7.4.1 Sample/RF Frequency Offset Estimation

Here we describe a novel SF offset estimator to estimate the SF offset coefficient β, and either the residual RF oscillator frequency offset ϵ_e (after coarse and fine frequency synchronization) or the phase noise ϑ, but not both. Starting with the dth block, the received sample matrix corresponding to the kth pilot subchannel position is

$$\mathbf{R}_k^d = A_k^d C_k \mathbf{\Lambda}_k \mathbf{S}_k^d \cdot \mathbf{H}_k^d + \mathbf{W}_k^d \ . \tag{7.4.5}$$

If we proceed under the assumption that the channel is sufficiently static for $2Q$ consecutive OFDM symbols (such as fixed wireless access), then the received sample matrix corresponding to the pilot subchannel positions for the next block of Q OFDM symbols is

$$\mathbf{R}_k^{d+Q} = A_k^{d+Q} C_k \mathbf{\Lambda}_k \mathbf{S}_k^{d+Q} \cdot \mathbf{H}_k^d + \mathbf{W}_k^{d+Q} \ . \tag{7.4.6}$$

Hence, we can correlate $\mathbf{R}_k^d$ and $\mathbf{R}_k^{d+Q}$ at the pth pilot tone position to obtain an initial per sub-carrier estimate of $D_p = k_p\beta + \Upsilon$ as,

$$\begin{aligned} \widehat{D_p} &= \frac{\angle \mathrm{Tr}[\mathbf{R}_{k_p}^{d\ \mathrm{H}} \mathbf{R}_{k_p}^{d+Q}] + f\left(P_d, P_{d+Q}\right)}{2\pi Q(N+G)/N} \\ \widehat{D_p} &= \frac{\angle \sum_{d=1}^{Q} \sum_{\ell=1}^{L} |R_{k_p}^{d,\ell}|^2 e^{j2\pi Q(k_p\beta+\Upsilon)(1+\frac{G}{N})} + f\left(P_d, P_{d+Q}\right)}{2\pi Q(1+\frac{G}{N})}, \end{aligned} \tag{7.4.7}$$

where $\widehat{\Upsilon} = \widehat{\epsilon_e}$ or $\widehat{\Upsilon} = [\widehat{\vartheta}^{d+Q} - \widehat{\vartheta}^{d}]/[2\pi Q(1+G/N)]$ and $f\left(P_d, P_{d+Q}\right) = \pi$ if $P_d \neq P_{d+Q}$ and 0 if $P_d = P_{d+Q}$. Combining, the results for all M pilot tone positions, the above measurements may be put in the matrix form

$$\underbrace{\begin{bmatrix} \widehat{D}_1 \\ \vdots \\ \widehat{D}_M \end{bmatrix}}_{\underline{\widehat{D}}} = (2\pi Q(1+G/N)) \underbrace{\begin{bmatrix} k_1 & 1 \\ \vdots & \vdots \\ k_M & 1 \end{bmatrix}}_{\mathbf{J}} \underbrace{\begin{bmatrix} \beta \\ \Upsilon \end{bmatrix}}_{\underline{\widehat{\theta}}} + \underbrace{\begin{bmatrix} \epsilon_1 \\ \vdots \\ \epsilon_M \end{bmatrix}}_{\underline{\epsilon}} . \quad (7.4.8)$$

The optimum set of weights $\{k_1, \ldots, k_M\}$ in matrix J depends on the pilot tone matrix structure as described in Appendix B.

Step E. Sample/RF Frequency Offset Estimation - The estimate of the sample frequency offset $\underline{\widehat{\theta}} = [\widehat{\beta}\ \widehat{\Upsilon}]^T$ is obtained using a least-square fit as

$$\underline{\widehat{\theta}} = \frac{1}{(2\pi Q(1+G/N))}(\mathbf{J}^H\mathbf{J})^{-1}\mathbf{J}^H\underline{\widehat{D}} \ . \quad (7.4.9)$$

At this stage, the system designer needs to choose whether to estimate the residual frequency offset ϵ_e or the phase noise difference $[\widehat{\vartheta}^{d+Q} - \widehat{\vartheta}^d]$. If the frame-size is large, then any residual frequency offset causes a linear increase in the phase of the received symbols with time which results in a BER floor. On the other hand, for short frame sizes, large fluctuations in the phase noise should not be ignored. Hence, either the residual frequency offset can be estimated as $\widehat{\epsilon_e} = \left[\underline{\widehat{\theta}}\right]_2$ or the phase noise difference may be estimated as $[\widehat{\vartheta}^{d+Q} - \widehat{\vartheta}^d] = [2\pi Q(1+\frac{G}{N})]\left[\underline{\widehat{\theta}}\right]_2$, but not both.

The residual frequency offset estimate is fed back to the time domain RF oscillator frequency offset correction circuit, whereas the estimates of the SF offset and the phase noise difference, are used to construct the terms $\widehat{A}_k^d\widehat{C}_k\widehat{\mathbf{\Lambda}}_k$. The new estimate of the received and demodulated sample matrix is then obtained using $\widehat{\mathbf{R}}_k = \left(\widehat{A}_k^d\widehat{C}_k\widehat{\mathbf{\Lambda}}_k\right)^{-1}\mathbf{R}_k$ which is used for decoding the data. The theoretical MSE performance of the above estimator is derived in Appendix C.

Tracking Mode

The sampling time T'_{sa} at the receiver suffers from jitter, which is reflected by changes in the SF offset coefficient β. At the same time, the RF phase noise may assume a new value. If we assume that β and ϑ assume new values every Q OFDM symbols, then new estimates of these parameters may be obtained using (7.4.9). If $\widehat{\beta}^{d-Q}$ and $\widehat{\beta}^d$ are the sampling frequency offset estimates for the MIMO-OFDM blocks at the $(d-Q)^{\text{th}}$ and d^{th} OFDM symbols, respectively, then the final estimate of the sampling frequency offset may be computed using a first-order frequency locked loop

$$\widehat{\beta}^{d} = \alpha_{\beta}\widehat{\beta}^{d-Q} + (1-\alpha_{\beta})\widehat{\beta}^{d} \ . \tag{7.4.10}$$

Similarly, phase noise difference estimates may be obtained using (7.4.9) for every Q OFDM symbols. The estimate $\widehat{\beta}^{d}$ is then used to compensate for sampling frequency offset by computing $\widehat{\mathbf{R}}_k = (\widehat{A}_k^d\widehat{C}_k\widehat{\mathbf{\Lambda}}_k)^{-1}\mathbf{R}_k$. where $(\widehat{A}_k^d\widehat{C}_k\widehat{\mathbf{\Lambda}}_k)$ is formed using $\widehat{\beta}^{d}$.

The above tracking scheme does not take into account the phase error built up due to error in the sampling frequency offset estimation. When this happens, compensation must be carried out partly in time domain and partly in the frequency domain as suggested in [305] and requires the use of a Phase Locked Loop (PLL).

7.4.2 There is a Time for Channel Estimation

The receiver should first estimate the sampling frequency offset using signal matrices transmitted at the pilot subchannels in the preamble and the next block of Q OFDM data symbols, before estimating the channel. This ensures that the sampling frequency offset has already been corrected before the channel is estimated. If statistics of the channel and noise are available at the receiver then they can be used to initially estimate the channel as [306]

$$\widehat{\mathbf{H}}_k = (\mathbf{B}_k^H\mathbf{C}_{HH,k}\mathbf{B}_k + \mathbf{C}_{WW,k})^{-1}(\mathbf{C}_{HH,k}\mathbf{B}_k^H)\mathbf{R}_k \ , \tag{7.4.11}$$

where $\mathbf{B}_k = \mathbf{\Lambda}_k\mathbf{S}_k$, $\mathbf{C}_{HH,k} = E[\eta_k\eta_k^H]$ is the channel correlation matrix of dimension $Q\times Q$ and $\mathbf{C}_{WW,k} = E[\mathbf{W}_k\mathbf{W}_k^H]$ is the noise covariance matrix of dimension $Q\times Q$. In the absence of the necessary correlation functions the LS estimate can be obtained as

$$\widehat{\mathbf{H}}_k = (\mathbf{B}_k^H\mathbf{B}_k)^{-1}\mathbf{B}_k^H\mathbf{R}_k. \tag{7.4.12}$$

The MSE of the channel estimation is equal to N_o as shown in Appendix D.

When $N_I \neq N$, the initial channel coefficients obtained from the preamble must be interpolated/ extrapolated to obtain the channel estimates for all the sub-carriers. For example, a training sequence of length $N_I = N/4$ corresponds to the excitation of every fourth subchannel of an N-point system. Here we use simple linear interpolation/ extrapolation to estimate the missing channel estimates followed by simple frequency domain smoothing $\widehat{\eta}_{ij,k} = (\bar{\eta}_{ij,k-1} + \bar{\eta}_{ij,k+1})/2$. Of course a variety of more complex algorithms may be employed to improve the channel estimates.

Table 7.1. SUI-4 Channel Model [308]

	Tap 1	Tap 2	Tap 3	Units
Delay	0	1.5	4.0	μs
Power (omni ant.)	0	-4	-8	dB
f_m	0.2	0.15	0.25	Hz

7.5 Simulation Results

Simulations have been carried out for a 4 × 4 MIMO-OFDM system very similar to IEEE802.16a [299] operating on a Stanford University Interim SUI-4 channel [307] for fixed wireless access. The SUI-4 model is appropriate for intermediate path loss with moderate to heavy tree densities and has the parameters listed in Table 7.1. The Doppler power spectrum for the channel taps is [308]

$$S(f) = \begin{cases} 1 - bf_0^2 + cf_0^4 & |f_0| \leq 1 \\ 0 & |f_0| > 1 \end{cases} \quad \text{where } f_0 = \frac{f}{f_m}$$

where $b = 1.72$, $c = 0.784$ and f_m represents the maximum Doppler frequency. Rayleigh faded tap coefficients for the MIMO channel are generated using a filtered Gaussian noise source as described in [2]. The fading tap coefficients are uncorrelated. The sampling rate is assumed to be $1/T_{\text{sa}} = 4.0$ MHz.

The preamble signal matrix with 4 transmit antennas shown in (7.3.2) is used with BPSK modulation. The OFDM block size is $N = 256$, and the guard interval is $G = N/4 = 64$. Out of 256 tones, the dc tone and 55 other tones at the band edges are set to zero [299]. This means that for $N = 256$, tones $[1 \ldots 100\ 156 \ldots 255]$ contain data and the number of used tones is N_u=200. The preamble of a system with $I = 2$, would have a known symbol at subchannels $[2\ 4 \ldots 100\ 156\ 158 \ldots 254]$ and the channel estimates for the remaining sub-channels must be obtained by interpolation or extrapolation. In the data mode, 8 pilot tones are inserted for each OFDM block in a manner described earlier. Each frame consists of a preamble followed by $F = PQ$ (typically 80) OFDM symbols.

The sampling frequency offset coefficient β is initialized at 10 parts per million (ppm) and fluctuates around that value every Q OFDM symbols, as $\beta = 10\text{ppm} + N(0, 10\% \text{ of } \beta \text{ ppm})$, characterizing the timing jitter. Assuming that the RF and SF clocks are derived from different sources, ϵ is initialized to 0.2 subchannel spacings, and is made to vary from frame to frame in a random walk as $\epsilon = 0.2 + N(0, 10\%$ of ϵ ppm), symbolizing the relatively slow RF oscillator frequency offset drift. Phase noise is assumed to have a zero-mean Gaussian distribution with variance (Var$[\vartheta]$) of 10π and 50π ppm. Phase noise is assumed to change every Q OFDM symbols. The integer frequency offset k_o is assumed to be one sub-carrier spacing.

Figure 7.4 shows the coarse and fine time synchronization performance for a 4×4 system with $N_I = 128, I = 2$, and E_s/N_0=10 dB. The symbol energy to noise ratio $[E_s/N_0]_\ell$ is defined as $\sum_{q=1}^{Q} \mathrm{E}[|H_k^{q,\ell}|^2]/(QN_0)$.

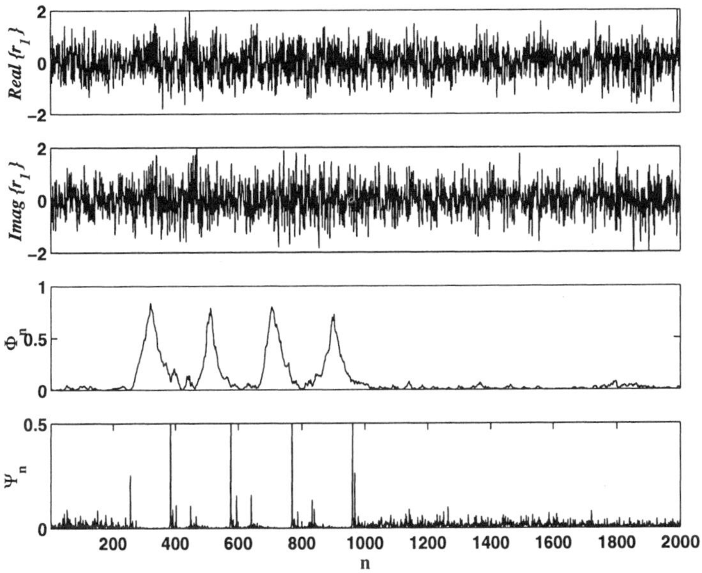

Figure 7.4. Coarse and fine time synchronization for a 4×4 system with $N_I = 128$ at an E_s/N_0 of 10 dB and frequency offset $k_o + \epsilon$ of 1.2 subchannel spacings. Steps A and D.

Figure 7.5 shows the performance of the channel estimator for various values of I (or N_I). Here we use the LS channel estimator in (7.4.12) with frequency domain linear interpolation extrapolation and smoothing. No further post-processing is performed on the channel estimates. As I increases, the channel estimation performance becomes poorer because of imperfect interpolation and extrapolation, especially at the band edges.

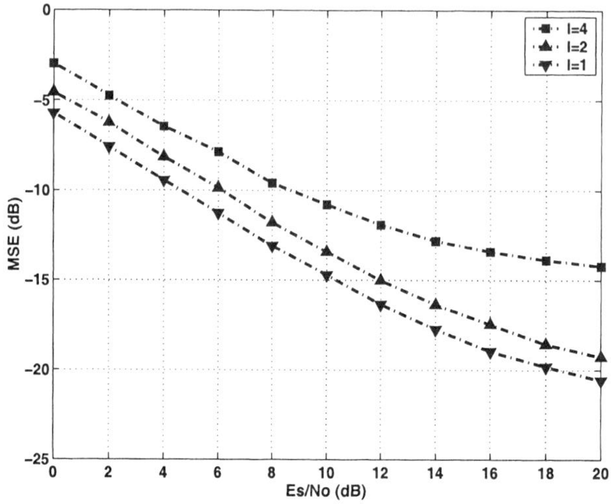

Figure 7.5. MSE in channel estimate as a function of E_s/N_0, $N = 256$, $N_u = 200$ with linear interpolation and frequency domain smoothing.

Fig. 7.6 shows the theoretical and simulated MSE performance of the sampling/RF frequency offset estimator. Simulated MSE performance closely follows the theoretical values, and performs well for the system under consideration. Fig. 7.7 shows the theoretical and simulated MSE of the proposed residual frequency offset estimator. Our analysis and simulation results confirm that this estimator yields a 13 dB performance improvement in MSE over the Simple RF Frequency Offset Estimator. Fig. 7.8 shows the theoretical and simulated MSE performance of the phase noise estimator for a perfect frequency offset estimation, and frequency offset estimation using *Steps A-C* defined earlier. As can be seen, presence of residual frequency offset affects the phase noise estimate.

7.6 Summary and Further Reading

Time synchronization, RF frequency offset estimation and tracking, channel estimation and tracking, and sampling frequency offset estimation and

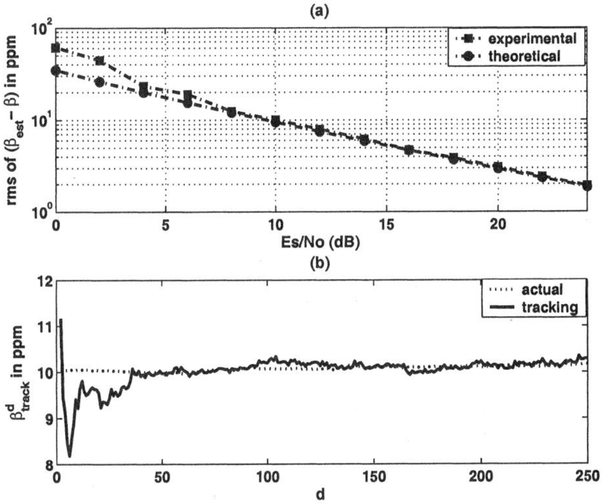

Figure 7.6. (a) Estimation performance - Analytical and experimental root mean square (rms) of the $(\widehat{\beta} - \beta)$ in ppm as a function of E_s/N_0 for a 4×4 system. (b) Tracking performance - SF offset variations and the tracking performance using a single pole frequency lock loop with $\alpha_\beta = 0.5$ at an $E_s/N_0 = 20$ dB.

tracking are essential functions in a MIMO-OFDM receiver. A large number of studies have addressed these issues, but usually individually and for single-input single-output (SISO) OFDM systems. Moose [303] proposed a popular technique for robust frequency offset estimation based on the transmission of two identical OFDM symbols. van de Beek *et al.* [159] suggest time synchronization based on the periodicity of the cyclic prefix (CP) and proved it to be a maximum likelihood (ML) estimator. In their method, the estimate of the frequency offset is obtained from the phase of the auto-correlation function and the estimation range is $\pm 1/2$ subchannel spacings. Other work on synchronization in SISO-OFDM systems is available in [117], [132], [137], [143], [309], [310], [311], [312].

Most approaches for OFDM channel estimation estimate the channel in the frequency domain followed by some post-processing for performance enhancement [159], [175], [306], [313], [314], [315]. Almost all this existing work assumes perfect time and frequency synchronization. Finally, some literature on SISO-OFDM system implementation includes the work by Speth [316] who has proposed a receiver architecture for digital video broadcasting (DVB-T) applications.

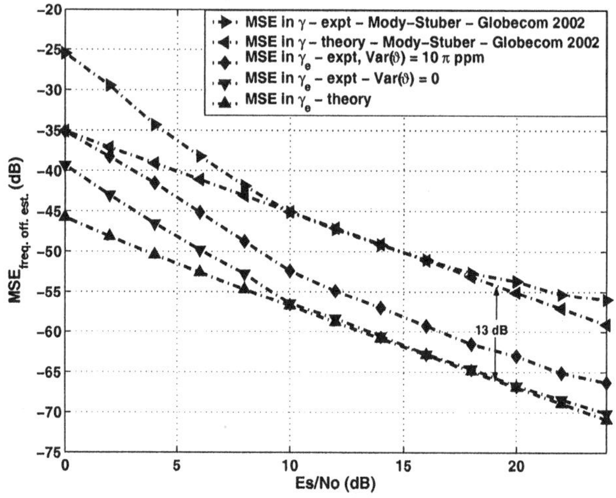

Figure 7.7. Theoretical and experimental MSE in frequency offset estimates as a function of E_s/N_0 for a 4×4 system using Mooses' method (7.4.2), and the residual sample/RF oscillator frequency offset estimation in the presence and the absence of phase noise. $\epsilon = 0.2$ subchannel spacings, Frame size $F = 80$ OFDM symbols. Our estimator gives approximately 13 dB gain as compared to Mooses' approach.

Appendix 7A: Gram-Schmidt Orthonormalization to Make S_ks Unitary

Given a matrix $\mathbf{S}_{in_k}$ whose rows are not orthonormal we can use the Gram-Schmidt procedure [317] to orthonormalize the rows. The matrix $\mathbf{S}_{out_k}$ that is formed using this procedure is unitary provided that $\mathbf{S}_{in_k}$ has full rank. The first row of $\mathbf{S}_{in_k}$ remains unchanged and based on the first, row Gram-Schmidt procedure is performed to make the remaining rows orthonormal. The following example shows the conversion of a 3×3 matrix $\mathbf{S}_{in_k}$ to $\mathbf{S}_{out_k}$.

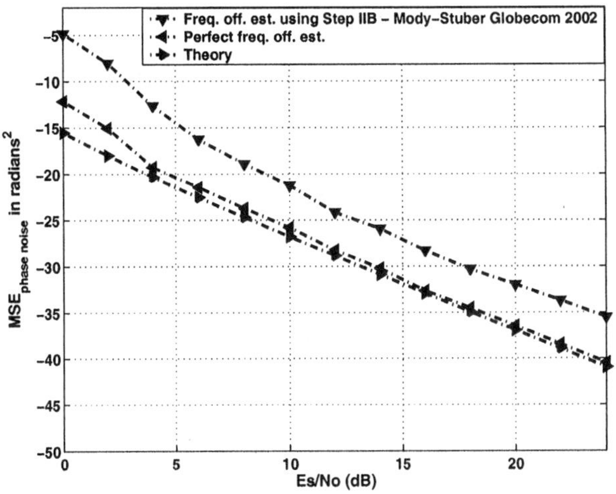

Figure 7.8. Analytical and experimental MSE in the phase noise estimates in radians^2 as a function of E_s/N_0 for a 4×4 system, $\text{Var}[\vartheta] = 50\pi$ ppm, Frame size $F = 40$ OFDM symbols. The graph shows that the presence of any residual RF oscillator frequency offset will affect the performance of the phase noise estimator.

Given $\mathbf{S}_{in_k}$ whose elements are chosen from a 16-PSK alphabet,

$$\mathbf{S}_{in_k} = \begin{bmatrix} .5774e^{0j} & .5774e^{1.5708j} & .5774e^{0j} \\ .5774e^{2.3562j} & .5774e^{0j} & .5774e^{0j} \\ .5774e^{0j} & .5774e^{-0.3926j} & .5774e^{0j} \end{bmatrix}, \qquad (7A.1)$$

after the Gram-Schmidt orthogonalization, we obtain a unitary $\mathbf{S}_{out_k}$ as follows:

$$\mathbf{S}_{out_k} = \begin{bmatrix} .5774e^{0j} & .5774e^{1.5708j} & .5774e^{0j} \\ .6634e^{2.3562j} & .5291e^{-0.1078j} & .5291e^{0.1078j} \\ .4760e^{-0.5890j} & .6219e^{-0.9818j} & .6219e^{1.3745j} \end{bmatrix}. \qquad (7A.2)$$

Appendix 7B: Optimum Weight Calculation for Sampling Frequency Offset Estimates

From ([318], page 57), the Cramer-Rao Bound (CRB) for estimation of the frequency of a sinusoid is $\mathrm{Var}(\widehat{\omega}) \geq 12/(E_s/N_0)M(M^2-1)$ where ω is the angular frequency, and M is the number of observations. Since $\omega = 2\pi k\widehat{\beta}(N+G)/N$, the CRB for $\widehat{\beta}_k$ may be simplified to $\mathrm{Var}(\widehat{\beta}_k) \geq 1/Ck^2$, where C is a constant. Let $\alpha_1, \alpha_2, \ldots, \alpha_M$ be the weights for M pilot tones $k_1, k_2, \ldots, k_M$ such that $\sum_{m=1}^{M} \alpha_m = 1$. Then the final estimate $\widehat{\beta}$ is

$$\widehat{\beta} = \alpha_1\widehat{\beta}_1 + \alpha_2\widehat{\beta}_2 + \ldots + \left(1 - \sum_{m=1}^{M-1} \alpha_m\right)\widehat{\beta}_M. \tag{7B.1}$$

To find the optimal weights we obtain the minimum

$$\begin{aligned}\min_{\alpha_i}\left\{E[(\beta-\widehat{\beta})^2]\right\} &= {\alpha_1}^2 E[(\beta-\widehat{\beta}_1)^2] + \ldots \\ &+ \left(1 - \sum_{m=1}^{M-1} \alpha_m\right)^2 E[(\beta-\widehat{\beta}_M)^2], \quad (7B.2)\end{aligned}$$

by taking derivative and setting it to zero as

$$\begin{aligned}\frac{d}{d\alpha_i}\left[\frac{\alpha_1^2}{Ck_1^2} + \frac{\alpha_2^2}{Ck_2^2} + \ldots + \left(1 - \sum_{m=1}^{M-1}\alpha_m\right)^2 \frac{1}{Ck_M^2}\right] &= 0 \quad (7B.3)\\ \Rightarrow \frac{\alpha_i}{k_1^2} + \frac{1}{k_M^2}\sum_{m=1}^{M-1}\alpha_m &= \frac{1}{k_M^2}.\end{aligned}$$

This system of linear equations can be put in a matrix form and solved for $\underline{\alpha}$ as

$$\begin{bmatrix} \frac{k_M^2}{k_1^2}+1 & 1 & 1 & \ldots & 1 \\ 1 & \frac{k_M^2}{k_2^2}+1 & 1 & \ldots & 1 \\ \vdots & & \ddots & & \vdots \\ 1 & 1 & \ldots & 1 & \frac{k_M^2}{k_{M-1}^2}+1 \end{bmatrix} \begin{bmatrix} \alpha_1 \\ \alpha_2 \\ \vdots \\ \alpha_{M-1} \end{bmatrix} = \begin{bmatrix} 1 \\ 1 \\ \vdots \\ 1 \end{bmatrix}, \tag{7B.4}$$

and $\alpha_M = \left(1 - \sum_{m=1}^{M-1}\alpha_m\right)$. For 8 subchannels placed at [12 36 60 84 172 196 220 244] as in IEEE 802.16a [299], the optimal weights are [0.0009 0.0072 0.0196 0.0380 0.1574 0.2042 0.2569 0.3158].

Appendix 7C: MSE Analysis for the Sampling Frequency/ Residual RF Oscillator Frequency Offset Estimator

For the MSE analysis of the sampling frequency offset estimator, we assume without loss of generality that $P_d = P_{d+Q}$, which implies that $f(P_d, P_{d+Q}) = 0$. The received OFDM-demodulated sample matrix at pilot tone p is

$$\mathbf{R}_p^d = \underbrace{A_p^d C_p \mathbf{\Lambda}_k \mathbf{S}_p^d \cdot \mathbf{H}_p^d}_{\mathbf{X_p}} + \mathbf{W}_p. \tag{7C.1}$$

The term $\angle \mathrm{Tr}[\mathbf{R}_p^{d\ \mathrm{H}} \mathbf{R}_p^{d+Q}]$ in the sampling and the residual frequency offset estimator can be expanded as

$$\begin{aligned} \angle \mathrm{Tr}[\mathbf{R}_p^{d\ \mathrm{H}} \mathbf{R}_p^{d+Q}] &= \angle \mathrm{Tr}[(\mathbf{X}_p^{d\ H} + \mathbf{W}_\mathbf{p}^{\mathbf{d}\ \mathbf{H}})((\mathbf{X}_p^{d+Q} + \mathbf{W}_p^{d+Q})] \quad (7C.2)\\ &= \angle \Big\{ \mathrm{Tr}[\mathbf{X}_p^{d\ H} \mathbf{X}_p^{d+Q}] + \mathrm{Tr}[\mathbf{X}_p^{d\ H} \mathbf{W}_p^{d+Q}] \\ &\quad + \mathrm{Tr}[\mathbf{W}_p^{d\ H} \mathbf{X}_p^{d+Q}] + \mathrm{Tr}[\mathbf{W}_p^{d\ H} \mathbf{W}_p^{d+Q}] \Big\} \quad (7C.3)\\ &= \angle \Big(A \cdot \exp\{j\chi\} + \epsilon_p^{'} \Big), \end{aligned}$$

where $A = \left| \mathrm{Tr}[\mathbf{X}_p^{d\ H} \mathbf{X}_p^{d+Q}] \right| = \sum_{q=1}^{Q} \sum_{\ell=1}^{L} |H_p^{q,\ell}|^2$, χ is the desired phase, and $\epsilon_p^{'}$ represents the contribution from the error terms. Assuming a small error analysis, the angle of the error terms may be expressed as

$$\epsilon_p \approx \mathrm{Im}\{\epsilon_p^{'} \exp\{-j\chi\}\}/A. \tag{7C.4}$$

Let $\underline{\theta} = [\beta\ \Upsilon]^T$ and $\widehat{\underline{\theta}} = [\widehat{\beta}\ \widehat{\Upsilon}]^T$. From (7.4.9),

$$\widehat{\underline{\theta}} = \frac{1}{(2\pi Q(1 + G/N))} (\mathbf{J}^H \mathbf{J})^{-1} \mathbf{J}^H \underline{D}. \tag{7C.5}$$

Substituting the value of $\underline{D}$ yields

$$\widehat{\underline{\theta}} = \frac{1}{(2\pi Q(1 + G/N))} (\mathbf{J}^H \mathbf{J})^{-1} \mathbf{J}^H ((2\pi Q(1 + G/N)) \mathbf{J} \underline{\theta} + \underline{\epsilon}). \tag{7C.6}$$

Hence,

$$\widehat{\underline{\theta}} = \underline{\theta} + \frac{1}{(2\pi Q(1 + G/N))} (\mathbf{J}^H \mathbf{J})^{-1} \mathbf{J}^H \underline{\epsilon}. \tag{7C.7}$$

The MSE in the estimates of $\underline{\theta}$ ($\text{MSE}_{\widehat{\underline{\theta}}}$) may be expressed as

$$\begin{aligned}
\mathrm{E}\left[(\widehat{\underline{\theta}}-\underline{\theta})(\widehat{\underline{\theta}}-\underline{\theta})^H\right] &= \mathrm{E}\left[\left(\frac{1}{(2\pi Q(1+G/N))}(\mathbf{J}^H\mathbf{J})^{-1}\mathbf{J}^H\underline{\epsilon}\right)\right.\\
&\quad \times\left.\left(\frac{1}{(2\pi Q(1+G/N))}(\mathbf{J}^H\mathbf{J})^{-1}\mathbf{J}^H\underline{\epsilon}\right)^H\right]\\
&= \frac{1}{(2\pi Q(1+G/N))^2}(\mathbf{J}^H\mathbf{J})^{-1}\mathbf{J}^H\mathrm{E}\left[\underline{\epsilon\epsilon}^H\right]\\
&\quad \times\left((\mathbf{J}^H\mathbf{J})^{-1}\mathbf{J}^H\right)^H. \qquad (7\text{C}.8)
\end{aligned}$$

The term $\mathrm{E}\left[\underline{\epsilon}\;\underline{\epsilon}^H\right]$ is a diagonal matrix since the transmitted symbols and noise at different sub-carriers are uncorrelated. Assuming that the $\epsilon_p^{'}$ are Gaussian distributed, the $(p,p)^{\text{th}}$ element is given by

$$\begin{aligned}
\mathrm{E}[\underline{\epsilon\epsilon}^H]_{(p,p)} &= \mathrm{E}\left[\mathrm{Im}(\epsilon_p^{'})\mathrm{Im}(\epsilon_p^{'})/(A)^2\right]\\
&= [N_0\bar{\mathbf{H}}^2+N_0^2/2Q^2]/(\bar{\mathbf{H}}^2)^2, \qquad (7\text{C}.9)
\end{aligned}$$

where $\bar{\mathbf{H}}^2=\mathrm{E}\left[\sum_{q=1}^{Q}\sum_{\ell=1}^{L}|H_p^{q,\ell}|^2\right]$. Hence,

$$\left[\text{MSE}_{\widehat{\underline{\theta}}_{(i)}}\right]=\left[\frac{\left[N_0+\frac{N_0^2}{2\bar{\mathbf{H}}^2}Q^2\right]}{\left(2\pi Q\left(1+\frac{G}{N}\right)\right)^2\bar{\mathbf{H}}^2}(\mathbf{J}^H\mathbf{J})^{-1}\right]_{(i,i)}. \qquad (7\text{C}.10)$$

Finally, we have the MSE of the sample clock offset

$$\text{MSE}_\beta = \left[\frac{\left[N_0+\frac{N_0^2}{2\bar{\mathbf{H}}^2}Q^2\right]}{\left(2\pi Q\left(1+\frac{G}{N}\right)\right)^2\bar{\mathbf{H}}^2}\left(\frac{P}{P\sum_{p=1}^{P}k_p^2-\left(\sum_{p=1}^{P}k_p\right)^2}\right)\right], \qquad (7\text{C}.11)$$

the MSE of the residual RF frequency offset

$$\text{MSE}_{\epsilon_e} = \left[\frac{\left[N_0+\frac{N_0^2}{2\bar{\mathbf{H}}^2}Q^2\right]}{\left(2\pi Q\left(1+\frac{G}{N}\right)\right)^2\bar{\mathbf{H}}^2}\left(\frac{\sum_{p=1}^{P}k_p^2}{P\sum_{p=1}^{P}k_p^2-\left(\sum_{p=1}^{P}k_p\right)^2}\right)\right], \qquad (7\text{C}.12)$$

and the MSE of the phase noise

$$\mathrm{MSE}_{\vartheta} = \left[\frac{\left[N_0 + \frac{N_0^2}{2\bar{\mathbf{H}}^2} Q^2\right]}{\bar{\mathbf{H}}^2} \left(\frac{\sum_{p=1}^{P} k_p^2}{P \sum_{p=1}^{P} k_p^2 - \left(\sum_{p=1}^{P} k_p\right)^2} \right) \right] . \tag{7C.13}$$

Appendix 7D: MSE Analysis of the Channel Estimator

The least-square channel estimator is given by (7.4.12)

$$\begin{aligned} \widehat{\mathbf{H}}_k &= (\mathbf{B}_k^H \mathbf{B}_k)^{-1} \mathbf{B}_k^H \mathbf{R}_k \\ &= (\mathbf{B}_k^H \mathbf{B}_k)^{-1} \mathbf{B}_k^H (\mathbf{B}_k \mathbf{H}_k + \mathbf{W}_k) \\ &= \mathbf{H}_k + (\mathbf{B}_k^H \mathbf{B}_k)^{-1} \mathbf{B}_k^H \mathbf{W}_k \end{aligned} \tag{7D.1}$$

Hence, the error in the channel estimates is

$$\widehat{\mathbf{H}}_k - \mathbf{H}_k = (\mathbf{B}_k^H \mathbf{B}_k)^{-1} \mathbf{B}_k^H \mathbf{W}_k, \tag{7D.2}$$

and the MSE of channel estimate is

$$\begin{aligned} \mathrm{MSE}_H &= \frac{1}{QL} \mathrm{E}\left[||\widehat{\mathbf{H}}_k - \mathbf{H}_k||^2 \right] \\ &= \frac{1}{QL} \mathrm{E}\left[\mathrm{Tr}\left[((\mathbf{B}_\mathrm{k}^\mathrm{H} \mathbf{B}_\mathrm{k})^{-1} \mathbf{B}_\mathrm{k}^\mathrm{H} \mathbf{W}_\mathrm{k}) \right. \right. \\ &\quad \left. \left. \times \, ((\mathbf{B}_k^H \mathbf{B}_k)^{-1} \mathbf{B}_k^H \mathbf{W}_k)^H \right] \right] \\ &= \frac{1}{QL} \mathrm{Tr}\left[((\mathbf{B}_k^H \mathbf{B}_k)^{-1} \mathbf{B}_k^H) \mathrm{E}\left[\mathbf{W}_k \mathbf{W}_k^H \right] \right. \\ &\quad \left. \times \, ((\mathbf{B}_k^H \mathbf{B}_k)^{-1} \mathbf{B}_k^H)^H \right] \\ &= \frac{1}{QL} \mathrm{Tr}\left[((\mathbf{B}_k^H \mathbf{B}_k)^{-1} \mathbf{B}_k^H)(L N_0 \mathbf{I}_{Q \times Q}) \right. \\ &\quad \left. \times \, ((\mathbf{B}_k^H \mathbf{B}_k)^{-1} \mathbf{B}_k^H)^H \right] \\ &= N_0 \end{aligned} \tag{7D.3}$$

BIBLIOGRAPHY

[1] R. van Nee, G. Awater, M. Morikura, M. Webster, and K. W. Halford, "New high-rate wireless LAN standards," *IEEE Commun. Mag.*, pp. 82–88, Dec. 1999.

[2] G. L. Stüber, *Principles of Mobile Communication, 2/e*, Norwell, MA: Kluwer 2001.

[3] S. B. Weinstein and P. M. Ebert, "Data transmission by frequency division multiplexing using the discrete Fourier transform," *IEEE Trans. Commun.*, vol. 19, no. 10, pp. 628–634, Oct. 1971.

[4] C. E. Shannon, "A Mathematical Theory of Communication," *Bell System Tech. J.*, vol. 27, pp. 379–423, 1948.

[5] E. Telatar, "Capacity of multi-antenna gaussian channels," *AT&T-Bell Labs Internal Tech. Memo.*, June 1995.

[6] G. J. Foschini and M. J. Gans, "On limits of wirelss communications in a fading enviroment when using multiple antennas," *Wireless Personal Commun.*, vol. 6, no. 3, pp. 311–335, 1998.

[7] G. J. Foschini, "Layered space-time architecture for wireless communication in a fading enviornment when using multi-element antennas," *Bell Labs Tech. J.*, pp. 41–59, Autumn 1996.

[8] J. Tan and G. Stüber, "Multicarrier delay diversity modulation for MIMO systems," *IEEE Trans. Wireless Commun.*, vol. 3, no. 5, pp. 1756–1763, Sept. 2004.

[9] R. W. Chang, "Synthesis of band-limited orthogonal signals for multichannel data," *Bell System Tech. J.*, pp. 1775–1797, Dec. 1966.

[10] B. R. Saltzberg, "Performance of an efficient parallel data transmission systems," *IEEE Trans. Commun. Tech.*, Dec. 1967.

[11] R. W. Chang, "Orthogonal frequency division multiplexing," U.S. Patent 3,488,445, filed Nov. 4 1966 issued Jan. 1970.

[12] R. W. Chang and R. A. Gibby, "A theoretical study of performance of an orthogonal multiplexing data transmission scheme," *IEEE Trans. Commun. Tech.*, Aug. 1968.

[13] M. L. Doelz, E. T. Heald, and D. L. Martin, "Binary data transmission techniques for linear systems," *Proc. IRE*, vol. 45, pp. 656–661, May 1957.

[14] G. C. Porter, "Error distribution and diversity performance of a frequency-differential PSK HF modem," *IEEE Trans. Commun. Tech.*, pp. 197–205, April 1967.

[15] M. S. Zimmerman and A. L. Kirsch, "The AN/GSC-10 (KATHRYN) variable rate data modem for HF radio," *IEEE Trans. Commun. Tech.*, pp. 197–205, April 1967.

[16] P. A. Bello, "Selective fading limitations of the Kathryn modem and some system design considerations," *IEEE Trans. Commun. Tech.*, vol. 13, pp. 320–333, Sept. 1965.

[17] W. E. Keasler, "Reliable data communications over the voice bandwidth telephone using orthogonal frequency division multiplexing," Ph.D. dissertation, Univ. of Illinois Urbana-Champaign, 1982.

[18] J. Holsinger, "Digital communication over fixed time-continuous channel with memory - with special application to telephone channels," Ph.D. dissertation, Massachusetts Institute of Technology, 1964.

[19] B. Hirosaki, S. Hasegawa, and A. Sabato, "Advanced groupband data modem using orthogonally multiplexed QAM techniques," *IEEE Trans. Commun.*, June 1986.

[20] A. Peled and A. Ruiz, "Frequency domain data transmission using reduced computational complexity algorithms," in *Proc. of 1980 IEEE International Conf. on Acoustics Speech and Signal Processing*, 1980, pp. 964–967.

[21] L. J. Cimini, Jr., "Analysis and simulation of a digital mobile channel using orthogonal frequency division multiplexing," *IEEE Trans. Commun.*, pp. 665–675, July 1985.

[22] E. F. Casas and C. Leung, "OFDM for data communication over mobile radio FM channels," *IEEE Trans. Commun.*, pp. 783–793, May 1991.

[23] T. J. Willink and P. H. Wittke, "Optimization and performance evaluation of multicarrier transmission," *IEEE Trans. Inform. Theory*, vol. 43, March 1997.

[24] I. Kalet, "The multitone channel," *IEEE Trans. Commun.*, pp. 119–124, Feb. 1989.

[25] J. A. C. Bingham, "Multicarrier modulation for data transmission: An idea whose time has come," *IEEE Commun. Mag.*, pp. 5–14, May 1990.

[26] L. J. Cimini, Jr. and N. R. Sollenberger, "OFDM with diversity and coding for high-bit-rate mobile data applications," in *Proc. of the 3rd International Workshop on Mobile Multimedia Commun.*, Sept. 1996.

[27] L. J. Cimini, Jr., B. Daneshrad, and N. R. Sollenberger, "Clustered OFDM with transmitter diversity and coding," in *Proc. of 1996 IEEE Global Telecommun. Conf.*, 1996, pp. 703–707.

[28] L. J. Cimini, Jr., J. C. Chuang, and N. R. Sollenberger, "Advanced Cellular Internet Service (ACIS)," *IEEE Commun. Mag.*, pp. 150–159, Oct. 1998.

[29] "Special issue on multi-carrier modulation," in *Kluwer Wireless Pers. Commun.*, Jan. 1996.

[30] R. van Nee and R. Prasad, *OFDM for Wireless Multimedia Communications.* Artech, 2000.

[31] L. Hanzo, W. Webb, and T. Keller, *Single- and Multi-carrier Quadrature Amplitude Modulation.* Wiley, 2000.

[32] J. A. C. Bingham, *ADSL VDSL and Multicarrier Modulation.* Wiley, 2000.

[33] J. C.-I. Chuang, "The effect of time delay spread on portable radio communications channels with digital modulation," *IEEE J. Select. Areas Commun. SAC-5*, pp. 879–889, June 1987.

[34] J. G. Proakis, *Digital Communications*, 4th ed. New York: McGraw-Hill, 2001.

[35] W. C. Jakes, Ed., *Microwave Mobile Communications.* Wiley, 1974.

[36] P. Robertson and S. Kaiser, "The effects of Doppler spreads in OFDM(A) mobile radio systems," in *Prof. of 1999-Fall IEEE Veh. Tech. Conf.*, 1999, pp. 329–333.

[37] Y. G. Li and L. J. Cimini, Jr., "Bound on the interchannel interference of OFDM in time-varying impairments," *IEEE Trans. Commun.*, vol. 49, pp. 401–404, March 2001.

[38] M. Russell and G. L. Stüber, "Interchannel interference analysis of OFDM in a mobile environment," in *Proc. of 1995 IEEE Veh. Tech. Conf.*, 1995, pp. 820–824.

[39] T. Pollet, P. Spruyt, and M. Moeneclaey, "The BER performance of OFDM systems using non-synchronized sampling," in *Proc. of 1994 IEEE Global Telecommun. Conf.*, vol. 1, San Francisco, CA, Nov. 1994, pp. 253–257.

[40] X. Li and L. J. Cimini, Jr., "Effects of clipping and filtering on the performance of OFDM," in *Proc. of 1997 IEEE Veh. Tech. Conf.*, 1634-1638 1997.

[41] B. Hirosaki, "An analysis of automatic equalizers for orthogonally multiplexed QAM systems," *IEEE Trans. Commun.*, Jan. 1980.

[42] R. Gallager, *Information Theory and Reliable Communication.* New York, NY: Wiley, 1968.

[43] A. Ruiz, J. Cioffi, and S. Kasturia, "Discrete multiple tone modulation with coset coding for the spectrally shaped channel," *IEEE Trans. Commun.*, vol. 40, no. 5, May 1992.

[44] J. Cioffi and G. Forney, Jr., "Generalized decision-feedback equalization for packet transmission with ISI and Gaussian noise," from *Communications, Computation, Control and Signal Processing*, edited by A. Paulraj, V. Roychowdhury, and C. D. Schaper, Kluwer: Boston, 1997.

[45] G. Cherubini, J. Cioffi, E. Eleftheriou, and S. Ölçer, "Filter-bank modulation techniques for very-high-speed digital subscriber lines," *IEEE Commun. Mag.*, May 2000.

[46] J. Chow, "Finite-length equalization for multicarrier transmission Systems," *PhD dissertation, Stanford University*, June 1992.

[47] N. Al-Dhahir and J. Cioffi, "Optimum finite-length equalization for multicarrier transceivers," *IEEE Trans. Commun.*, vol. 44, no. 1, pp. 56–64, Jan. 1996.

[48] J. Cioffi, G. Dudevoir, M. V. Eyuboglu, and G. Forney, "MMSE decision-feedback equalizers and coding – Parts I and II," *IEEE Trans. Commun.*, vol. 43, no. 10, pp. 2582–2604, Oct. 1995.

[49] D. Pal, G. Iyengar, and J. Cioffi, "A new method of channel shortening with applications to discrete multi-tone (DMT) systems," in *Proc. of 1998 IEEE International Conf. on Commun.*, June 1998, pp. 763–768.

[50] C. Berrou, A. Glavieux, and P. Thitimajshima, "Near Shannon limit error-correcting coding and decoding: Turbo codes," in *Proc. of 1993 IEEE International Conf. on Commun.*, Geneva, Switzerland, May 1993, pp. 1064–1070.

[51] D. Hughes-Hartog, "Ensemble modem structuire for imperfect transmission Media," *U.S. Patent 4,833,706, Telebit Corp., see also 4,679,227, and 4,731,816 by same author and same title*, May 23 1989.

[52] J. C. de Souza, "Discrete Bit Loading for Multicarrier Modulation Systems," *PhD dissertation, Stanford University*, June 1999.

[53] H. Levin, "A complete and optimal data allocation method for practical discrete multitone systems," in *Proc. of 2001 IEEE Global Telecommun. Conf.* , vol. 1, pp. 369–374, Nov. 2001.

[54] P. Chow, "Bandwidth Optimized Digital Transmission Techniques for Spectrally Shaped Channels," *PhD dissertation, Stanford University*, May 1993.

[55] P. Chow, J. Cioffi, and J. Bingham, "A practical discrete multitone transceiver loading algorithm for data transmission over spectrally shaped channels," *IEEE Trans. Commun.*, vol. 43, pp. 773–775, Feb.-March 1995.

[56] T. Cormen, C. Leiserson, and R. Rivest, *Introduction To Algorithms.* Cambridge, Massachusetts: McGraw-Hill, 1990.

[57] J. Campello, "Practical bit loading for DMT," *Proc. of 1999 IEEE International Conf. on Commun.*, June 1999, pp. 801–805.

[58] S. Pradhan and K. Ramchandran, "Optimized eembedded multicarrier modulation for efficient delivery of layered video data," *Proc. of 1998 IEEE International Conf. on Commun.* , June 1998, pp. 1009–12.

[59] L. Hoo, J. Tellado, and J. Cioffi, "Multiuser loading algorithms for multicarrier systems with embedded constellations," in *Proc. of 2000 IEEE International Conf. on Commun.*, June 2000, pp. 1115–1119.

[60] R. Ahlswede, "Multi-way communication channels," in *Proc. 2nd. Int. Symp. Inf. The.*, Sept. 1971, pp. 23–52.

[61] H. Liao, "Multiple Access Channels," *PhD dissertation, University of Hawaii, Honolulu*, 1972.

[62] T. Cover and J. Thomas, *Elements of Information Theory.* New York: Wiley & Sons, 1991.

[63] S. Verdú, "Multiple-access channels with memory with and without frame synchronism," *IEEE Trans. Inform. Theory*, vol. 35, no. 3, pp. 605–619, May 1989.

[64] ——, "The capacity region of the symbol-asynchronous Gaussian multiple-access channel," *IEEE Trans. Inform. Theory*, vol. 35, no. 4, pp. 733–751, July 1989.

[65] R. Cheng and S. Verdú, "Gaussian multiaccess channels with ISI: Capacity region and multiuser water-filling," *IEEE Trans. Inf. The.*, vol. 39, no. 3, pp. 773–785, May 1993.

[66] S. Diggavi, "Multiuser DMT: a multiple access modulation scheme," in *Proc. of 1996 IEEE Global Telecommun. Conf.* , Nov. 1996, pp. 1566–1570.

[67] C. Zeng, L. Hoo, and J. Cioffi, "Optimal water-filling algorithms for a Gaussian multiaccess channel with intersymbol interference," in *Proc. of 2001 IEEE International Conf. on Commun.*, June 2001, pp. 2421–2427.

[68] ——, "Efficient water-filling algorithms for a Gaussian multiaccess channel with ISI," in *Proc. of 2000 IEEE Veh. Tech. Conf.*, Sept. 2000, pp. 1072–1077.

[69] W. Yu and J. Cioffi, "FDMA capacity of Gaussian multiple-access channels with ISI," *IEEE Trans. Commun.*, vol. 50, no. 1, pp. 102–111, Jan. 2002.

[70] T. Cover, "Broadcast channels," *IEEE Trans. Inform. Theory*, vol. 18, no. 1, pp. 2–14, Jan. 1972.

[71] P. Bergmans, "Random coding theorem for broadcast channels with degraded components," *IEEE Trans. on Inf. The.*, vol. 19, pp. 197–207, March 1973.

[72] ——, "A simple converse for broadcast channels with additive white Gaussian noise," *IEEE Trans. Inform. Theory*, vol. 20, pp. 279–280, March 1974.

[73] R. Gallager, "Capacity and coding for degraded broadcast channels," *Probl. Pered. Inform.*, vol. 10, no. 3, pp. 185–193, July-Sept. 1974.

[74] A. Goldsmith and M. Effros, "The capacity region of broadcast channels with intersymbol interference and colored gaussian noise," *IEEE Trans. Inform. Theory*, vol. 47, no. 1, pp. 219–240, Jan. 2001.

[75] W. Hirt and J. Massey, "Capacity of the discrete-time gaussian channel with intersymbol interference," *IEEE Trans. Inform. Theory*, vol. 34, pp. 380–388, May 1988.

[76] L. Hoo, "Multiuser Transmit Optimization For Multicarrier Modulation Systems," *Ph.D. dissertation, Stanford University*, Dec. 2000.

[77] L. Hoo, B. Halder, J. Tellado, and J. Cioffi, "Asymptotic FDMA capacity region for broadcast channels with ISI," in *Proc. of 2001 IEEE International Conf. on Commun.*, June 2001, pp. 1648–1653.

[78] ——, "Multiuser transmit optimization for multicarrier broadcast channels: asymptotic FDMA capacity region and algorithms," *IEEE Trans. Commun.*, vol. 52, no. 6, pp. 922–930, June 2004.

[79] C. Wong and R. Cheng, "Multiuser OFDM with adaptive subcarrier, bit and power allocation," *IEEE J. Select. Areas Commun.*, vol. 17, no. 10, pp. 1747–1757, Oct. 1999.

[80] C. Wong, C. Tsui, R. Cheng, and K. Letaief, "A real-time sub-carrier allocation scheme for multiple access downlink OFDM transmission," in *Proc. of 1999 IEEE Veh. Tech. Conf.*, Sept. 1999, pp. 1124–1128.

[81] W. Rhee and J. Cioffi, "Increase in capacity of multiuser OFDM system using dynamic subchannel allocation," *Proc. of 2000 IEEE Veh. Tech. Conf.*, May 2000, pp. 1085–1089.

[82] L. Hoo, J. Tellado, and J. Cioffi, "FDMA-based multiuser transmit optimization for broadcast channels," in *Proc. of 2004 IEEE Wireless Commun. and Networking Conf.*, vol. 2, Sept. 2000, pp. 597–602.

[83] J.-B. Hiriart-Urruty and C. Lemarâechal, *Convex Analysis and Minimization Algorithms.* Berlin ; New York: Springer-Verlag, 1993.

[84] L. Hoo, J. Tellado, and J. Cioffi, "Dual QoS loading algorithms for multicarrier systems offering different CBR services," in *Proc. of IEEE 2005 International Symp. on Personal, Indoor and Mobile Radio Commun.*, Sept. 1998, pp. 278–282.

[85] ——, "Dual QoS loading algorithms for DMT systems offering CBR and VBR services," in *Proc. of 1998 IEEE Global Telecommun. Conf.* , Nov. 1998, pp. 25–30.

[86] ——, "Discrete dual QoS loading algorithms for multicarrier systems," in *Proc. of 1999 IEEE International Conf. on Commun.*, June 1999, pp. 796–800.

[87] B. Schein and M. Trott, "Sub-optimal power spectra for colored Gaussian channels," *IEEE Int. Symp. Information Theory*, p. 340, 1997.

[88] F. Kozamernik, "Eureka 147 - Digital audio broadcasting system for Europe and worldwide," in *Digital Consumer Electronics Handbook*, R. K. Jurgen, Ed. McGraw-Hill: New York, 1997, ch. 14, pp. 14.1–14.23.

[89] M. Alard and R. Lassalle, "Principles of modulation and channel coding for digital broadcasting for mobile receivers," *EBU Review - Technical*, vol. 1, no. 224, pp. 168–190, Aug. 1987.

[90] B. Marti, P. Bernard, N. Lodge, and R. Schafer, "European activities on digital television broadcasting - from company to cooperative projects," *EBU Technical Review*, no. 256, pp. 20–29, 1993.

[91] G. Stüber and M. Russell, "Terrestrial digital video broadcasting for mobile reception using OFDM," in *Proc. of 1995 IEEE Global Telecommun. Conf.* , Nov. 1995, pp. 2049–53.

[92] B. Sklar, *Digital Communications - Fundamentals and Applications.* Prentice Hall, 1988.

[93] J. Kim, L. C. Jr., and J. Chuang, "Coding strategies for OFDM with antenna diversity for high-bit-rate mobile data applications ," in *Proc. of 1998 IEEE Veh. Tech. Conf.*, May 1998, pp. 763–767.

[94] S. Lin and D. C. Jr., *Error Control Coding, Fundamentals and Applications.* Englewood Cliffs, NJ: Prentice-Hall, Inc., 1983.

[95] S. Wicker, *Error Control Systems for Digital Communication and Storage.* Upper Saddle River, NJ: Prentice-Hall, 1995.

[96] T. Starr, J. Cioffi, and P. Silverman, *Understanding Digital Subscriber Lines.* Upper Saddle River, NJ: Prentice-Hall, 1999.

[97] T. Rappaport, *Wireless Communications: Principles and Practice.* Upper Saddle River, NJ: Prentice-Hall, 1996.

[98] J. Nicolas and J. Lim, "On the performance of multicarrier modulation in a broadcast multipath environment," in *Proc. of 1994 IEEE International Conf. on Acoustics, Speech and Signal Processing*, 1994, pp. 245–248.

[99] V. Tarokh, N. Seshadri, and A. Calderbank, "Space-time codes for high data rate wireless communication: performance criterion and code construction," *IEEE Trans. Inform. Theory*, vol. 44, pp. 744–765, March 1998.

[100] D. Agrawal, V. Tarokh, A. Naguib, and N. Seshadri, "Space-time coded OFDM for high data-rate wireless communication over wideband channels ," in *Proc. of 1998 IEEE Veh. Tech. Conf.*, May 1998, pp. 2232 – 2236.

[101] R. Wesel, "Trellis Code Design for Correlated Fading and Achievable Rates for Tomlinson-Harashima Precoding," *PhD dissertation, Stanford University*, Aug. 1996.

[102] A. Leke and J. Cioffi, "Dynamic bandwidth optimization for wireline and wireless channels ," in *Conf. Rec. of the Thirty-second Asilomar Conf.*, vol. 2, Nov. 1998, pp. 1753–1757.

[103] G. Raleigh and J. Cioffi, "Spatio-temporal coding for wireless communications," *IEEE Trans. Commun.*, vol. 46, no. 3, pp. 357–366, March 1998.

[104] A. Ruiz and J. Cioffi, "A frequency-domain approach to combined spectral shaping and coding," in *Proc. of 1997 IEEE International Conf. on Commun.*, Seattle, WA, June 1987, pp. 1711–1715.

[105] T. N. Zogakis, J. T. Aslanis, and J. M. Cioffi, "A coded and shaped discrete multitone system," *IEEE Trans. Commun.*, vol. 43, no. 12, pp. 2941–2949, Dec. 1996.

[106] C. Berrou, A. Glavieux, and P. Thitimajshima, "Near Shannon limit error-correcting coding and decoding: turbo codes," in *Proc. of 1993 International Conf. on Commun.*, Geneva, Switzerland, May 1993, pp. 1064–1070.

[107] C. Douillard, M. Jezequel, C. Berrou, A. Picart, and P. Didier, "Iterative Correction of Intersymbol Interference: Turbo-Equalization," *European Trans. Telecommun.*, vol. 6, no. 5, pp. 507–511, Sept.-Oct. 1995.

[108] J. Lauer and J. Cioffi, "Turbo coding for discrete multitone transmission systems," in *Proc. of 1998 IEEE Global Telecommun. Conf.*, Sydney, Australia, Nov. 1998, pp. 3256–3260.

[109] L. Lin, L. Cimini, Jr., and J. Chuang, "TURBO codes for OFDM with antenna diversity," in *Proc. of 1999 IEEE Veh. Tech. Conf.*, May 1999, pp. 1664–1668.

[110] L. Hazy and M. El-Tanany, "Synchronization of ofdm systems over frequency selective fading channels," in *Proc. of 1997 IEEE Veh. Tech. Conf.*, vol. 3, Phonex, AZ, May 1997, pp. 2094–2098.

[111] J.-J. van de Beek, P. Börjesson, M.-L. Boucheret, D. Landström, J. M. Arenas, P. Ödling, C. Östberg, M. Wahlqvist, and S. K. Wilson, "A time and frequency synchronization scheme for multiuser ofdm," *IEEE JSAC: Wireless Commun. Series*, vol. 17, no. 11, pp. 1900–1914, Nov. 1999.

[112] J.-S. Oh, Y.-M. Chung, and U.-S. Lee, "A carrier synchronization technique for OFDM on the frequency-selective fading environment," in *Proc. of 1996 IEEE Veh. Tech. Conf.*, vol. 3, Atlanta, GA, May 1996, pp. 1574–1578.

[113] H. Meyr and G. Ascheid, *Synchronization in Digital Communications : Phase-, Frequency-Locked Loops, and Amplitude Control*, 1st ed., ser. Wiley Series in Telecommunications. John Wiley and Sons, 1999.

[114] "Radio broadcasting systems; Digital Audio Broadcasting (DAB) to mobile, portable and fixed receivers," ETSI – European Telecommunications Standards Institute, Valbonne, France, ETS 300-401, Feb. 1995.

[115] A. Eyadeh, P. Nobles, F. Halsall, and T. Davies, "OFDM fine time synchronisation for indoor wireless data communications," in *IEE Colloq. on Wireless Tech.*, vol. 7. London: IEE, Nov. 1996, pp. 1–5.

[116] B. L. Floch, R. Halbert-Lassalle, and D. Castelain, "Digital sound broadcasting to mobile receivers," *IEEE Trans. Consumer Electronics*, vol. 35, no. 3, pp. 493–503, Aug. 1989.

[117] T. Schmidl and D. Cox, "Robust frequency and timing synchronization for OFDM," *IEEE Trans. Commun.*, vol. 45, no. 12, pp. 1613–1621, Dec. 1997.

[118] P. H. Moose, "A technique for orthogonal frequency-division multiplexing frequency offset correction," *IEEE Trans. Commun.*, vol. 42, no. 10, pp. 2908–2914, Oct. 1994.

[119] H. L. van Trees, *Detection, Estimation and Modulation Theory.* New York, USA: John Wiley & Sons, 1968, vol. 1.

[120] J.-J. van de Beek, M. Sandell, M. Isaksson, and P. O. Börjesson, "Low-complex frame synchronization in OFDM systems," Tokyo, Japan, Nov. 1995, pp. 982–986.

[121] J.-J. van de Beek, M. Sandell, and t. . P. O. Börjesson, "ML estimation of time and frequency offset in OFDM systems," *IEEE Trans. Signal Processing* vol. 45, no. 7, pp. 1800 – 1805, July 1997

[122] D. Lee and K. Cheun, "A new symbol timing recovery algorithm for OFDM systems," *IEEE Trans. Consumer Electronics*, vol. 43, no. 3, pp. 767–75, June 1997.

[123] M. Speth, Michael, F. Classen, and H. Meyr, "Frame synchronization ofdm systems in frequency selective fading channels," in *Proc. 1997 IEEE Veh. Tech. Conf.*, vol. 3, Phoenix, AZ, May 1997, pp. 1807–1811.

[124] U. Lambrette, M. Speth, and H. Meyr, "A new symbol timing recovery algorithm for OFDM systems," *IEEE Commun. Letters*, vol. 1, no. 2, pp. 46–48, March 1997.

[125] M. Luise and R. Reggiannini, "Carrier frequency acquisition and tracking for OFDM systems," *IEEE Trans. Commun.*, vol. 44, no. 11, pp. 1590–1598, Nov. 1996.

[126] F. Classen and H. Meyr, "Frequency synchronization algorithms for OFDM systems suitable for communication over frequency-selective fading channels," in *IEEE Veh. Tech. Conf.*, vol. 3, Stockholm, Sweden, 1994, pp. 1655–1659.

[127] R. Ziemer and W. Tranter, *Principles of Communications*, 4th ed. Wiley, 1995.

[128] F. Daffara and O. Adami, "A new frequency detector for orthogonal multicarrier transmission techniques," in *Proc. of 1995 IEEE Veh. Tech. Conf.*, vol. 2, Chicago, IL, 1995, pp. 804–809.

[129] ——, "A novel carrier recovery technique for orthogonal multicarrier systems," *European Trans. Telecommun.*, vol. 7, no. 4, pp. 323–334, July-August 1996.

[130] B. Yang, K. B. Letaief, R. S. Cheng, and Z. Cao, "Timing recovery for ofdm transmission," *IEEE J. Select. Areas Commun.*, vol. 50, no. 9, pp. 1525–1534, Sept. 2002.

[131] A. I. Bo, G. E. Jian-hua, and Y. Wang, "Symbol synchronization technique in cofdm systems," *IEEE Trans. Broadcasting*, vol. 50, no. 1, pp. 56–62, March 2004.

[132] A. J. Coulson, "Maximum likelihood synchronization for ofdm using a pilot symbol: algorithms," *IEEE J. Select. Areas Commun.*, vol. 19, no. 12, pp. 2486–2494, Dec. 2001.

[133] D. Lee and K. Cheun, "Coarse symbol synchronization algorithms for ofdm systems in multipath channels," *IEEE Commun. Letter*, vol. 6, no. 10, pp. 446–448, Oct. 2002.

[134] D. Landstrom, S. K. Wilson, J.-J. van de Beek, P. Odling, and P. O. Borjesson, "Symbol time offset estimation in coherent ofdm systems," *IEEE Trans. Commun.*, vol. 50, no. 4, pp. 545–549, April 2002.

[135] K. W. Kang, J. Ann, and H. S. Lee, "Decision-directed maximum-likelihood estimation of OFDM frame synchronization offset," *Electronics Letters*, vol. 30, no. 25, pp. 2153–2154, Dec. 1994.

[136] L. Jing and T.-S. Ng, "A consistent ofdm carrier frequency offset estimator based on distinctively spaced pilot tones," *IEEE Trans. Wireless Commun.*, vol. 3, no. 2, pp. 588–599, March 2004.

[137] T. Keller, L. Piazzo, P. Mandarini, and L. Hanzo, "Orthogonal frequency division multiplex synchronization techniques for frequency-selective fading channels," *IEEE J. Select. Areas Commun.*, vol. 19, no. 6, pp. 999–1008, June 2001.

[138] B. Ai, J.-H. Ge, Y. Wang, S.-Y. Yang, P. Liu, and G. Liu, "Frequency offset estimation for ofdm in wireless communications," *IEEE Trans. Consumer Elect.*, vol. 50, no. 1, pp. 73–77, Feb. 2004.

[139] N. Lashkarian and S. Kiaei, "Class of cyclic-based estimators for frequency-offset estimation of ofdm systems," *IEEE Trans. Commun.*, vol. 48, no. 12, pp. 2139–2149, Dec. 2000.

[140] S. Giovanni and F. Andrea, "Description of a frequency and symbol synchronization system for ofdm signals and simulation results in a multipath channel," in *Proc. of 1997 International Conf. on Commun.*, Montreal, Canada, June 1997, pp. 939–943.

[141] N. Morelli, A. N. D'Andrea, and U. Mengali, "Feedback frequency synchronization for ofdm applications," *IEEE Trans. Commun. Letter*, vol. 5, no. 1, pp. 28–30, Jan. 2001.

[142] E. Chiavaccini and G. M. Vitetta, "Maximum-likelihood frequency recovery for ofdm signals transmitted over multipath fading channels," *IEEE Trans. Commun.*, vol. 52, no. 2, pp. 244–251, Feb. 2004.

[143] M. H. Minn, V. K. Bhargava, and K. K. B. Letaief, "A robust timing and frequency synchronization for ofdm systems," *IEEE Trans. Wireless Commun.*, vol. 2, no. 4, pp. 822–839, July 2003.

[144] K. Shi and E. Serpedin, "Coarse frame and carrier synchronization of ofdm systems: a new metric and comparison," *IEEE Trans. Wireless Commun.*, vol. 3, no. 4, pp. 1271–1284, July 2004.

[145] G. L. Stüber, J. R. Barry, S. W. McLaughlin, Y. G. Li, M. A. Ingram, and T. G. Pratt, "Broadband mimo-ofdm wireless communications," *Proc. of the IEEE*, vol. 92, no. 2, pp. 271–294, Feb. 2004.

[146] M. Morelli, "Timing and frequency synchronization for the uplink of an ofdma system," *IEEE Trans. Commun.*, vol. 52, no. 2, pp. 296–306, Feb. 2004.

[147] P.-Y. Tsai, H.-Y. Kang, and T.-D. Chiueh, "Joint weighted least-squares estimation of carrier-frequency offset and timing offset for ofdm systems over multipath fading channels," *IEEE Trans. Veh. Tech.*, vol. 54, no. 1, pp. 211–223, Jan. 2005.

[148] M.-H. Hsieh and C.-H. Wei, "A low-complexity frame synchronization and frequency offset compensation scheme for ofdm systems over fading channels," *IEEE Trans. Veh. Tech.*, vol. 48, no. 5, pp. 1596–1609, Sept. 1999.

[149] V. Lottici, M. Luise, M. Marselli, and R. Reggiannini, "Blind subcarrier frequency ambiguity resolution for ofdm signals over selective channels," *IEEE Trans. Commun.*, vol. 52, no. 9, pp. 1532–1537, Sept. 2004.

[150] H. Bolcskei, "Blind estimation of symbol timing and carrier frequency offset in wireless ofdm systems," *IEEE Trans. Commun.*, vol. 49, no. 6, pp. 988–999, June 2001.

[151] M. Ghogho and A. Swami, "Blind frequency-offset estimator for ofdm systems transmitting constant-modulus symbols," *IEEE Commun. Letter*, vol. 6, no. 8, pp. 343–345, Aug. 2002.

[152] B. Park, H. Cheon, E. Ko, C. Kang, and D. Hong, "A blind ofdm synchronization algorithm based on cyclic correlation," *IEEE Signal Processing Letter*, vol. 11, no. 2, pp. 83–85, Feb. 2004.

[153] R. Negi and J. M. Cioffi, "Blind ofdm symbol synchronization in isi channels," *IEEE Trans. Commun.*, vol. 50, no. 9, pp. 1525–1534, Sept. 2002.

[154] T. Pollet and M. Peters, "Symbol synchronization in multicarrier systems," in *Copper Wire Access Systems (CWAS) '97 Proc.*, Budapest, Hungary, Oct. 1997, pp. 125–126.

[155] H. Nogami and T. Nagashima, "A frequency and timing period aquisition techique for ofdm systems," *IEICE Trans. Commun.*, vol. E79-B, no. 8, pp. 1135–1146, Aug. 1996.

[156] Z.-W. Zheng, Z.-X. Yang, C.-Y. Pan, , and Y.-S. Zhu, "Novel synchronization for tds-ofdm-based digital television terrestrial broadcast systems," *IEEE Trans. Broadcasting*, vol. 50, no. 2, pp. 148–153, June 2004.

[157] B. Ai, J.-H.Ge, Y. Wang, S.-Y. Yang, and P. Liu, "Decimal frequency offset estimation in cofdm wireless communications," *IEEE Trans. Broadcasting*, vol. 50, no. 2, pp. 154–158, June 2004.

[158] J.-J. van de Beek, O. Edfors, M. Sandell, S. K. Wilson, and P. O. Börjesson, "On channel estimation in OFDM systems," in *Proc. of 45th IEEE Veh. Tech. Conf.*, Chicago IL, July 1995, pp. 815–819.

[159] ——, "OFDM channel estimation by singular value decomposition," *IEEE Trans. Commun.*, vol. 46, pp. 931–939, July 1998.

[160] A. Papoulis, *Probability Random Variables and Stochastic Processes*, 3rd ed. McGraw-Hill, 1991.

[161] Y. G. Li, L. J. Cimini, Jr., and N. R. Sollenberger, "Robust channel estimation for OFDM systems with rapid dispersive fading channels," *IEEE Trans. Commun.*, vol. 46, pp. 902–915, July 1998.

[162] P. Hoeher, S. Kaiser, and P. Robertson, "Two-dimensional pilot-symbol-aided channel estimation by Wiener filtering," in *Proc. of 1997 IEEE International Conf. on Acoustics Speech and Signal Processing*, Munich Germany, April 1997, pp. 1845–1848.

[163] ——, "Pilot-symbol-aided channel estimation in time and frequency," in *Proc. of 1997 IEEE Global Telecommun. Conf.*, Phoenix AZ, Nov. 1997, pp. 90–96.

[164] R. M. Mersereau and T. C. Speake, "The processing of periodically sampled multidimensional signals," *IEEE Trans. Signal Processing*, vol. 31, pp. 188–194, Feb. 1983.

[165] P. P. Vaidyanathan, *Multirate Systems and Filter Banks*. New Jersey: Prentice Hall Inc., 1993.

[166] V. Mignone and A. Morello, "CD3-OFDM: a novel demodulation scheme for fixed and mobile receivers," *IEEE Trans. Commun.*, vol. 44, pp. 1144–1151, Sept 1996.

[167] A. V. Oppenheim and R. W. Schafer, *Discrete-Time Signal Processing.* New Jersey: Prentice Hall Inc., 1989.

[168] X. Wang and K. J. R. Liu, "Channel estimation for multicarrier system using a time-frequency polynomial model," *IEEE Trans. Commun.*, vol. 50, pp. 1045–1048, July 2002.

[169] H. Zhang and Y. G. Li, "Clustered OFDM with adaptive antenna arrays for interference suppression," *IEEE Trans. Wireless Commun.*, vol. 3, no. 6, pp. 2189-2197, Nov. 2004 / also in *Proc. of 2003 IEEE International Conf. Commun.*, June 2003.

[170] H. H. H'mimy, "Channel estimation based on coded pilot for OFDM," in *Proc. of 47th IEEE Veh. Tech. Conf.*, Phoenix AZ, May 1997, pp. 1375–1379.

[171] F. Tufvesson and T. Maseng, "Pilot assisted channel estimation for OFDM in mobile cellular systems," in *Proc. of 47th IEEE Veh. Tech. Conf.*, Phoenix AZ, May 1997, pp. 1639–1643.

[172] Y. G. Li, "Pilot-symbol-aided channel estimation for OFDM in wireless systems," *IEEE Trans. Veh. Tech.*, vol. 49, pp. 1207–1215, July 2000.

[173] Y. G. Li and N. Sollenberger, "Interference suppression in OFDM systems using adaptive antenna arrays," in *Proc. of 1998 IEEE Global Telecommun. Conf.*, Sydney Australia, Nov. 1998, pp. 213–218.

[174] ——, "Adaptive antenna arrays for OFDM systems with co-channel interference," *IEEE Trans. Commun.*, vol. 47, pp. 217–229, Feb. 1999.

[175] Y. G. Li, N. Seshadri, and S. Ariyavisitakul, "Channel estimation for transmitter diversity in OFDM systems with mobile wireless channels," *IEEE J. Selected Areas Commun.*, vol. 17, pp. 461–471, March 1999.

[176] Y. G. Li, "Simplified channel estimation for OFDM systems with multiple transmit antennas ," *IEEE Trans. Wireless Commun.*, vol. 1, pp. 67–75, Jan. 2002.

[177] Y. G. Li, J. H. Winters, and N. R. Sollenberger, "MIMO-OFDM for wireless communications: signal detection with enhanced channel estimation," *IEEE Trans. Commun.*, vol. 50, pp. 1471–1477, Sept. 2002.

[178] A. Leke and J. M. Cioffi, "Impact of imperfect channel knowledge on the performance of multicarrier systems," in *Proc. of 1998 IEEE Global Telecommun. Conf.*, Nov. 1998, pp. 951–955.

[179] L. Jarbot, "Combined decoding and channel estimation of OFDM systems in mobile radio networks," in *Proc. of 47th IEEE Veh. Tech. Conf.*, May 1997, pp. 1601–1604.

[180] M. Bossert, A. Donder, and V. Zyablov, "Improved channel estimation with decision feedback for OFDM systems," *Electronics Letters*, vol. 34, pp. 1064–1065, May 1998.

[181] T. Onizawa, M. Mizoguchi, T. Sakata, and M. Morikura, "A simple adaptive channel estimation scheme for OFDM systems," *Proc. of 50th IEEE Veh. Tech. Conf.*, pp. 279–283, Sept. 1999.

[182] E. Al-Susa and R. F. Ormondroyd, "A predictor-based decision feedback channel estimation method for COFDM with high resilience to rapid time-variations," *Proc. of 50th IEEE Veh. Tech. Conf.*, pp. 273–278, Sept. 1999.

[183] J. K. Cavers, "An analysis of pilot symbol assisted modulation for Rayleigh fading channels," *IEEE Trans. Veh. Tech.*, pp. 686–693, Nov. 1991.

[184] R. Negi and J. Cioffi, "Pilot tone selection for channel estimation in a mobile OFDM system," *IEEE Trans. Consumer Electronics*, vol. 44, pp. 1122–1128, June 1998.

[185] M. J. Fernandez-Getino, J. M. Paez-Borrallo, and S. Zazo, "Efficient pilot patterns for channel estimation in OFDM systems over HF channels," in *Proc. of 50th IEEE Veh. Tech. Conf.*, Sept. 1999, pp. 2193–2197.

[186] M.-H. Hsieh and C.-H. Wei, "Channel estimation for OFDM systems based on comb-type pilot arrangement in frequency selective fading channels," *IEEE Trans. Consumer Electronics*, vol. 44, pp. 217–225, Feb. 1998.

[187] A. Huang and Y. Zhao, "Estimating channel response from pilot subcarrier pairs for OFDM systems," in *Proc. of the 40th Midwest Symp. on Circuits and Systems*, Aug. 1997, pp. 774–777.

[188] J. K. Moon and S. I. Choi, "Performance of channel estimation methods for OFDM systems in a multipath fading channels," *IEEE Trans. Consumer Electronics*, vol. 46, pp. 161–170, Feb. 2000.

[189] B. Yang, K. B. Letaief, R. S. Cheng, and Z. Cao, "Robust and improved channel estimation for OFDM systems in frequency selective fading channels," in *Proc. of 1999 IEEE Global Telecommun. Conf.*, Dec. 1999, pp. 2499–2503.

[190] V.-P. Kaasila, "Performance analysis of an OFDM system using data-aided channel estimation," in *Proc. of 49th IEEE Veh. Tech. Conf.*, May 1999, pp. 2303–2307.

[191] C.-S. Yeh and Y. Lin, "Channel estimation using pilot tones in OFDM systems," *IEEE Trans. Broadcasting*, vol. 45, pp. 400–409, Dec. 1999.

[192] S.-Y. Park and C.-G. Kang, "Performance of pilot-assisted channel estimation for OFDM system under time-varying multi-path Rayleigh fading with frequency offset compensation," in *Proc. of 51st IEEE Veh. Tech. Conf.*, May 2000, pp. 1245–1249.

[193] F. Said and A. H. Aghvami, "Two dimensional pilot assisted channel estimation for turbo coded OFDM systems," in *Proc. of IEE Colloquium on Turbo Codes in Digital Broadcasting - Could It Double Capacity?*, 1999, pp. 19/1–19/6.

[194] A. Chini, Y. Wu, M. El-Tanany, and S. Mahmoud, "Filtered decision feedback channel estimation for OFDM-based DTV terrestrial broadcasting system," *IEEE Trans. Broadcasting*, vol. 44, pp. 2–11, March 1998.

[195] K. Fazel and M. J. Ruf, "A hierarchical digital HDTV transmission scheme for terrestrial broadcasting," in *Prof. of 1993 IEEE Global Telecommun. Conf.*, Nov. 1993, pp. 12–17.

[196] A. A. Hutter, R. Hasholzner, and J. S. Hammerschmidt, "Channel estimation for mobile OFDM systems," in *Proc. of 50th IEEE Veh. Tech. Conf.*, Sept. 1999, pp. 305–309.

[197] N. Chotikakamthorn and H. Suzuki, "On identifiability of OFDM blind channel estimation," in *Proc. of 50th IEEE Veh. Tech. Conf.*, Sept. 1999, pp. 2358–2361.

[198] B. Muquet, M. de Courville, P. Duhamel, and V. Buzenac, "A subspace based blind and semi-blind channel identification method for OFDM systems," in *Proc. of 1999 2nd IEEE Workshop on Signal Processing Advances in Wireless Commun.*, May 1999, pp. 170–173.

[199] C. Tepedelenlioglu, "'deterministic blind estimation of time- and frequency-selective fading channels using filterbank precoders," in *Proc. of 1999 2nd IEEE Workshop on Signal Processing Advances in Wireless Commun.*, May 1999, pp. 74–77.

[200] Y. Song, S. Roy, and L. A. Akers, "Joint blind estimation of channel and data symbols in ofdm," in *Proc. of 51st IEEE Veh. Tech. Conf.*, May 2000, pp. 46 – 50.

[201] Z. Liu, G. B. Giannakis, and S. Barbarossa, "Decoding and equalization of unknown multipath channels based on block precoding and transmit-antenna diversity," in *Proc. of the 33rd Asilomar Conf.on Sign. Syst. and Comp.*, Oct. 1999, pp. 1557–1561.

[202] C. Tepedelenlioglu and G. B. Giannakis, "Transmitter redundancy for blind estimation and equalization of time- and frequency-selective channels," *IEEE Trans. Signal Processing*, vol. 48, pp. 2029 – 2043, July 2000.

[203] J. Rinne and M. Renfors, "An equalization method for orthogonal frequency division multiplexing systems in channels with multipath propagation frequency offset and phase noise," in *Proc. of 1996 IEEE Global Telecommun. Conf.*, Nov. 1996, pp. 1443–1446.

[204] A. Czylwik, "Degradation of multicarrier and single carrier transmission with frequency domain equalization due to pilot-aided channel estimation and frequency synchronization," in *Proc. of 1997 IEEE Global Telecommun. Conf.*, Nov. 1997, pp. 27–31.

[205] R. F. Ormondroyd and E. A. Al-Susa, "A high efficiency channel estimation and equalization strategy for a broadband COFDM system," *1998 URSI International Sym. on Signals Systems and Electronics*, pp. 471–475, Oct. 1998.

[206] G.-T. Ye, G.-A. Bi, and L.-Q. Fang, "Time domain channel estimation for the uplink of multicarrier DS-CDMA systems," in *Proc. of the 32nd Asilomar Conf. on Sign. Syst. & Comp.*, Nov. 1998, pp. 1878–1881.

[207] S. Kaiser, "MC-FDMA and MC-TDMA versus MC-CDMA and SS-MC-MA: performance evaluation for fading channels," in *Proc. of 1998 IEEE 5th International Symp. on Spread Spectrum Tech. and Appl.*, Sept. 1998, pp. 200–204.

[208] Y. G. Li and N. R. Sollenberger, "Clustered OFDM with channel estimation for high rate wireless data," in *Proc. of IEEE MoMuc'99*, Nov. 1999, pp. 43–50.

[209] Y. G. Li, N. Seshadri, and S. Ariyavisitakul, "Transmitter diversity of OFDM systems with dispersive fading channels," in *Proc. of 1998 IEEE Global Telecommun. Conf.*, Sydney Australia, Nov. 1998, pp. 968–973.

[210] V. K. Jones and G. C. Raleigh, "Channel estimation for wireless OFDM systems," in *Proc. of 1998 IEEE Global Telecommun. Conf.*, Sydney Australia, Nov. 1998, pp. 980–985.

[211] Y.-H. Zeng and T.-S. Ng, "A semi-blind channel estimation method for multiuser multiantenna OFDM systems ," *IEEE Trans. Signal Processing*, vol. 52, no. 5, pp. 1419–1429, May 2004.

[212] Z.-J. Wang, Z. Han, and K. J. R. Liu, "A MIMO-OFDM channel estimation approach using time of arrivals ," *IEEE Trans. Wireless Commun.*, vol. 4, no. 3, pp. 1207–1213, May 2005.

[213] X.-L. Ma, M.-K. Oh, G. B. Giannakis, and D.-J. Park "Hopping pilots for estimation of frequency-offset and multiantenna channels in MIMO-OFDM ," *IEEE Trans. on Commun.*, vol. 53, no. 1, pp. 162–172, Jan. 2005.

[214] W. Zhu and M. P. Fitz, "Adaptive channel estimation for trained MIMO-OFDM," in *Proc. of the 38th Asilomar Conf. on Signal, Systems, and Computers*, Nov. 2004, pp. 697–701.

[215] Y. Jing, D.-M. Wang, M. Chen, S.-X. Cheng, and H.-F. Wang, "Reduced complexity MIMO-OFDM channel estimation based on parametric channel model," in *Proc. of 2004 IEEE 60th Veh. Tech. Conf.*, Sept. 2004, pp. 1372–1376.

[216] G. L. I. Barhumi and M. Moonen, "Optimal training design for MIMO OFDM systems in mobile wireless channels," *IEEE Trans. on Signal Processing*, vol. 51, no. 6, pp. 1615–1624, June 2003.

[217] C. K. H. S.-M. Sun, I. Wiener and T. T. Tjhung, "Training sequence assisted channel estimation for MIMO OFDM," in *Proc. of 2003 IEEE Wireless Commun. and Networking Conf.*, March 2003, pp. 38–43.

[218] H. Minn and N. Al-Dhahir, "Training signal design for MIMO OFDM channel estimation in the presence of frequency offsets," in *Proc. of 2005 IEEE Wireless Commun. and Networking Conf.*, March 2005, pp. 13–17.

[219] F. Horlin and L. V. der Perre, "Optimal training sequences for low complexity ML multi-channel estimation in multi-user MIMO OFDM-based communications," in *Proc. of 2004 International Conf. on Commun.*, June 2004, pp. 2427–2431.

[220] J. Salz, "Optimum mean-square decision feedback equalization," *Bell System Tech. J.*, pp. 1341–1371, Oct. 1973.

[221] S. H. Müller and J. B. Huber, "A novel peak power reduction scheme for OFDM," in *Proc. of IEEE 1997 International Symp. on Personal, Indoor and Mobile Radio Commun.*, Sept. 1997, pp. 1090–1094.

[222] J. Tellado and J. M. Cioffi, *PAR reduction in multicarrier transmission systems.* ANSI T1E1.4 committee contribution number 97-367, Dec. 1997.

[223] M. Friese, "OFDM signals with low crest-factor," in *Proc. of 1997 IEEE Global Telecommun. Conf.*, 1997, pp. 290–294.

[224] R. Dinis, P. Montezuma, and A. Gusmao, "Performance trade-offs with quasi linearly amplified OFDM through a two-branch combining technique," in *Proc. of 1996 IEEE Veh. Tech. Conf.*, April 1996, pp. 899–903.

[225] J. Jong, K. Yang, W. E. Stark, and G. I. Hadded, "Power optimization of OFDM systems with dc bias controlled nonlinear amplifiers," in *Proc. of 1999 IEEE Veh.Tech. Conf.*, Sept. 1999, pp. 268–272.

[226] J. Bingham and J. M. Cioffi, *Dynamic scaling for clip mitigation in ADSL standard issue 2.* ANSI T1E1.4 committee contribution number 96-019, January 1996.

[227] C. Röbing and V. Tarokh, "A construction of OFDM 16-QAM sequences having low peak powers," *IEEE Trans. Inform. Theory*, vol. 47, pp. 2091–2094, July 2001.

[228] C. V. Chong and V. Tarokh, "A simple encodable/decodable OFDM QPSK code with low peak-to-mean," *IEEE Trans. Inform. Theory*, vol. 47, pp. 3025–3029, Nov. 2001.

[229] T. Ginige, N. Rajatheva, and K. M. Ahmed, "Dynamic spreading code selection method for PAPR reduction in OFDM-CDMA systems with 4-QAM modulation," *IEEE Commun. Lett.*, vol. 5, pp. 39–43, Oct. 2001.

[230] M. Park, H. Jun, J. Cho, N. Cho, D. Hong and C. Kang, "PAPR reduction in OFDM transmission using Hadamard transform," in *Proc. of 2000 IEEE Veh. Tech. Conf.*, 2000, pp. 430–433.

[231] L. J. Cimini, Jr. and N. R. Sollenberger, "Peak-to-average power ratio reduction of an OFDM signal using partial transmit sequences," in *Proc. of 1999 IEEE International Conf. on Commun.*, Dec. 1999, pp. 511–515.

[232] H. Ochiai, "Analysis and reduction of peak-to-avarage power ratio in OFDM systems," Ph.D. dissertation, The University of Tokyo, March 2001.

[233] B. H. Sharif and M. Khalaj., "Peak to mean envelope power ratio of oversampled OFDM signals: An analytical approach," in *Proc. of 2001 IEEE International Conf. on Commun.*, 2001, pp. 1476 –1480.

[234] S. Boyd, "Multitone signals with low crest factors," *IEEE Trans. Circuits Syst.*, vol. 33, pp. 1018–1022, Oct. 1986.

[235] G. Benke, "On the maximum modulus for a certain class of unimodular," in *Recent advances in Fourier analysis and its applications*, Kluwer Academic Publishers, 1990, pp. 83–100.

[236] B. M. Popovic, "Synthesis of power efficient multitone signals with flat amplitude spectrum," *IEEE Trans. Commun.*, vol. 39, pp. 1031–1033, July 1991.

[237] M. Friese, "Multitone signals with low crest factor," *IEEE Trans. Commun.*, vol. 45, pp. 1338–1344, Oct. 1997.

[238] A. Brajal and A. Chouly, "Compensation of nonlinear distortions for orthogonal multicarrier," in *Proc. of 1994 IEEE Global Telecommun. Conf.*, 1994, pp. 1909–1914.

[239] H. W. Kang, Y. S. Cho and D. H. Youn, "On compensating nonlinear distortions of an OFDM system using an efficient adaptive predistorter," *IEEE Trans. Commun.*, vol. 47, pp. 522–526, Apr. 1999.

[240] G. Santella and F. Mazzenga, "A model for performance evaluation in M-QAM-OFDM scheme in presence of nonlinear distortion," in *Proc. of 1995 IEEE Veh. Tech. Conf.*, 1995, pp. 830–834.

[241] C. Rapp, "Effects of HPA-nonlinearity on a 4-DPSK/OFDM-Signal for a digital sound broadcasting system," in *Proc. of Second European Conf. on Satellite Commun.*, Oct. 1991, pp. 179–184.

[242] R. O'Neill and L. N. Lopes, "Envelope variations and spectrual splatter in clipped multicarrier signals," in *Proc. of IEEE 1995 International Symp. on Personal, Indoor and Mobile Radio Commun.*, 1995, pp. 71–75.

[243] M. Friese, "On the degradation of OFDM-signals due to peak-clipping in optimally predistorted power amplifiers," in *Proc. of 1998 IEEE Global Telecommun. Conf.*, Nov. 1998, pp. 939–944.

[244] E. Costa, M. Midrio, and S. Pupolin, "Impact of amplifier nonlinearities on OFDM transmission system," *IEEE Commun. Lett.*, vol. 3, pp. 37–39, Feb. 1999.

[245] X. Li and L. J. Cimini Jr., "Effects of clipping and filtering on the performance of OFDM," *IEEE Commun. Lett.*, vol. 2, pp. 131–133, May 1998.

[246] M. Pauli and H.-P. Kuchenbebecker, "Minimization of the intermodulation distortion of a nonlinearly amplified OFDM signal," *Wireless Personal Commun.*, vol. 4, pp. 93–101, Jan. 1997.

[247] R. van Nee and A. D. Wild, "Reducing the peak-to-average power ratio of OFDM," in *Proc. of 1998 IEEE Veh. Tech. Conf.*, 1998, pp. 2072–2076.

[248] D. J. G. Mestdagh and P. M. P. Spruyt, "A method to reduce the probability of clipping in DMT-based transceivers," *IEEE Trans. Commun.*, vol. 44, pp. 1234–1238, Oct. 1996.

[249] R. W. Bäuml, R. F. H. Fischer and J. B. Huber, "Reducing the peak-to-average power ratio of multicarrier modulation by selected mapping," *IEE Elect. Lett.*, vol. 32, pp. 2056–2057, Oct. 1996.

[250] P. V. Eetvelt, G. Wade, and M. Thompson, "Peak to average power reduction for OFDM schemes by selected scrambling," *IEE Elect. Lett.*, vol. 32, pp. 1963–1964, Oct. 1996.

[251] R. Verbin, *Efficient algorithm for clip probability reduction.* ANSI T1E1.4 committee contribution number 97-323, Sept. 1997.

[252] A. Gatherer and M. Polly, *Controlling clipping probability in DMT transmission.* ANSI T1E1.4 committee contribution number 97-367, Dec. 1997.

[253] M. Friese, "Mehrträgermodulation mit kleinem Crest-Faktor (in german)," Ph.D. dissertation, Technical University of Darmstadt, Nov. 1996.

[254] J. Tellado, Louise M. C. Hoo, and J. M. Cioffi, "Maximum likelihood detection of nonlinearly distorted multicarrier symbols by iterative decoding," in *Proc. of 1999 IEEE Global Telecommun. Conf.*, 1999, pp. 2493–2498.

[255] J. Tellado and J. M. Cioffi, "PAR reduction in multicarrier transmission systems," Stanford University, Tech. Rep., 1998.

[256] J. Tellado, "Peak to average power reduction for multicarrier modulation," Ph.D. dissertation, Stanford University, Sept. 1999.

[257] J. Tellado and J. M. Cioffi, "Efficient algorithms for reducing PAR in muticarrier systems," in *Proc. of IEEE Int. Symp. Information Theory*, 1998, p. 191.

[258] M. J. Fernandez-Getino Garcia, J. M. Paez-Borrallo, and O. Edfors, "Orthogonal pilot sequences for peak-to-average power reduction in OFDM," in *Proc. of 2001 IEEE Veh. Tech. Conf.*, Atlantic City, NJ, USA, 2001, pp. 650–654.

[259] O. M. T. Takada and Y. Akaiwa, "Peak power suppression with parity carrier for multi-carrier transmission," in *Proc. of 1999 IEEE Veh. Tech. Conf.*, 1999, pp. 2903–2907.

[260] H. Schmidt and K. D. Kammeyer, "Reducing the peak to average power ratio of multicarrier signals by adaptive subcarrier selection," in *Proc. of IEEE ICUPC*, 1998, pp. 933–937.

[261] A. E. Jones and T. A. Wilkinson, "Combined coding for error control and increased robustness to system in OFDM," in *Proc. of 1996 IEEE Veh. Tech. Conf.*, 1996, pp. 904–907.

[262] M. Friese, "Multicarrier modulation with low peak-to-average power ratio," *IEE Elect. Lett.*, vol. 32, pp. 713–712, Apr. 1996.

[263] S. H. Müller and J. B. Huber, "OFDM with reduced peak-to-average power ratio by optimum combination of partial transmit sequences," *IEE Elect. Lett.*, vol. 33, pp. 368–369, Feb. 1997.

[264] E. van der Ouderaa, J. Schoukens and J. Renneboog, "Peak factor minimization of input and output signals on linear systems," *IEEE Trans. Instrum. Meas.*, vol. 37, pp. 207–212, June 1988.

[265] L. J. Cimini, Jr. and N. R. Sollenberger, "Peak-to-average power ratio reduction of an OFDM signal using partial transmit sequences," *IEEE Commun. Lett.*, vol. 4, pp. 511–515, Mar. 1999.

[266] P. K. Frenger and N. A. B. Sevensson, "Parallel combinatory OFDM signalling," in *Proc. of IEEE 1996 International Symp. on Personal, Indoor and Mobile Radio Commun.*, 1996, pp. 1069–1073.

[267] ——, "Parallel combinatory OFDM signalling," *IEEE Trans. Commun.*, vol. 47, pp. 558–567, Apr. 1999.

[268] J. Tellado and J. M. Cioffi, "Peak power reduction for multicarrier transmission," in *Proc. of 1998 IEEE Global Telecommun. Conf.*, 1998, pp. 219–224.

[269] R. W. Bauml, R. F. H. Fischer, and J. B. Huber, "Reducing the peak-to-average power ratio of multicarrier modulation," *IEE Elect. Lett.*, pp. 2056–2057, Oct. 1996.

[270] S. H. Müller, R. W. Bäuml, R. F. H. Fisher and J. B. Huber, "OFDM with reduced peak-to-average power ratio by multiple signal representation," *Annals of Telecommun.*, vol. 52, pp. 58–67, Feb. 1997.

[271] A. D. S. Jayalath and C. Tellambura, "Reducing the peak-to-average power ratio of an OFDM signal through bit or symbol interleaving," *IEE Elect. Lett.*, vol. 36, pp. 1161–1163, June 2000.

[272] G. A. Awater and R. D. J. van Nee., "Implementation of the magic WAND wireless ATM modem." in *Proc. of 1999 IEEE Global Telecommun. Conf.*, 1999, pp. 1–6.

[273] B. Tarokh, H. R. Sadjadpour, "Construction of OFDM M-QAM sequences with low peak-to-average," *IEEE Trans. Commun*, vol. 51, pp. 25–28, Jan. 2003.

[274] J. A. Davis and J. Jedwab, "Peak-to-mean power control in OFDM, Gloay complementary sequences and Reed-Muller codes," in *Proc. of International Symposium on Information Theory '98.* IEEE, 1998, p. 190.

[275] ——, "Peak-to-mean power control in OFDM Golay complementary sequences and Reed-Muller codes." HPL-97-158, HP Laboratories, Tech. Rep., 1997.

[276] ——, "Peak-to-mean power control in OFDM, Golay complementary sequences, and Reed-Muller codes," *IEEE Trans. Inform. Theory*, vol. 45, pp. 2397–2417, Nov. 1999.

[277] T. Hoholdt, H. E. Jensen and J. Justesen, "Aperiodic correlations and the merit factor of a class of binary sequences," *IEEE Trans. Inform. Theory*, vol. 31, pp. 549–552, July 1985.

[278] R. L. Frank, "Polyphase complementary codes," *IEEE Trans. Inform. Theory*, vol. 26, pp. 641–647, Nov. 1980.

[279] R. D. J. van Nee, "OFDM codes for peak-to-average power reduction and error correction," in *Proc. of 1996 IEEE Global Telecommun. Conf.*, 1996, pp. 740–744.

[280] H. Ochiai and H. Imai, "Bloack coding scheme based on complementary sequences for multicarrier signals," *IEICE Trans. Fundamentals*, vol. 11, pp. 2136–2143, Nov. 1997.

[281] A. J. Grant and R. van Nee, "Efficient maximum likelihood decoding of Q-ary modulated codes," *IEEE Commun. Lett.*, vol. 2, pp. 134–136, May 1998.

[282] P. C. Fossorier and S. Lin, "Soft-decision decoding of linear block codes based on ordered statistics," *IEEE Trans. Inform. Theory*, vol. 41, pp. 1379–1396, Sept. 1995.

[283] H. Ochiai, Marc P. C. Fossorier and H. Imai, "On decoding of block codes with peak reduction in OFDM systems," *IEEE Commun. Lett.*, vol. 4, pp. 226–228, July 2000.

[284] K. G. Paterson and A. E. Jones, "Efficient decoding algorithms for generalized Reed-Muller codes," *IEEE Trans. Commun.*, vol. 48, pp. 1272–1285, Aug. 2000.

[285] J. Park, J. Kim and S. Choi, "Performance of MC-CDMA systems in non-independant Rayleigh fading," in *Proc. of 1999 IEEE International Conf. on Commun.*, 1999, pp. 506–510.

[286] H. Ochiai and H. Imai, "OFDM-CDMA with peak power reduction based on the spreading sequences," in *Proc. of 1998 IEEE International Conf. on Commun.*, 1998, pp. 1299–1303.

[287] L. Freiberg, A. Annamalai, and K. Bhargaava., "Crest factor reduction using orthogonal spreading codes in multi-carrier CDMA systems," in *Proc. of IEEE 1997 International Symp. on Personal, Indoor and Mobile Radio Commun.*, New York, NY, USA, 1997, pp. 1220–1224.

[288] K. Laird, N. Whinnett and S. Buljore, "A peak-to-average power reduction method for third generation CDMA reverse links," in *Proc. 1999 of IEEE Veh. Tech. Conf.*, 1999, pp. 551 –555,.

[289] B.-J Choi, E.-L Kuan, and L. Hanzo, "Crest factor study of MC-CDMA and OFDM," in *Proc. of 1999 IEEE Veh.Tech. Conf.*, 1999, pp. 233–237.

[290] Chih-Lin I and R. D. Gitlin, "Multi-code CDMA wireless personal communications networks," in *Proc. of 1995 IEEE International Conf. on Commun.*, 1995, pp. 1060–1064.

[291] T. Wada, "Charatecristic of bit sequence applicable to constant amplitude orthogonal multicode systems," *IEICE Trans. Fundamentals*, vol. 11, pp. 2160–2164, Nov. 2000.

[292] T. Wada, T. Yamazato and A. Ogawa, "A constant amplitude coding for orthogonal multi-code CDMA systems," *IEICE Trans. Fundamentals*, vol. 12, pp. 2477–2484, Dec. ember 1997.

[293] ——, "Error correcting capability of constant amplitude coding for orthogonal multi-code CDMA systems," *IEICE Trans. Fundamentals*, vol. 10, pp. 2166–2169, Oct. 1998.

[294] N. Guo and L. B. Milstein, "Uplink performance evaluation of multicode DS-CDMA systems in the presence of nonlinear distortions," *IEEE J. Select. Areas Commun.*, vol. 18, no. 2000, pp. 1418–1428, Aug.

[295] K. G. Paterson, "On codes with low peak-to-average power ratio for multi-code CDMA," vol. 50, pp. 550–559, March 2004.

[296] S. L. Miller and R. J. O'Dea, "Peak power and bandwidth efficient linear modulation," *IEEE Trans. Commun.*, vol. 46, pp. 1639–1648, Dec. 1998.

[297] A. G. Armada and M. Calvo, "Phase noise and sub-carrier spacing effects on the performance of an OFDM communication system ," *IEEE Commun. Lett.*, vol. 2, pp. 11–13, Jan. 1998.

[298] S. Alamouti, "A Simple Transmit Diversity Technique for Wireless Communications, ," *IEEE J. Select. Areas Commun.*, pp. 1451–1458, Oct. 1998.

[299] "IEEE Standard 802.16a, for Local and Metropolitan Area Networks - Part 16, Air Interface for Fixed Broadband Wireless Access Systems - Medium Access Control and Additional Physical Layer Specifications for 2-11 GHz ," iEEE Std. 802.16a, 2003.

[300] C. Tellambura, M. G. Parker, Y. J. Gao, S. J. Shepherd, and S. K. Barton, "Optimal Sequences for channel estimation using discrete Fourier transform techniques ," *IEEE Trans. Commun.*, vol. 47, no. 2, Feb. 1999.

[301] V. Tarokh and H. Jafarkhani, " On Computation and Reduction of the Peak-to-Average Power Ratio in Multicarrier Communications," *IEEE Trans. Commun.*, vol. 48, pp. 37–44, Jan. 2000.

[302] S. H. Müller-Weinfurtner, "On Optimality of Metrics for Coarse Frame Synchronization in OFDM: A Comparison ," in *Proc. of IEEE 1998 International Symp. on Personal, Indoor and Mobile Radio Commun.*, Boston, MA, 1998, pp. 533–537.

[303] P. H. Moose, " A Technique for Orthogonal Frequency Division Multiplexing Frequrncy Offset Correction," *IEEE Trans. Commun.*, vol. 42, no. 10, pp. 2908–2914, Oct. 1994.

[304] H. Zou, B. McNair, and B. Daneshrad, "An integrated OFDM receiver for high-speed mobile data communications ," in *Proc. of 2001 IEEE Global Telecommun. Conf.*, San Antonio, TX, Nov. 2001, pp. 3090–3094.

[305] B. G. Yang, K. B. Letaief, R. S. Cheng, and Z. Cao, "Timing recovery for OFDM transmission ," *IEEE J. Select. Areas Commun.*, vol. 18, no. 11, pp. 2278–2291, Nov. 2000.

[306] J. H. Kotecha and A. M. Sayeed, "Optimal signal design for estimation of correlated MIMO channels ," in *Proc. of 2003 IEEE International Conf. on Commun.*, Anchorage, Alaska, May 2003, pp. 3170–3174.

[307] V. Erceg *et al.*, "An empirically based path loss model for wireless channels in suburban environments ," *IEEE J. Select. Areas Commun.*, vol. 17, no. 7, pp. 1205–1211, July 1999.

[308] "Contribution to the IEEE 802.16a Working Group ," *80216a-03_01.pdf. [Online]. Available: http://ieee802.org/16/tga/docs/80216a-03_01.pdf*, June 2003.

[309] W. D. Warner and C. Leung, " OFDM/FM Frame Synchronization for Mobile Radio Data Communication." *IEEE Trans. Veh. Tech.*, vol. 42, no. 3, pp. 302–313, Aug. 1993

[310] G. Santella, "A frequency and symbol synchronization system for OFDM signals:architecture and simulation results," *IEEE Trans. Veh. Technol.*, vol. 49, no. 1, pp. 245–275, Jan. 2000.

[311] J. Li, G.-Q. Liu and G. B. Giannakis, "Carrier frequency offset for OFDM based WLANs ," *IEEE Signal Processing Lett.*, vol. 8, no. 3, pp. 80–82, March 2001.

[312] E. G. Larsson, G.-Q. Liu, J. Li, and G. B. Giannakis, "Joint Symbol Timing and Channel Estimation for OFDM Based WLANs ," *IEEE Commun. Lett.*, vol. 5, no. 8, pp. 325–327, Aug. 2001.

[313] V. Mignone and A. Morello, "CD3-OFDM: A Novel Demodulation Scheme for Fixed and Mobile Receivers ," *IEEE Trans. Commun.*, vol. 44, no. 9, pp. 1144–1151, Sept. 1996.

[314] G. G. Raleigh and V. K. Jones, "Multivariate modulation and coding wireless communication ," *IEEE J. Select. Areas Commun.*, vol. 17, no. 5, pp. 851–886, May 1999.

[315] Z. C. B. Yang and K. B. Letaief, "Analysis of low-complexity windowed DFT-based MMSE estimator for OFDM systems ," *IEEE J. Select. Areas Commun.*, vol. 49, no. 11, pp. 1977–1987, Nov. 2001.

[316] M. Speth, S. A. Fechtel, G. Fock, and H. Meyr, "Optimum Receiver Design for Wireless OFDM-Based Broadband Transmission-Part II: A Case Study," *IEEE Trans. Commun.*, vol. 49, no. 4, pp. 571–578, April 2001.

[317] G. Strang, *Linear Algebra and its Applications, (3rd ed).* New York, NY: Saunders, Harcourt College Publishing, 1986.

[318] S. M. Kay, *Fundamentals of Statistical Signal Processing: Estimation Theory*. New Jersey: Prentice Hall, 1993.

SUBJECT INDEX